HAZARDOUS MATERIALS RESPONSE HANDBOOK

HAZARDOUS MATERIALS RESPONSE HANDBOOK

First Edition

Based on the 1989 Editions of NFPA 472, *Standard for Professional Competence of Responders to Hazardous Materials Incidents*, and NFPA 471, *Recommended Practice for Responding to Hazardous Materials Incidents*.

Martin F. Henry, Editor
Director, Public Fire Protection Division

National Fire Protection Association
NFPA® Quincy, Massachusetts

This first edition of the *Hazardous Materials Response Handbook* is an essential reference for everyone who responds to hazardous materials incidents. It covers the spectrum of hazardous materials preparedness, from recognizing risks to decontamination. State-of-the-art, international guidelines are included, together with the latest NFPA haz mat standards.

This handbook features the complete texts of NFPA's new hazardous materials documents, *Standard for Professional Competence of Responders to Hazardous Materials Incidents* (NFPA 472) and *Recommended Practice for Responding to Hazardous Materials Incidents* (NFPA 471), along with explanatory commentary, which is integrated between requirements.

All NFPA codes and standards are processed in accordance with NFPA's *Regulations Governing Committee Projects*. The commentary in this handbook is the opinion of the author(s), recognized experts in the field of hazardous materials. It is not, however, processed in accordance with the NFPA *Regulations Governing Committee Projects*, and therefore shall not be considered to be, nor relied upon as, a Formal Interpretation of the meaning or intent of any specific provision or provisions of NFPA 471 or NFPA 472. The language contained in NFPA 471 and 472, rather than this commentary, represents the official position of the Committee on Hazardous Materials Response Personnel and the NFPA.

This handbook is a must for all public, military, industrial, and volunteer fire departments that respond to hazardous materials incidents, as well as those responsible for the preparedness and training of the response community.

Project Manager: Jennifer Evans
Project Editor: Julie Mason Eksteen
Composition: Louise Grant, Cathy Ray, and Elizabeth Turner
Illustrations: Paula Eissmann and Nancy Maria
Cover Design: Greenwood Associates
Production: Donald McGonagle and Stephen Dornbusch

NFPA No. 472HB89
ISBN No. 0-87765-358-5
Library of Congress No. 89-60577
Printed in the United States of America

First Printing, September 1989

Dedication

This edition of the *Hazardous Materials Response Handbook* is dedicated to the late Percy Bugbee, who will always be "Mr. NFPA."

Contents

Recommended Practice for Responding to Hazardous Materials Incidents, NFPA 471

Supplements

Foreword

There are many who think that concern about hazardous materials is a recent development, which is not an accurate assessment. The National Fire Protection Association has been involved for many decades in developing standards that relate to various hazardous materials. Consider, as examples, the several codes that deal with flammable liquids, oxidizers, gases, pesticides, dangerous chemicals, chemical reactions, and explosives, as well as other documents that pertain to identification, classification, and proper handling of some of the above.

It is true, however, that there is a great deal of recent interest and emphasis on the subject, and for several reasons.

People have come to realize, more than ever before, that the world's resources are both precious and limited. Concern for the welfare of our environment has heightened. Environmentalists, who were treated with scorn not too many years ago, are now give the respect that is their due. In the United States, the Environmental Protection Agency has become a major regulator of hazardous substances, and enforcement of their regulations is becoming increasingly efficient.

Several incidents have also served to heighten concern about hazardous materials, perhaps none more so than Bhopal, India. Consider also the Mexican LP Gas explosions that killed an estimated five hundred people; the chemical fire in Basel, Switzerland that did prolonged damage to the Rhine river; the Kansas City, Kansas explosion that cost the lives of six fire fighters; and the costly and damaging Valdez oil tanker spill. Similar incidents are likely to recur, and they invariably result in demands for more and better regulations.

Another major driving force has been concern for worker safety and health. The Occupational Safety and Health Administration has developed rules that have helped to focus attention on the needs of employees as well as on the obligations of employers. Labor unions and litigation have helped bring about appropriate observance of the many and varied OSHA regulations.

Interestingly, concern for worker safety has also resulted in a growing realization that emergency responders need, and are indeed entitled to, adequate protection against hazardous materials. They are no longer considered expendable, however noble their calling. OSHA's Fire Brigade Standard, and more recently NFPA 1500, have highlighted the need for protection of responders.

In 1984, two separate requests to NFPA expressed a need for standards relating to responders to hazardous materials incidents. One was from the International Society of Fire Service Instructors (ISFSI), the other from the International Fire Service Training Association (IFSTA). NFPA sought public support for the project, and the responses indicated substantial agreement that there was indeed a need for standards. Both ISFSI and IFSTA should be commended for their foresight.

Establishing a technical committee became a labor of love for me. NFPA anticipated an eager response from the fire service, but other key interest groups had to be involved. The committee would need the expertise of regulators, transporters, labor, manufacturers, educators, environmentalists, and so on. Warren Isman, Fire Chief of

Fairfax County Fire and Rescue, and Gerry Grey of the San Francisco Fire Department agreed to serve as chairman and vice chairman, respectively. The committee process benefited greatly from their leadership.

As anticipated, requests for membership poured in from fire service members and in such numbers that many had to be placed on a waiting list for committee membership. That list numbered over 60 at one time. This speaks well of fire service willingness to participate in the committee process.

I cannot adequately express my admiration and gratitude to all members of the technical committee. They constituted a "blue ribbon" panel of willing experts, and they can be proud of the work they have done. My gratitude also extends to the several non-members who attended most of our meetings and contributed substantially to the final product.

Finally, allow me at this stage to extol the NFPA standards making process. It is a wondrous procedure to observe, a rewarding process in which to participate, and it works well.

Acknowledgements

The editor acknowledges with gratitude the following people:

- Members of the Technical Committee on Hazardous Materials Response Personnel who faced with aplomb the monumental task of finding a starting point in a diverse subject,
- All non-members who attended our committee meetings and lent their expertise and elbow grease to the effort,
- Those who participated in the committee process by submitting their constructive comments, and
- The many members of the NFPA staff who contributed to the production of this Handbook.

Note:

The text, illustrations, and captions to photographs that make up the commentary on various sections of this Handbook are printed in color and in a column narrower than the standard text. The photographs, all of which are part of the commentary, are printed in black for clarity. The text of the standard and the recommended practice is printed in black.

NFPA 472

Standard for

Professional Competence of Responders to Hazardous Materials Incidents

1989 Edition

NOTICE: An asterisk (*) following the number or letter designating a paragraph indicates explanatory material on that paragraph in Appendix A. Material from Appendix A is integrated with the text and is identified by the letter A preceding the subdivision number to which it relates.

Information on referenced publications can be found in Chapter 5 and Appendix C.

1 Administration

Editor's Note

It is most important to note at the very outset that this document does not limit its application to the fire service. It is the intent of the Standard, as clearly indicated in the Committee Scope approved by NFPA's Standards Council; and it is likewise the intent of the Committee, as illustrated in its responses to the many comments viewed on the Technical Committee Report; that the provisions of the Standard be applied to all persons who might be called on to respond to or take action at a hazardous materials incident. Fire service personnel come immediately to mind as the premier front line emergency responders. Police departments will also frequently find themselves the first at the scene of a hazardous materials transportation incident. Many other groups, however, will also become party to remediation of an incident. These might include government and industrial responders, private fire brigades, and workers who are engaged in operations, both mobile and fixed, that involve everyday dealing with hazardous materials in the broadest use of that term.

1-1 Scope. This standard identifies the levels of competence required of responders to hazardous materials incidents. It specifically covers the requirements for first responder, hazardous materials technician, and hazardous materials specialist.

The key to achieving a given level of competence is training. A specified number of hours of training will not achieve equal results for every individual, since there are many variables involved. These include the abilities of the trainer and trainee and the quality, intensity, frequency, and recency of the instruction. For this reason, the Standard does not prescribe a number of training hours. Instead, it deals with certain objectives and abilities that should be attained by the responder.

Note that three principal levels of response are indicated. In Chapter 2, you will see that the first level is subdivided into Awareness and Operational. The levels of response are somewhat arbitrary and subjective. Further, the lines of demarcation that separate one level from another are not always clear.

Another aspect of the levels of response that is worth discussing is that the competency levels might be chemical-specific or site-specific.

For example, a response team can be trained to handle just one hazardous chemical, such as chlorine. The training, experience, competency, and equipment possessed by members of the response team would lead one to consider them at the specialist level, but only for an incident that involves chlorine. At accidents where other chemicals are encountered, the response team might well be designated at a lower level of response competency.

At a given fixed facility where just a few chemicals are handled, an on-site response team might be trained to a specialist level for the particular chemicals that are present. Once the response team, or its individual members, leaves the confines of the fixed facility, their response competency level might be lowered.

It is also important to point out that a response team might be comprised of members who find themselves at different levels. It is not inconceivable that a team would have some members at the specialist level and others who are technicians. Some team members could even fall into the first responder operational level.

It is apparent, then, that the levels of competence refer primarily to individuals rather than to response teams. However, a response team can be made up entirely of members in one category and can be designated as belonging to that response category. We should not let the designations obstruct us in any way.

1-2 Purpose.   The purpose of this standard is to specify minimum requirements of competence for those who will respond to hazardous materials incidents. It is not the intent of this standard to restrict any jurisdiction from exceeding these minimum requirements.

Hazardous materials are a pervasive part of the world in which we live, and incidents involving hazardous materials are inevitable. Anyone who is likely to deal with an incident must be trained for that eventuality. This would include the fixed site worker as well as the person involved in transportation. It most certainly includes every emergency responder. Employers (including federal, state, and local governments) have an obligation to provide training that is adequate for the anticipated degree of involvement of their employees. One of the principal reasons underlying that obligation is safety. The more competent the responder, the less likely will be the chance of injury from the mishandling of an incident.

Figure 1.1   Hazardous materials incidents can happen at any time. Preplanning and training allow the safe and effective handling of such situations.

1-2.1   One of the purposes of the qualification requirements contained herein is to reduce the numbers of accidents, injuries, and illnesses during response to hazardous materials incidents and to help prevent exposure to hazardous materials to reduce the probability of fatalities, illnesses, and disabilities affecting emergency response personnel.

Providing protection to the responder and assuring enhanced safety at the scene of an incident have to be the principal purposes of this standard. Safety at operations must always be observed. To some extent, every responder must be his own safety officer and must use good judgement in the performance of necessary duties. Adherence to the provisions of this Standard at each of the levels of competency should provide a high degree of safety, despite the hazards encountered.

1-3*   Definitions.

Approved.   Acceptable to the "authority having jurisdiction."

NOTE: The National Fire Protection Association does not approve, inspect or certify any installations, procedures, equipment, or materials nor does it approve or evaluate testing laboratories. In determining the acceptability of installations or procedures, equipment or materials, the authority having jurisdiction may base acceptance on compliance with NFPA or other appropriate standards. In the absence of such standards, said authority may require evidence of proper installation, procedure or use. The authority having jurisdiction may also refer to the listings or labeling practices of an organization concerned with product evaluations which is in a position to determine compliance with appropriate standards for the current production of listed items.

Acceptance by an authority having jurisdiction is usually based on tests or experience rather than on an arbitrary decision.

Authority Having Jurisdiction.   The "authority having jurisdiction" is the organization, office or individual responsible for "approving" equipment, an installation or a procedure.

NOTE: The phrase "authority having jurisdiction" is used in NFPA documents in a broad manner since jurisdictions and "approval" agencies vary as do their responsibilities. Where public safety is primary, the "authority having jurisdiction" may be a federal, state, local or other regional department or individual such as a fire chief, fire marshal, chief of a fire prevention bureau, labor department, health department, building official, electrical inspector, or others having statutory authority. For insurance purposes, an insurance inspection department, rating bureau, or other insurance company representative may be the "authority having jurisdiction." In many circumstances the property owner or his designated agent assumes the role of the "authority having jurisdiction"; at government installations, the commanding officer or departmental official may be the "authority having jurisdiction."

In a document where one of the subject areas is the very broad concept of hazardous materials, there will be many authorities exercising jurisdiction. This will include, at least in some cases and to some extent, the manufacturer and shipper and the local community. Every level of government will also play a role.

Cold Zone.   This area contains the command post and such other support functions as are deemed necessary to control the incident. This is also referred to as the clean zone or support zone in other documents.

It might appear that there is no outer boundary to the cold zone, but this is not the case. One might equate the outer boundary at a hazardous materials incident with the fire lines that are often established at a major fire or emergency and that are usually controlled by the police department. The public at large would not have access to the cold zone under most circumstances.

Competence.   Possessing knowledge, skills, and judgment needed to perform indicated objectives satisfactorily.

While knowledge and skills are capable of measurement, one's judgment is not easily evaluated. Under varying circumstances, the proper exercise of judgment or decision-making can change substantially. Nonetheless, training and experience can effectively improve the emergency decision making process.

Since the outcome of an incident will be determined by the decisions that are made in handling it, the need for proper and adequate training becomes paramount.

Confinement.   Those procedures taken to keep a material in a defined or local area.

Confinement is but one technique that can be employed in achieving control of an incident. An assumption is made that the material that is involved is outside of its normal contained configuration.

Container.   Any bag, barrel, bottle, box, can, cylinder, drum, reaction vessel, storage tank, or the like that contains a hazardous material.

Some codes define a container by placing limitations on its capacity. The meaning in this document is not affected by size. In effect, a container is anything that is designed or intended to hold a hazardous material.

Containment.   Those procedures taken to keep a material in its container.

Containment is a second technique that can be employed in controlling an incident. An assumption is made that the container in question is capable of holding or receiving the material.

Contaminant/Contamination.   A substance or process that poses a threat to life, health, or the environment.

Inherent to the definition of contaminant is the concept that the offending material is present where it should not be and does not belong. A second aspect essential to the meaning is that the offending material is somehow toxic or harmful.

Control.   The procedures, techniques, and methods used in the mitigation of a hazardous materials incident, including containment, extinguishment, and confinement.

Control can be used interchangeably with the word mitigation. Every measure taken to control a hazardous materials incident is part of the mitigation process. Limiting the degree of contamination, by whatever measure, is also part of the control or mitigation process.

Control Zones.   The designation of areas at a hazardous materials incident based upon safety and the degree of hazard. Many terms are used to describe the zones involved in a hazardous materials incident. For purposes of this standard, these zones shall be defined as the hot, warm, and cold zones.

Other terms that are commonly used include "site control zones" and "work zones." The Committee's choice of basic terms like "hot," "warm," and "cold" was based on the fact that the words are simple and easily understood and that they clearly suggest the nature of the situation one would expect to encounter in an area so designated.

Coordination.   The process used to get people, who may represent different agencies, to work together harmoniously in a common action or effort.

One of the more difficult problems at major incidents can be the determination of who is in overall charge of the incident; another more remote concern involves ultimate responsibility for the consequences associated with the incident. The first problem is one that can be solved by good local emergency planning, which is an essential element in proper handling of incidents. The second problem is one that may need judicial determination.

Decontamination (Contamination Reduction).   The physical and/or chemical process of reducing and preventing the spread of contamination from persons and equipment used at a hazardous materials incident.

The broad concept of decontamination includes actions taken at an incident that will serve to reduce or even prevent contamination. The narrower meaning suggests those steps that are taken to remove contaminants that may have accumulated on equipment or personnel at an incident. Good decontamination practices should incorporate and emphasize the notion of prevention as a first step.

Decontamination Area.   The area, usually located within the warm zone, where decontamination takes place.

The purpose of a decontamination area is to reduce the likelihood of the spreading of contamination to the cold zone. It is usually located in the warm zone because that is the transition area between the clean or uncontaminated area and the hot zone.

Degradation.   A chemical action involving the molecular breakdown of a protective clothing material due to contact with a chemical. The term degradation may also refer to the molecular breakdown of the spilled or released material to render it less hazardous.

Physical degradation of protective clothing can also take place, such as might occur from rubbing against a rough surface, or even from normal usage. Chemical degradation can be minimized by avoiding unnecessary contact with chemicals. The result of degradation will be increased likelihood of permeation and penetration.

Figure 1.2 Normally, when a suit is contaminated in the field you would decontaminate initially with a soap and water wash; saving the drain-off water as contaminated waste. Once this is done you may consider some type of neutralization or decontaminating agent other than soap and water.

Demonstrate.   To show by actual use. This may be supplemented by simulation, explanation, illustration, or a combination of these.

Describe.   To explain verbally or in writing using standard terms recognized in the hazardous materials response community.

Hazard/Hazardous.   Capable of posing an unreasonable risk to health, safety, or the environment; capable of doing harm.

Hazard Sector.   That function of an overall Incident Command System that deals with the actual mitigation of a hazardous materials incident. It is directed by a sector officer and principally deals with the technical aspects of the incident.

The hazard sector, though crucial to the successful outcome of an incident, is not the only sector at work. At major incidents, there could be several sectors operating.

Hazard Sector Officer.   The person responsible for the management of the hazard sector.

The hazard sector officer, although directly responsible for the "hands-on" mitigation of the incident, is not the incident commander. He reports to the incident commander, as do those in command of other sectors that might be operating at the scene of the incident.

Hazardous Material.* A substance (gas, liquid, or solid) capable of creating harm to people, property, and the environment. See specific regulatory definitions in Appendix A.

Class. The general grouping of hazardous materials into nine categories identified by the United Nations Hazard Class Number System, including:

Explosives

Gases (compressed, liquefied, dissolved)

Flammable Liquids

Flammable Solids

Oxidizers

Poisonous Materials

Radioactive Materials

Corrosive Materials

Other Regulated Materials

A-1-3 Hazardous Materials Definitions. There are many definitions and descriptive names being used for the term hazardous materials, each of which depends on the nature of the problem being addressed.

Unfortunately, there is no one list or definition that covers everything. The United States agencies involved, as well as state and local governments, have different purposes for regulating hazardous materials that, under certain circumstances, pose a risk to the public or the environment.

(a) *Hazardous Materials.* The United States Department of Transportation (DOT) uses the term *hazardous materials*, which covers eight hazard classes, some of which have subcategories called classifications, and a ninth class covering other regulated materials (ORM). DOT includes in its regulations hazardous substances and hazardous wastes as an ORM-E, both of which are regulated by the Environmental Protection Agency (EPA), if their inherent properties would not otherwise be covered.

(b) *Hazardous Substances.* EPA uses the term *hazardous substance* for the chemicals which, if released into the environment above a certain amount, must be reported and, depending on the threat to the environment, federal involvement in handling the incident can be authorized. A list of the hazardous substances is published in 40 CFR Part 302, Table 302.4.

(c) *Extremely Hazardous Substances*. EPA uses the term *extremely hazardous substance* for the chemicals which must be reported to the appropriate authorities if released above the threshold reporting quantity. Each substance has a threshold reporting quantity. The list of extremely hazardous substances is identified in Title III of Superfund Amendments and Reauthorization Act (SARA) of 1986 (40 CFR Part 355).

(d) *Toxic Chemicals*. EPA uses the term *toxic chemical* for chemicals whose total emissions or releases must be reported annually by owners and operators of certain facilities that manufacture, process, or otherwise use a listed toxic chemical. The list of toxic chemicals is identified in Title III of SARA.

(e) *Hazardous Wastes*. EPA uses the term *hazardous wastes* for chemicals that are regulated under the Resource, Conservation and Recovery Act (40 CFR Part 261.33). Hazardous wastes in transportation are regulated by DOT (49 CFR Parts 170-179).

(f) *Hazardous Chemicals*. The United States Occupational Safety and Health Administration (OSHA) uses the term *hazardous chemical* to denote any chemical that would be a risk to employees if exposed in the work place. Hazardous chemicals cover a broader group of chemicals than the other chemical lists.

(g) *Hazardous Substances*. OSHA uses the term *hazardous substance* in 29 CFR Part 1910.120, which resulted from Title I of SARA and covers emergency response. OSHA uses the term differently than EPA. Hazardous substances, as used by OSHA, cover every chemical regulated by both DOT and EPA.

The United Nations Committee of Experts on the Transport of Dangerous Goods established recommendations that would help to standardize the regulations governing transportation of hazardous materials. The recommendations covered classification and definitions of classes, listing of the principal hazardous materials, and packing, labeling, and shipping papers. The recommended regulations do not apply to bulk shipments, since these are generally subject to special regulations in most countries. The purpose of the United Nations system is to eliminate or reduce risks and at the same time to establish some uniformity in worldwide transportation of hazardous materials.

The UN system classifies materials by type of risk presented. The classes are not presented in any order or degree of danger.

In the United States, the UN class number may be displayed at the bottom of a placard or label, or on a shipping paper after the listed shipping name. The class number is used in place of the written name of the hazard class in the shipping paper description. The UN classes have specified subdivisions.

Explosives

Division 1.1	Explosives with a mass explosion hazard
Division 1.2	Explosives with a projection hazard
Division 1.3	Explosives with predominantly a fire hazard
Division 1.4	Explosives with no significant blast hazard
Division 1.5	Very insensitive explosives

Gases (compressed, liquefied, dissolved)

Division 2.1	Flammable gases
Division 2.2	Nonflammable gases
Division 2.3	Poison gases
Division 2.4	Corrosive gases (Canadian)

Flammable Liquids

Division 3.1	Flash point below −18 °C (O °F)
Division 3.2	Flash point −18 °C and above, but less than 23 °C (73 °F)
Division 3.3	Flash point of 23 °C and up to 61 °C (141 °F)

Flammable Solids

Division 4.1	Flammable solids
Division 4.2	Spontaneously combustible materials
Division 4.3	Materials that are dangerous when wet

Oxidizers

Division 5.1	Oxidizers
Division 5.2	Organic peroxides

Poisonous Materials

Division 6.1	Poisonous materials
Division 6.2	Etiologic (infectious) materials

Radioactive Materials

A radioactive material is defined as any substance with a specific activity greater than 0.002 microcurie per gram.

Corrosive Materials

These are substances that, by chemical action, will cause severe damage when in contact with living tissue, or, in the case of leakage, will materially damage or even destroy other freight or the means of transport; they may also cause other hazards.

Other Regulated Materials

These are substances that during transport present a danger not covered by other classes.

Classification.   The individual divisions of hazardous materials called "hazard classes" in the United States and "divisions" in the United Nations system, including:

Explosives A

Class A explosives are sensitive to heat and shock, will detonate, and present a maximum hazard. There are nine types identified in the U.S. Department of Transportation (DOT) regulations. Class A explosives include dynamite, TNT, black powder, and some types of military ammunition.

Explosives B

Class B explosives function by rapid combustion rather than by detonation. They possess a high flammability hazard and include most propellant materials like rocket motors, display fireworks, and some military ammunition.

Explosives C

Class C explosives include manufactured articles that contain limited amounts of Class A or Class B explosives. This class includes fireworks, explosive rivets, detonating fusee, small arms ammunition, etc. They will not normally mass detonate under fire conditions.

Blasting Agents

Blasting agents present little probability of accidental ignition. Blasting agents are used primarily in demolition, mining, and quarrying. They are not cap sensitive and require a strong primer.

Nonflammable Gases

Nonflammable gases are compressed gases that will not normally burn, but that may support combustion. Examples of nonflammable gases include carbon dioxide, oxygen, and nitrogen.

Flammable Gases

Flammable gases are compressed gases that will burn. Examples include hydrogen, acetylene, vinyl chloride, and propane.

Flammable Liquids

A flammable liquid is any liquid with a flash point below 100 °F (38 °C). The flash point of a liquid is the lowest temperature at which enough vapor is given off to form an ignitible mixture with air near the surface of the liquid.

Combustible Liquids

A combustible liquid is one with a flash point at or above 100 °F (38 °C) and below 200 °F (93 °C). It is important to note that the NFPA definition of combustible liquid does not establish an upper limit flash point.

One should not be deceived into thinking that liquids with flash points above 200 °F (93 °C) will not burn.

Flammable Solids

A flammable solid is any solid material other than an explosive that ignites readily and burns vigorously. Under conditions that are normally incident to transportation, such as friction or retained heat from manufacturing or processing, a flammable solid can ignite, burn furiously, and present a hazard to transportation. Some flammable solids are air-reactive or pyrophoric, some react with water, and some are spontaneously combustible.

Oxidizers

Oxidizers are materials that contain large amounts of chemically bound oxygen that is easily released, especially when heated, and that will stimulate the burning of combustible material.

Organic Peroxides

Organic peroxides were given a separate hazard class because of their dangerously destructive potential. Nearly all of the organic peroxides are used in the plastics industry. They are combustible and will also increase the intensity of a fire. The decomposition process releases a great deal of heat and gaseous products that are often toxic.

Poison Gases

Poison gases are of such a nature that a very small amount is dangerous to life.

Poison B

Class B poisons are less hazardous but still present a significant hazard if released during transportation. This class also includes gases that, in the absence of adequate data on human toxicity, are presumed to be toxic to humans based on testing performed on laboratory animals.

Irritating Materials

Irritating materials give off dangerous or intensely irritating fumes. Tear gas is an example of an irritating material.

Etiologic Agents

Etiologic agents have properties similar to those of poisons. An etiologic agent is a living microorganism that may cause human disease. They might include biological specimens and some specimens used for testing and research.

Radioactive Materials

Radioactive materials are widely used in medicine, industry, research, and the generation of electric power. The majority of radioactive shipments involve small carton-type packages used in medical applications.

Corrosive Materials

Corrosive materials are liquids or solids that can destroy human skin tissue or severely corrode steel. Examples include sulfuric acid, hydrochloric acid, and sodium hydroxide.

ORM A

Other Regulated Materials (ORM) are divided into five classes, A through E. ORMs do not meet the definitions of any of the DOT hazard classes, but possess enough hazardous characteristics in transport that they require some regulation.

ORM A is a material that has an anesthetic, irritating, noxious, toxic, or other similar property and that can cause extreme annoyance or discomfort to passengers and crew in the event of leakage. An example would be chloroform or carbon tetrachloride.

ORM B

ORM B is a material capable of causing significant damage to a transport vehicle or vessel from leakage during transportation; an example would be quicklime.

ORM C

ORM C is a material that is unsuitable for shipment unless properly identified and prepared for transportation. An example would be excelsior.

ORM D

ORM D is a material, such as a consumer commodity, that meets the definition of a hazardous material but presents a limited hazard during transportation because of its form, quantity, and packaging. An example would be household cleaning supplies.

ORM E

ORM E is a material that is not included in any other hazard class but is subject to the requirements of this subchapter. Materials in this class include hazardous waste and hazardous substances as defined by DOT and EPA regulations. An example would be waste acetone.

Hazardous Materials Response Team. A group of trained response personnel operating under an emergency response plan and appropriate standard operating procedures to control or otherwise minimize or eliminate the hazards to people, property, or the environment from a released hazardous material.

In most cases, a constituted response team will have their duties and responsibilities clearly listed and described in a standard operating procedure manual. A typical fire department situation might have designated members of a hazardous materials response unit perform in other capacities, such as engine company, ladder company, rescue, or squad; and respond to hazardous materials when specially called to do so. The members so designated will have been adequately trained for the hazardous material response.

In some few cases, there are "dedicated" response teams whose only function is to handle hazardous materials incidents.

There is no requirement that a fire department have a designated hazardous materials response team. However, all members are required to be trained to some degree to respond to incidents that might involve hazardous materials.

High Temperature Protective Clothing.   Protective clothing designed to protect the wearer for short-term high temperature exposures. This type of clothing is usually of limited use in dealing with chemical commodities.

This definition of high temperature protective clothing comes from the Chemical Manufacturers Association. It should be noted that NFPA does not have a standard on this category of protective clothing or equipment. It is most important that the user be aware of the stated limitations in the CMA definition, namely "short-term" and "limited use."

Hot Zone.   Area immediately surrounding a hazardous materials incident, which extends far enough to prevent adverse effects from hazardous materials

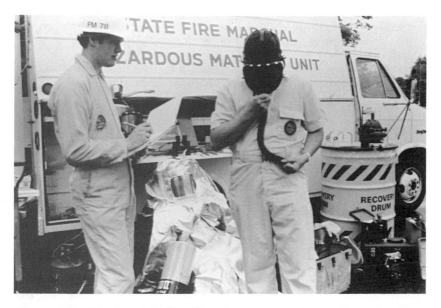

Figure 1.3   Whether a dedicated unit or a team that performs in other capacities, a hazardous materials response team should be properly equipped and fully trained.

Figure 1.4   Approach suits can be used in applications where
temperatures reach as high as 2000 °F. However, they are meant for
brief exposures of three minutes or less. These suits will provide
limited protection against both hot and cold liquids, mildly corrosive
chemicals as well as steam.

releases to personnel outside the zone. This zone is also referred to as the
exclusion zone or restricted zone in other documents.

The hot zone is the area where contamination does or could take
place. It is also the area where cleanup operations will be performed. The boundary between the hot zone and the warm zone
should be clearly indicated by some physical means, such as lines,
hazard tape, equipment barriers, and the like. Movement of personnel from one zone to another must be tightly regulated and
supervised in order to minimize contamination. This will allow for
greater control of the operations within the zone.

Identify.   To physically select, indicate, or explain verbally or in writing
using recognized standard terms.

Incident.   A fire involving a hazardous material or a release or potential
release of a hazardous material.

There is a wide difference of opinion on what constitutes a
hazardous materials incident when fire is involved. At one extreme are those who would like to categorize every fire as an
incident, on the basis that every fire will produce toxic products of

combustion that are clearly hazardous. A more reasonable approach is to limit the definition to the actual involvement in fire of an acknowledged hazardous material. Frequently the line of demarcation is blurred. Is a fire in a fuel storage tank to be considered a fire, or should it be classified as a hazardous materials incident? The answer may vary from one agency to another. A fire department might list it as a fire incident; an environmental agency might see it as a hazardous materials incident. There are many other examples that can be used to illustrate the problem, and it should not be taken lightly. There are many more complicating factors involved in operating at a designated hazardous materials incident than there are in operating at a fire, just from the standpoint of personnel.

Incident Command System.   An organized system of roles, responsibilities, and standard operating procedures used to manage and direct emergency operations.

Incident Commander.   The person responsible for all decisions relating to the management of the incident. The incident commander is in charge at the incident.

The incident commander can be compared to a fireground commander, who is responsible for overall control of operations. It is important that one person be in responsible command of an

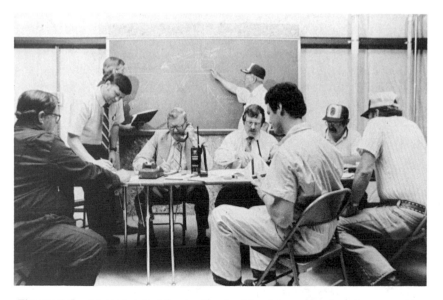

Figure 1.5   At major incidents, the incident command system may be at a remote location.

incident. The actual command may change from one person to another as an incident develops and intensifies.

Listed.   Equipment or materials included in a list published by an organization acceptable to the "authority having jurisdiction" and concerned with product evaluation, that maintains periodic inspection of production of listed equipment or materials and whose listing states either that the equipment or material meets appropriate standards or has been tested and found suitable for use in a specified manner.

NOTE: The means for identifying listed equipment may vary for each organization concerned with product evaluation, some of which do not recognize equipment as listed unless it is also labeled. The "authority having jurisdiction" should utilize the system employed by the listing organization to identify a listed product.

Material Safety Data Sheet (MSDS).   Provided by manufacturers and compounders (blenders) of chemicals, with minimum information about chemical composition, physical and chemical properties, health and safety hazards, emergency response, and waste disposal of the material as required by OSHA 1910.1200.

A typical material safety data sheet (MSDS) is available from the U.S. Department of Labor, Occupational Safety and Health Administration (OSHA). (Form OSHA-20.)

Monitoring Equipment.   Instruments and devices used to identify and quantify contaminants.

The use of monitoring equipment at an incident is important for several reasons. It can be used to determine the personal protective equipment that will be needed, it determines where the protection is needed, it can assist in evaluating the severity of the situation in terms of exposure, and it dictates the need for medical monitoring and surveillance. Monitoring equipment ranges from inexpensive and very simple to use, to costly and highly sophisticated.

Objective.   A goal that is achieved through the attainment of a skill, knowledge, or both, which can be observed or measured.

Packaging.   Any container that holds a material (hazardous or nonhazardous). Packaging includes nonbulk and bulk packaging.

Nonbulk Packaging.   Any packaging having a capacity meeting one of the following criteria:

(a)   Liquid—internal volume of 118.9 gallons (450 L) or less;

(b)   Solid—capacity of 881.8 pounds (400 kg) or less; or

(c) Compressed gas — water capacity of 1000 pounds (453.6 kg) or less.

Bulk Packaging.   Any packaging, including transport vehicles, having a capacity greater than described above under nonbulk packaging. Bulk packaging for transportation can be either placed on or in a transport vehicle or vessel or is constructed as an integral part of the transport vehicle.

Penetration.   The movement of a material through a suit's closures, such as zippers, buttonholes, seams, flaps, or other design features of chemical protective clothing, and through punctures, cuts, and tears.

A regular and routine program of inspection can help uncover conditions that lend themselves to penetration.

Permeation.   A chemical action involving the movement of chemicals, on a molecular level, through intact material.

Information on the resistance of types of clothing material to hundreds of chemicals is available. Suppliers of the new materials and manufacturers and vendors of protective clothing will make data on resistance to permeation available on request. No one material will protect against all chemicals or combination of chemicals, and no material will protect against prolonged exposure to certain chemicals. In any event, extreme caution on the part of the user is advised. Available data is not capable of dealing with every situation that might be encountered.

Figure 1.6   Railroad cars are an example of bulk packaging.

The rate of permeation is a function of clothing type, thickness, quality of the finished product, concentration of the chemical, ambient conditions, solubility of the chemical in the material, and its diffusion rate in the material. The rate of permeation will vary depending on these several conditions.

Personal Protective Equipment. The equipment provided to shield or isolate a person from the chemical, physical, and thermal hazards that may be encountered at a hazardous materials incident. Adequate personal protective equipment should protect the respiratory system, skin, eyes, face, hands, feet, head, body, and hearing. Personal protective equipment includes both personal protective clothing and respiratory protection.

The purpose of personal protective equipment is to protect the responder against every anticipated hazard. Protective equipment has improved dramatically and substantially in recent years, but the nature of the hazards to which emergency personnel must respond have also become exceedingly complex. Protective equipment will not prevent injuries or illnesses if improper selection is made, if the equipment is not used, or if it is not used properly. Federal regulations (OSHA and EPA) require the use of personal protective equipment.

It is important to stress repeatedly that no combination of protective equipment will protect the responder against every conceivable hazard. Also, the equipment itself can cause injuries if it is not used appropriately and intelligently. The more sophisticated the equipment, the greater is the need for training and experience.

Protective Clothing. Equipment designed to protect the wearer from heat and/or hazardous materials contacting the skin or eyes. Protective clothing is divided into three types:

(a) structural fire fighting protective clothing;

NFPA 1971 (1986), *Standard on Protective Clothing for Structural Fire Fighting*, defines structural fire fighting as the activities of rescue, fire suppression, and property conservation in buildings, enclosed structures, vehicles, vessels, or like properties that are involved in a fire or emergency situation. Protective clothing is defined as the protective garments configured as a coat and trousers, or as a coverall, and designed to provide protection to the fire fighter's body.

(b) chemical protective clothing; and

Chemical protective clothing is designed and intended to provide protection, generally, against chemical exposure. There are many types of chemical protective clothing available, including those intended to protect the full body, the head, eyes and face, ears, hands and arms, and the feet. Table 1-1 (reprinted from NIOSH/OSHA/USCG/EPA Occupational Safety and Health Guidance Manual for Hazardous Waste Site Activities) will illustrate the range of protective clothing.

Clothing appropriate for an incident is greatly dependent on the types of hazards that are present.

(c) high temperature protective clothing.

It should be pointed out that there is no NFPA standard on high temperature protective clothing.

Qualified.   Having satisfactorily completed the requirements of the objectives.

Respiratory Protection.   Equipment designed to protect the wearer from the inhalation of contaminants. Respiratory protection is divided into three types:

(a) positive pressure self-contained breathing apparatus;

(b) positive pressure self-contained air respirators; and

(c) air purifying respirators.

Inhalation of toxics is one of the principal causes of serious injury to responders, so respiratory protection is of the utmost importance.

A positive pressure self-contained breathing apparatus (SCBA) is one in which the pressure inside the facepiece, in relation to the immediate environment, is positive during both inhalation and exhalation when tested in accordance with 30CFR, Part II, Subpart H by NIOSH and using NIOSH test equipment. (SCBA is a respirator worn by the user that supplies a respirable atmosphere that is either carried in or generated by the apparatus and is independent of the ambient environment.) See NFPA 1981 (1987) for additional information on the subject.

A positive pressure self-contained air respirator is one that supplies air in positive pressure, during both inhalation and exhalation, from a source located at some distance from the user and connected to the user by an air-line hose.

An air purifying respirator passes ambient air through a filtering device prior to inhalation by the user. They do not have a separate air supply. These can be of the demand or negative pressure type

Table 1-1 Protective Clothing and Accessories

BODY PART PROTECTED	TYPE OF CLOTHING OR ACCESSORY	DESCRIPTION	TYPE OF PROTECTION	USE CONSIDERATIONS
Full Body	Fully-encapsulating suit	One-piece garment. Boots and gloves may be integral, attached and replaceable, or separate.	Protects against splashes, dust, gases, and vapors.	Does not allow body heat to escape. May contribute to heat stress in wearer, particularly if worn in conjunction with a closed-circuit SCBA; a cooling garment may be needed. Impairs worker mobility, vision, and communication
	Non-encapsulating suit	Jacket, hood, pants, or bib overalls, and one-piece coveralls.	Protects against splashes, dust, and other materials but not against gases and vapors. Does not protect parts of head and neck.	Do not use where gas-tight or pervasive splashing protection is required. May contribute to heat stress in wearer. Tape-seal connections between pant cuffs and boots and between gloves and sleeves.
	Aprons, leggings, and sleeve protectors	Fully sleeved and gloved apron. Separate coverings for arms and legs. Commonly worn over non-encapsulating suit.	Provides additional splash protection of chest, forearms, and legs.	Whenever possible, should be used over a non-encapsulating suit (instead of using a fully-encapsulating suit) to minimize potential for heat stress. Useful for sampling, labeling, and analysis operations. Should be used only when there is a low probability of total body contact with contaminants.
	Firefighters' protective clothing	Gloves, helmet, running or bunker coat, running or bunker pants (NFPA No. 1971, 1972, 1973), and boots.	Protects against heat, hot water, and some particles. Does not protect against gases and vapors, or chemical permeation or degradation. NFPA Standard No. 1971 specifies that a garment consist of an outer shell, an inner liner, and a vapor barrier with a minimum water penetration of 25 lbs/in^2 (1.8 kg/cm^2) to prevent the …	Decontamination is difficult. Should not be worn in areas where protection against gases, vapors, chemical splashes, or permeation is required.

Table 1-1   Protective Clothing and Accessories (cont.)

BODY PART PROTECTED	TYPE OF CLOTHING OR ACCESSORY	DESCRIPTION	TYPE OF PROTECTION	USE CONSIDERATIONS
Full Body (cont.)	Proximity garment (approach suit)	One- or two-piece overgarment with boot covers, gloves, and hood of aluminized nylon or cotton fabric. Normally worn over other protective clothing, such as chemical-protective clothing, firefighters' bunker gear, or flame-retardant coveralls.	Protects against brief exposure to radiant heat. Does not protect against chemical permeation or degradation. Can be custom-manufactured to protect against some chemical contaminants.	Auxiliary cooling and an SCBA should be used if the wearer may be exposed to a toxic atmosphere or needs more than 2 or 3 minutes of protection.
	Blast and fragmentation suit	Blast and fragmentation vests and clothing, bomb blankets, and bomb carriers.	Provides some protection against very small detonations. Bomb blankets and baskets can help redirect a blast.	Does not provide hearing protection.
	Radiation-contamination protective suit	Various types of protective clothing designed to prevent contamination of the body by radioactive particles.	Protects against alpha and beta particles. *Does NOT protect against gamma radiation.*	Designed to prevent skin contamination. If radiation is detected on site, consult an experienced radiation expert and evacuate personnel until the radiation hazard has been evaluated.
	Flame/fire retardant coveralls	Normally worn as an undergarment.	Provides protection from flash fires.	Adds bulk and may exacerbate heat stress problems and impair mobility.
	Flotation gear	Life jackets or work vests. (Commonly worn underneath chemical protective clothing to prevent flotation gear degradation by chemicals.)	Adds 15.5 to 25 lbs (7 to 11.3 kg) of buoyancy to personnel working in or around water.	Adds bulk and restricts mobility. Must meet USCG standards (46 CFR Part 160).

Table 1-1 Protective Clothing and Accessories (cont.)

BODY PART PROTECTED	TYPE OF CLOTHING OR ACCESSORY	DESCRIPTION	TYPE OF PROTECTION	USE CONSIDERATIONS
Full Body (cont.)	Cooling garment	One of three methods: (1) A pump circulates cool dry air throughout the suit or portions of it via an air line. Cooling may be enhanced by use of a vortex cooler, refrigeration coils, or a heat exchanger. (2) A jacket or vest having pockets into which packets of ice are inserted. (3) A pump circulates chilled water from a water/ice reservoir and through circulating tubes, which cover part of the body (generally the upper torso only).	Removes excess heat generated by worker activity, the equipment, or the environment.	(1) Pumps circulating cool air require 10 to 20 ft^3 (0.3 to 0.6 m^3) of respirable air per minute, so they are often uneconomical for use at a waste site. (2) Jackets or vests pose ice storage and recharge problems. (3) Pumps circulating chilled water pose ice storage problems. The pump and battery add bulk and weight.
Head	Safety helmet (hard hat)	For example, a hard plastic or rubber helmet.	Protects the head from blows.	Helmet shall meet OSHA standard 29 CFR Part 1910.135.
	Helmet liner		Insulates against cold. Does not protect against chemical splashes.	
	Hood	Commonly worn with a helmet.	Protects against chemical splashes, particulates, and rain.	
	Protective hair covering		Protects against chemical contamination of hair. Prevents the entanglement of hair in machinery or equipment. Prevents hair from interfering with vision and with the functioning of respiratory protective devices.	Particularly important for workers with long hair.
Eyes and Face[a]	Face shield	Full-face coverage, eight-inch minimum.	Protects against chemical splashes. Does not protect adequately against projectiles.	Face shields and splash hoods must be suitably supported to prevent them from shifting and exposing portions of the face or obscuring vision. Provides limited eye protection.

[a]All eye and face protection must meet OSHA standard 29 CFR Part 1910.133.

Table 1-1 Protective Clothing and Accessories (cont.)

BODY PART PROTECTED	TYPE OF CLOTHING OR ACCESSORY	DESCRIPTION	TYPE OF PROTECTION	USE CONSIDERATIONS
Eyes and Face (cont.)	Splash hood		Protects against chemical splashes. Does not protect adequately against projectiles.	
	Safety glasses		Protect eyes against large particles and projectiles.	If lasers are used to survey a site, workers should wear special protective lenses.
	Goggles		Depending on their construction, goggles can protect against vaporized chemicals, splashes, large particles, and projectiles (if constructed with impact-resistant lenses).	
	Sweat bands		Prevents sweat-induced eye irritation and vision impairment.	
Ears	Ear plugs and muffs		Protect against physiological damage and psychological disturbance.	Must comply with OSHA regulation 29 CFR Part 1910.95. Can interfere with communication. Use of ear plugs should be carefully reviewed by a health and safety professional because chemical contaminants could be introduced into the ear.
	Headphones	Radio headset with throat microphone.	Provide some hearing protection while enabling communication.	Highly desirable, particularly if emergency conditions arise.

Table 1-1   Protective Clothing and Accessories (cont.)

BODY PART PROTECTED	TYPE OF CLOTHING OR ACCESSORY	DESCRIPTION	TYPE OF PROTECTION	USE CONSIDERATIONS
Hands and Arms	Gloves and sleeves	May be integral, attached, or separate from other protective clothing.	Protect hands and arms from chemical contact.	Wear jacket cuffs over glove cuffs to prevent liquid from entering the glove. Tape-seal gloves to sleeves to provide additional protection.
		Overgloves.	Provide supplemental protection to the wearer and protect more expensive undergarments from abrasions, tears, and contamination.	
		Disposable gloves.	Should be used whenever possible to reduce decontamination needs.	
Foot	Safety boots	Boots constructed of chemical-resistant material.	Protect feet from contact with chemicals.	All boots must at least meet the specifications required under OSHA 29 CFR Part 1910.136 and should provide good traction.
		Boots constructed with some steel materials (e.g., toes, shanks, insoles).	Protect feet from compression, crushing, or puncture by falling, moving, or sharp objects.	
		Boots constructed from nonconductive, spark-resistant materials or coatings.	Protect the wearer against electrical hazards and prevent ignition of combustible gases or vapors.	
	Disposable shoe or boot covers	Made of a variety of materials. Slip over the shoe or boot.	Protect safety boots from contamination. Protect feet from contact with chemicals.	Covers may be disposed of after use, facilitating decontamination.
General	Knife		Allows a person in a fully-encapsulating suit to cut his or her way out of the suit in the event of an emergency or equipment failure.	Should be carried and used with caution to avoid puncturing the suit.

Table 1-1 Protective Clothing and Accessories (cont.)

BODY PART PROTECTED	TYPE OF CLOTHING OR ACCESSORY	DESCRIPTION	TYPE OF PROTECTION	USE CONSIDERATIONS
General (cont.)	Flashlight or lantern		Enhances visibility in buildings, enclosed spaces, and the dark.	Must be intrinsically safe or explosion-proof for use in combustible atmospheres. Sealing the flashlight in a plastic bag facilitates decontamination. Only electrical equipment approved as intrinsically safe, or approved for the class and group of hazard as defined in Article 500 of the National Electrical Code, may be used.
	Personal dosimeter		Measures worker exposure to ionizing radiation and to certain chemicals.	To estimate actual body exposure, the dosimeter should be placed inside the fully-encapsulating suit.
	Personal locator beacon	Operated by sound, radio, or light.	Enables emergency personnel to locate victim.	
	Two-way radio		Enables field workers to communicate with personnel in the Support Zone.	
	Safety belts, harnesses, and lifeline		Enable personnel to work in elevated areas or enter confined areas and prevent falls. Belts may be used to carry tools and equipment.	Must be constructed of spark-free hardware and chemical-resistant materials to provide proper protection. Must meet OSHA standards in 29 CFR Part 1926.104.

that depends on the user's inhalation to bring air into the face-piece, or powered air purifying respirators that deliver a continuous flow of air into the facepiece in a positive pressure configuration. With the latter type, a negative pressure can be created within the facepiece if the user is at a maximal breathing rate.

Response.   That portion of incident management in which personnel are involved in controlling a hazardous materials incident.

Safely.   To perform the objective without injury to self or others, property, or the environment.

Shall.   Indicates a mandatory requirement.

Should.   Indicates a recommendation or that which is advised but not required.

Stabilization.   The period of an incident where the adverse behavior of the hazardous material is controlled.

Stabilization is roughly equivalent to the "under control" terminology associated with fire fighting. It does not mean that the incident is over, but rather that there should be no further escalation or intensifying of the hazardous incident. It could well be that operations following stabilization can indeed be prolonged, and that specialized types of equipment and expertise will be needed before the operations are completed.

Termination.   That portion of incident management in which personnel are involved in documenting safety procedures, site operations, hazards faced, and lessons learned from the incident. Termination is divided into three phases: debriefing the incident, post-incident analysis, and critiquing the incident.

This definition differs from what one would expect it to mean at a fire incident, where termination usually is marked by the departure of the fire department from the fire scene.

Debriefing the incident involves the collection of all pertinent information relating to the nature of the incident itself, and all emergency actions and operations that transpired in achieving mitigation and control. Post incident analysis is concerned with evaluating the collected information. Critiquing the incident brings the operation to a fitting and appropriate conclusion, and is important as a learning mechanism. It affords all responders with valuable "lessons learned" so that future operations at hazardous materials incidents will be enhanced.

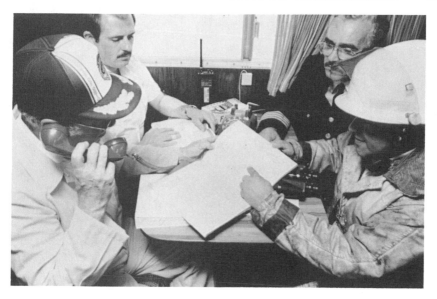

Figure 1.7   Post incident analysis includes debriefing, critiquing, and lessons learned.

Understanding.   The process of gaining or developing the meaning of various types of materials or knowledge.

Warm Zone.   The area where personnel and equipment decontamination and hot zone support take place. It includes control points for the access corridor and thus assists in reducing the spread of contamination. This is also referred to as the decontamination, contamination reduction, or limited access zone in other documents.

One of the purposes of the warm zone is to reduce the likelihood of contaminating the cold zone. To some extent, the warm zone serves as a buffer area between the hot and cold zones. The intensity of contamination in the warm zone should decrease the closer one approaches the cold zone, since decontamination procedures are taking place, and because of the intervening space that the warm zone affords between the other two. An access corridor refers to a defined path between the hot and cold zones where decontamination of personnel and equipment takes place. There may be a need for several access corridors at very large incidents. The access corridors must be tightly controlled and supervised so that movement between zones is regulated. Persons entering the warm zone from the cold zone must be wearing appropriate personal protective equipment.

2 First Responder

2-1 General. First responders are divided into two levels of competency: first responder awareness and first responder operational. First responders at the awareness level shall be trained to meet all of the requirements of Section 2-2 of this chapter and first responders at the operational level shall be trained to meet all of the requirements of Section 2-2 and Section 2-3 of this chapter. All first responders shall receive training to meet federal Occupational Safety and Health Administration (OSHA) or Environmental Protection Agency (EPA) requirements, whichever is appropriate for their jurisdiction.

When the Committee first began its work on levels of response, three major divisions were considered. It soon became apparent, however, that the primary or first level dealt with a very large population of responders whose anticipated competencies varied substantially. Qualifications ran the gamut from the very basic that would require relatively little training to quite complex abilities that one would only expect to find among organized groups of emergency personnel already possessing response training and experience.

The logical approach to resolving the dilemma was to divide the first response level in two: awareness and operational. The titles chosen to designate the two component parts suggest a good deal of what might be expected of each.

It would seem safe to conclude, then, that most people who find themselves involved as participants at hazardous materials incidents will fall into the first responder category.

It may seem unnecessary to make the following point, but it is important: persons who are in the first responder awareness category should know that they hold that designation. For example, a truck driver who routinely transports a hazardous material should be trained to the awareness level. It is not intended that one find oneself operating at the first responder awareness level only by accident.

It would seem logical that every police officer should be trained to at least the awareness level. Similarly, state police and highway

31

patrol officers should likely possess operational level abilities. Fire department personnel should also be at the operational level, since they are also likely to be summoned to most hazardous materials incidents.

It is important to point out that the lines separating the levels of response are not always clear and precise. This fact should not deter anyone from seeking as much training as possible so that they can operate safely and efficiently.

Federal regulations (OSHA and EPA) are far-reaching in terms of the populations that are affected. Every emergency response person in the country is covered in some way, be they fire service, police, EMS, or civil defense. All civilians who work with hazardous materials are also involved by virtue of the federal regulations.

2-2 First Responder Awareness Level.

2-2.1 Goal. The goal at the first responder awareness level shall be to provide those persons, who in the course of their normal duties may be the first on the scene of a hazardous materials incident, with the following competencies to respond in a safe manner when confronted with a hazardous materials incident. These personnel are not expected to take any actions other than to recognize that a hazard exists, call for trained personnel, and secure the area.

It is important to compare the language in the goal statements for awareness level and for operational level. There are significant differences. The awareness level responder *may* be first on the scene because of his normal duties. This could suggest a driver of hazardous material cargo, a train crew, or a local police officer. The operational level responder's duties *include* responding to the scene of emergencies. This category includes the fire service, EMS, and state police, among others. Further, the anticipated actions of the two categories of responders differ. The awareness level is not expected to take actions that would require a great deal of training and experience. Instead, the actions are basic and limited.

(a) An understanding of what hazardous materials are, and the risks associated with them in an incident;

This wording suggests a general awareness that there are such things as hazardous materials, that they can be dangerous, and that they do substantial harm when things go wrong. There is an implication that caution and care must be exercised in order to avoid making the situation even worse than it is.

(b) An understanding of the potential outcomes associated with an emergency created when hazardous materials are present;

Where one is dealing with a limited number of known substances, this competency can be relatively easy to acquire. Some awareness level responders might enjoy the luxury of such a situation. However, given the total array of substances available, the clear implication of this competency to a first on the scene police officer, for example, is to exercise extreme caution.

(c) The ability to recognize the presence of hazardous materials in an emergency;

This competency speaks to very basic recognition and identification. Indications of the presence of a hazardous material would include placards, markings and labels, the shape of containers, shipping papers, the occupancy involved, and the like. There is a clear indication that training in recognition and identification is a given, and that there are responsibilities and obligations that must be assumed by the employer, whether that employer is in the private or public sector.

(d) The ability to identify the hazardous materials and determine basic hazard and response information;

In addition to recognition and identification of the presence of a hazardous material, this subsection requires an ability to go a step further and begin deciding on an appropriate first course of action, dependent on an assessment of the basic hazard.

Figure 2.1 Improperly stored waste may lack accurate means of identification, making determination of their contents difficult.

(e) An understanding of the role of the first responder on the scene of a hazardous materials incident as identified in the local contingency plan for hazardous materials incidents;

There is a federal National Contingency Plan (NCP) that was created by a Presidential Executive Order in 1987. It provides for the establishment of a National Response Team (NRT), composed of representatives of several federal departments and agencies, which is responsible for national planning and coordination of response actions. The NRT is also responsible for the establishment of Regional Response Teams (RRTs), and these latter groups include representatives from state and local governments. Within each state there are regional or local contingency plans for responding to hazardous materials incidents.

The burden for implementation of local plans, and for the training of first responders, falls to those responsible for operation of the local contingency plan. This is especially true for training of persons at the awareness level. Response personnel at the operational level have generally been exposed to an organized training program that has prepared them for response to emergencies of various kinds.

(f) The ability to recognize the need for additional resources and make appropriate notifications; and

This is indeed an important competency for the awareness level responder. There are instances where a single notification to an appropriate emergency agency will determine the outcome of an incident. Even if no other action is taken beyond a prompt and appropriate call for assistance (and there may be situations where this is all that can be done), the emergency response plan will have been set in motion.

(g) The ability to initiate scene management (i.e., implement the Incident Command System, isolate the immediate site, deny entry to unauthorized persons, and evacuate).

It is standard procedure that the first emergency responder at the scene of an incident is in command until the arrival of a responder with higher authority. By virtue of first arrival, an awareness level responder may well be in charge. When he initiates any action, he is, in effect, implementing the basis of an incident command system. Important fundamental and immediate actions can keep the incident from causing unnecessary harm to those nearby. Keeping people away from the near environs of the incident accomplishes isolation and, to a lesser degree, evacuation.

2-2.2 Safety.   The first responder at the awareness level shall be capable of the following.

Safety must become a philosophy of emergency response life and an overriding consideration in all decision making. This is especially true in the world of hazardous materials response. There will be instances where a conscious decision to take no action, based on concerns for safety, will indeed be the most appropriate course to follow. However, decision making on the basis of safety will not always be simple and clear cut, and this is especially so in emergency operations. Time is not generally one of the luxuries afforded to the responder. Nonetheless, every emergency responder must incorporate the philosophy of safety in all of their actions.

2-2.2.1   Describe how hazardous materials incidents are different from other emergencies.

The adverse consequences of exposure to a hazardous material can be far reaching and severe. It is probably a truism for the responder that every emergency is different, or that no two are the same. However, the category of "hazardous materials emergency" stands aside from the rest in that its potential for doing harm is so great, and the responder must be specially trained and equipped to perform properly.

Figure 2.2   What may seem like a simple truck fire could become a
hazardous materials incident.

2-2.2.2*   Describe at least six ways hazardous materials are harmful to people at hazardous materials incidents.

A-2-2.2.2   The six ways are: thermal, radioactive, asphyxiation, chemical, etiologic, and mechanical. There may also be psychological harm.

The ways listed are not always a threat at every hazardous materials incident. The responder may be exposed to only one of the threats at a given emergency, or there may be a combination of several. Psychological harm might be considered as an ever-present phenomenon for the responder.

An etiologic agent is a living microorganism that may cause human disease. The other harmful mechanisms need no defining.

2-2.2.3*   Describe the general routes of entry for human exposure to hazardous materials.

A-2-2.2.3   These are: contact, absorption, inhalation, and ingestion. Absorption includes entry through the eyes and through punctures.

• *Contact* – A corrosive material will cause damage upon contact with skin or body tissue. Acids and alkalis can cause severe burns. If the skin is broken or if there is an open wound, another entry route exists in the presence of a hazardous material.

• *Absorption* – This is a process whereby one substance penetrates the inner structure of another. Hydrogen cyanide is an example of a substance that can be absorbed through the skin with fatal results.

• *Inhalation* – There are many substances that will cause severe damage if inhaled. Chlorine is an example of such a substance.

• *Ingestion* – Toxic substances can be present in drinking water and food.

The importance of proper personal protective equipment becomes obvious.

2-2.2.4   Describe the limitations of street clothes or work uniforms at the scene of hazardous materials incidents.

Given the six ways that hazardous materials can be harmful and given the several routes of entry for human exposure, it becomes patently obvious that ordinary clothing will not provide protection against hazardous materials. Specialized work uniforms, such as the station work uniform worn by some fire fighters, is also neither designed nor intended to protect the wearer in the presence of a hazardous material.

2-2.2.5 Describe the threats posed to property and the environment by hazardous materials releases.

Considering the major classes of hazardous materials (e.g., explosives, poisons, radioactive materials, and the like), and in light of the vulnerability of property and the environment when exposed to an incident, it is an easy task to imagine the countless possibilities for harm.

2-2.2.6* Describe the precautions necessary when rendering emergency medical care to victims of hazardous materials incidents.

A-2-2.2.6 These precautions are intended to apply to the responder as well as to the victim.

It is not intended that the awareness level responder go into substantial detail on this matter. It suffices to know that hazards exist for both the victim and the responder. The victim may well be contaminated, and decontamination measures must be considered. NFPA's Standards Council has approved the development of a document that will deal with emergency medical response to hazardous materials incidents. This resulted from instances where ambulances and even operating rooms became unnecessarily contaminated.

It should not be assumed that contamination is the only threat. The exposures to both victim and responder can be both numerous and severe, limited only by the specific circumstances involved in the incident.

One should not overlook the existence of the problem of communicable disease transmission when medical care is being provided.

2-2.2.7 Identify typical ignition sources found at the scene of hazardous materials incidents.

Sources of ignition could include all open flames, smoking, cutting and welding, heated surfaces, frictional heat, spontaneous ignition including heat-producing chemical reactions or the presence of pyrophoric materials, radiant heat, and static, electrical, and mechanical sparks. Lightning should not be overlooked as a source of ignition. Some of these sources of ignition can be controlled or eliminated by the responder, and some cannot. If flammable vapor is present, measures should be taken to reduce or eliminate the continued evolution of those vapors, or to disperse them into the atmosphere.

2-2.3 Resources and Planning. The person with a first responder awareness level shall be capable of the following.

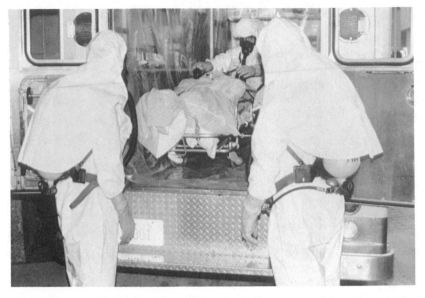

Figure 2.3   Ambulances should be specially prepared for transporting contaminated patients. Otherwise expensive medical equipment that becomes contaminated may have to be disposed of.

Resources will vary greatly from one area to another, and planning will have to be based in part on the availability of resources.

2-2.3.1   Describe the local procedures for requesting additional resources for dealing with hazardous materials incidents.

This may take no more than a single notification to an emergency dispatch center, where the local response plan will then be initiated.

2-2.3.2   Describe the role of the first responder at the scene of a hazardous materials incident, as identified in the local contingency plan for hazardous materials incidents.

See commentary following 2-2.1(e).

2-2.4 Incident Management.   The first responder at the awareness level shall be capable of the following.

2-2.4.1   Describe the purpose, need, and benefits of an Incident Command System at the scene of a hazardous materials incident.

Successful operations at an incident depend on proper use of the available resources at hand. An incident command system is

Figure 2.4   An incident command system enables the incident commander to manage and direct operations.

intended to enable the person in command to manage both personnel and equipment in an effective and efficient manner. The commander is burdened with the responsibility for overall control of operations at every incident, including strategy, tactics, and assignments. A major incident will involve a fairly complex command structure.

NFPA's Technical Committee on Fire Service Occupational Safety and Health is in the process of developing a standard for an incident management system that will be of use at hazardous materials incidents as well as at fire operations. The intent of the document, and indeed of any incident command system, will be to bring coordination and structure to the management of an operation. While such a system provides for effectiveness, it also assures safer operations.

2-2.4.2   Describe the process for implementing the Incident Command System at hazardous materials emergencies.

One person must assume responsibility for command at the very outset of operations. The person may be at the first responder awareness level. The command is transferred, perhaps several times at an incident, as additional resources arrive.

The system must be a structured one in which procedures for the transfer of command are clearly indicated. Involvement at the first responder awareness level will usually be minimal and of short duration.

2-2.4.3 Describe the basic techniques used to deny site entry.

Early site control is vital at an incident. In reporting an incident, or in summoning assistance, the awareness level responder should indicate the need for site control measures, if such are needed. The purpose is to minimize unnecessary contamination and to protect people from the harmful consequences of exposure. Whatever physical devices are at hand can be used, at least temporarily. These could include hazard tape, rope, warning cones, and the like.

2-2.4.4 Describe the basic techniques used to isolate the immediate site.

Even the awareness level responder must make an initial assessment of the situation at the scene of an incident. If a hazardous material is indeed involved, he must make this fact known to persons at the scene, as well as to incoming emergency responders. Subsequent actions may be guided by the information he or she provides. If the scene should be isolated, he should take whatever means are at hand to make that fact known to everyone in the vicinity.

2-2.4.5 Describe the basic techniques for evacuation in hazardous materials incidents.

The awareness level responder can indicate the need for evacuation in his notification to an emergency dispatch center or to incoming emergency response personnel. The actions already taken on denial of entry and on isolating the immediate area are the beginning steps in an evacuation process. Other means might be available, depending on circumstances pertaining to the incident, and the variables would be quite numerous. One is limited only by imagination.

2-2.5 Recognition of Hazardous Materials.   The first responder at the awareness level shall be capable of the following.

By being able to recognize hazardous materials, a responder can help to reduce the harmful consequences associated with an incident.

2-2.5.1 List the nine hazardous materials classes, describe the primary hazards of each class, and give examples of each class.

Class	Hazards	Examples
1. Explosives	Primary purpose is to function by explosion; overpressure or shock wave; fires; fragmentation	dynamite, TNT, display fireworks
2. Gases	Under considerable pressure at normal temperatures; may be flammable or corrosive; possibility of BLEVE	propane, chlorine, oxygen
3. Flammable Liquids	Ability to burn	gasoline, toluene
4. Flammable Solids	Ignites readily and burns vigorously	white phosphorus, metallic sodium
5. Oxidizers and Organic Peroxides	Readily releases oxygen, stimulates burning	sodium chlorate, nitric acid, sodium nitrate, benzoyl peroxide
6. Poisonous and Etiologic Materials	High degree of health hazard; disease causing	hydrogen cyanide, tetra-ethyl lead, rabies virus
7. Radioactive Materials	Radioactive exposure; contamination; radiation poisoning	enriched uranium, pluto-nium, radioactive waste
8. Corrosive Materials	Can destroy human tissue; some are oxidizers; some are toxic or unstable	sulfuric acid, sodium hydroxide
9. Other Regulated Materials (ORMs)	Possess hazardous characteristics in transport	chloroform, household cleaners

Figure 2.5   First responders need to know labels and placards associated with hazardous materials.

2-2.5.2* Use the six groups of clues to detect the presence of hazardous materials.

A-2-2.5.2 Six groups of clues for detecting the presence of hazardous materials are: occupancy and/or location, container shapes, markings and colors, placards and labels, shipping papers and material safety data sheets (MSDS), and senses.

(1) *Occupancy and/or location*: This applies chiefly to fixed sites and facilities. It could include manufacturing and processing plants, industrial operations, storage facilities, and railroad freight yards. While many occupancies and locations are obvious repositories of hazardous materials, others are more subtle. Local planning for response should be based in part on knowledge of the presence of hazardous materials at fixed locations.

(2) *Container shapes*: The configuration of some containers is so unusual that the presence of some hazardous materials can be determined. In many other cases, this is not so. Containers that provide clues include those that are used for radioactive materials, pressurized products, cryogenics, and corrosives, among others.

(3) *Container markings and colors*: This could include the four digit identification number, the NFPA 704 System of Identification, and military markings. Compressed gas cylinders are color coded to indicate contents.

(4) *Placards and labels*: In transportation, the placard would conform to the U.S. Department of Transportation regulations. At fixed facilities, one might find the NFPA 704 label being used on storage containers.

(5) *Shipping papers and material safety data sheets*: Shipping papers would provide information on the shipping name of the product, the identification number, the hazard class, and the quantity. A material safety data sheet provides information on the manufacturer, the hazardous ingredients, physical data, fire and explosion hazard, health and reactivity data, spill procedures, special protection, and special precautions.

(6) *Senses*: Colors and placards can be seen from a substantial distance, and sight should be used to advantage. Listening to the change in pitch or sound of an escaping pressurized gas can serve as a warning of container failure. All of our senses, with good common sense thrown in, should be used to assist us in detecting the presence of hazards.

2-2.5.3 Identify typical locations in the community or facility where hazardous materials are manufactured, transported, stored, used, or disposed of.

This will vary from one community to another, but there should be little difficulty in locating the more obvious fixed hazards. Most communities have service stations, school laboratories, waste disposal sites, medical facilities, and storage warehouses. Some communities have other sources of hazardous materials that are more obvious. Given the presence of hazardous materials at fixed sites, one can determine transportation routes to and from the sites. Many communities also contain major roadways or rail systems that can be used for passage through the community.

2-2.5.4 Describe placards, labels, container markings, and shipping papers used in the transportation of hazardous materials and explain their advantages and limitations in recognizing hazardous materials.

Advantages of placards, labels, container markings, and shipping papers are obvious. The system provides a national standard. The use of these devices assists in the ready identification of hazardous materials. Among the disadvantages or limitations are included the following:

• The Dangerous placard can be misleading in that it provides no specific information about the hazardous material. It would be possible to have 999 pounds total weight of materials such as flammable liquids, oxidizers, and flammable solids packed together with only a Dangerous placard.

• For many hazardous materials, quantities under 1000 pounds do not require a placard, and 990 pounds can be as dangerous as 1000 pounds of the same material.

• Placards and labels cannot always be seen at the time of an accident. However, the background color, if visible, can be a clue to the hazards involved.

• Some materials are not classified as dangerous by DOT and will not have a placard or label. Yet their properties can be hazardous enough to cause problems if an accident occurs.

• Identification by class, instead of by specific chemical, means that the multiple hazard problem is not recognized except in cases where multiple labels are required. It should be remembered that most hazardous materials have multiple hazardous characteristics.

• Human error is a factor in the effectiveness of labeling and placarding. For this reason, caution must be exercised in every incident.

2-2.5.5 Identify the shipping papers found in various modes of transportation, the individuals responsible for the papers, and the location where carried and found during an incident.

Highway transportation will have a bill of lading or manufacturer's data sheets in the cab of the vehicle, and the truck driver is the responsible person. On railroads, there will be a waybill, consist, or wheel report with the conductor, who is responsible for them. Shipping will have a dangerous cargo manifest in the wheelhouse and the Captain or Master is responsible. Air shipments require an air bill with shippers certification for restricted articles. It will be found in the cockpit and the pilot is the responsible person.

2-2.5.6 Given various examples of containers and packaging, identify the containers and packages by name and give an example of the materials that may typically be found inside.

There are several excellent programs available on the subject of recognition and identification, including one developed by the National Fire Academy.

2-2.5.7 Describe the types of specialized marking systems found at fixed facilities (such as military, special hazard communication markings, and NFPA 704, *Standard System for the Identification of the Fire Hazards of Materials*).

See the commentary for 2-2.5.4.

2-2.6 Classification, Identification, and Verification.

The first responder at the awareness level shall be capable of the following.

Figure 2.6   Information identifying the materials should also be available from the shipping papers with the vehicle.

2-2.6.1* Define the following terms:

A-2-2.6.1 Hazardous materials (known as dangerous goods in Canada) are substances capable of creating harm to people, property, or the environment. Although certain materials are classified differently in the two countries, they are still hazardous and should be dealt with accordingly. (*See A-1-3.*)

(a) Hazardous materials.

See Appendix A-1-3 for a definition of hazardous materials.

(b) Dangerous goods (in Canada).

In Canada, the Transportation of Dangerous Goods Acts divide dangerous goods into nine classes according to the type of hazard involved. The classes are the same as found in the United Nations Classification System.

2-2.6.2 Identify the specific name of a hazardous material involved in an emergency, or at least classify the material by its primary hazard using container markings, placards, and labels, pesticide labeling, shipping papers, material safety data sheets (MSDS), or personal contacts.

Examples:

• *Container marking*: A container with a three blade propeller type label would indicate a radioactive material, Class 7.

• *Placard*: A placard with a number 4 at the bottom of the placard would mean that the contents are in the flammable solid class.

• *Label*: A label with a skull and crossbones indicates a poison, Class 6.

• *Pesticide labeling*: A label with the word "Danger," a skull and crossbones, and an EPA registration number at the bottom would indicate pesticidal contents.

• *Shipping papers*: A waybill on a railroad would contain a description of articles carried. It might, for example, indicate liquefied petroleum gas, along with the major class category, flammable gas.

• *Material safety data sheet (MSDS)*: An MSDS would provide the name of the chemical and the chemical family. The description would provide information that would help to classify the chemical by its primary hazard.

• *Personal contact*: Communicating with the person in responsible charge, such as the driver of a truck, would enable you to determine the specific name of the material involved.

2-2.6.3* Identify three sources for obtaining hazard response information about hazardous materials and describe the types of information provided in each.

A-2-2.6.3 Sources of specific information to help identify hazardous materials include: MSDS's, reference guidebooks, hazardous materials data bases, technical information centers, technical information specialists, and monitoring equipment.

2-2.6.4 Demonstrate the use of the DOT *Emergency Response Guidebook (ERG)* in assessing hazards and response actions and determining isolation and evacuation distances.

As an example, assume that a placard displays the number 2188. When that number is found in the ERG, the chemical indicated, in bold type and with an asterisk, is arsine. The guide number is 18. An examination of Guide 18 explains the hazards. The asterisk and bold type indicate the user should also consult the Table of Evacuation Distances in the ERG for information on isolation from spills or leaks.

2-2.6.5 Demonstrate the use of a Material Safety Data Sheet (MSDS) in obtaining hazard and response information and determining isolation and evacuation distances.

A sample MSDS will contain information on the following major categories: manufacturer's name, location, and chemical name and family; hazardous ingredients; physical data; fire and explosion hazard data; health hazard data; reactivity data; spill or leak procedures; special protection information; and special precautions.

2-2.6.6 Explain the difficulties encountered in identifying the specific name of hazardous materials and their hazard and response information in an emergency.

There can be many difficulties, including some of the following:

• The placard is misleading or does not provide adequate information, as with the Dangerous placard.

• The placard or label cannot be seen.

• The placard or label is missing or is not required.

- There are multiple hazards associated with the chemical, but only one is indicated.

- The container is improperly placarded.

- There is no responsible person available for consultation.

- There are many chemicals involved at the same incident.

2-2.7 Chemistry of Hazardous Materials.   Reserved.

2-2.8 Hazard and Risk Assessment.   The first responder at the awareness level shall be capable of the following.

2-2.8.1   Describe the risk associated with hazardous materials located and transported through the community or facility and their potential threat to people, property, or the environment.

In order to describe your community's risks, you must be in possession of basic information regarding fixed facilities and the hazardous materials transportaion situation that prevails. The potential threat posed by the material would be dependent on some variables, such as the degree of hazard of the chemicals, the quantities involved, location, resources available to handle emergencies, density of population, nearby occupancies, evacuation problems, and so on.

Figure 2.7   A threat to the community must be identified quickly.

2-2.9 Personal Protective Equipment.   Reserved.

2-2.10 Hazardous Materials Control.   Reserved.

2-2.11 Decontamination.   Reserved.

2-2.12 Termination Procedures.   Reserved.

2-2.13 Educational and Medical Requirements.   Reserved.

2-3 First Responder Operational Level.

The person at this level will possess all of the competencies required of the awareness level and all of the additional competencies that are indicated in this section.

2-3.1 Goal.   The goal at the first responder operational level shall be to provide those persons, whose duties include responding to the scene of emergencies that may involve hazardous materials, with the following competencies to respond safely to hazardous materials incidents. The first responder at the operational level is not expected to use specialized chemical-protective clothing or special control equipment.

This wording indicates clearly that the person at this level belongs to some emergency response service or organization. It can be public or private, including such groups as the fire service, EMS, police, special chemical response teams, industrial brigades, or regional response teams. A certain level of training, experience, and skill in dealing with emergencies is assumed. However, they are not in the category of hazardous materials response teams, since they are not expected to use special chemical-protective clothing.

(a)   The ability to make initial basic hazard and risk assessments;

This requirement is related somewhat to the requirement in 2-2.8.1 for the awareness level, where a community's risk assessment was addressed. However, it is not identical with that section. This competency would pertain to a risk assessment at the site of an incident or at a training exercise. Knowing how to do a basic community risk assessment will certainly assist in performing an on-site emergency or training exercise risk assessment.

(b)   The ability to determine when the personal protective equipment provided to the first responders by the authority having jurisdiction for use in their normal response activities is adequate for a particular hazardous materials incident, and the ability to use that equipment properly;

The ability to use assigned equipment properly suggests that repeated periodic training takes place. Since hazardous materials

Figure 2.8   Risk assessment should be started
before entering the scene.

are steadily increasing in frequency, it is imperative that effective
training programs are in place.

This section indicates that not every incident is of such severity
that specialized clothing or equipment is needed. The incident
may have been remediated prior to the arrival of the responder, or
it may in fact be of such a nature that special clothing is unnecessary. A small spill of home heating fuel in a nonthreatening
location might serve as an appropriate example.

(c) An understanding of basic hazardous materials terms;

The definition section in Chapter 1 contains some of these basic
definitions, and 2-3.7.1 lists additional terms.

(d) The ability to perform hazardous materials control operations within
the capabilities of the resources and personal protective equipment available;

It is conceivable that this response level can achieve complete
control if the incident is relatively minor. At more severe incidents, they would merely perform to the extent that their training
and equipment allowed, and would have already summoned appropriate assistance. Control operations include containment, extinguishment, and confinement.

(e) An understanding of decontamination procedures;

See Supplement VIII on Decontamination. Note that a key element is the prevention of contamination, and this is especially significant for the operational level responder. The first step in decontamination is the establishment of procedures that reduce the potential for contamination.

(f) The ability to perform basic record keeping tasks; and

These tasks are likely within the present ability of the operational responder and should not require a great deal of additional training. At a hazardous materials incident, however, there will be additional and different record keeping tasks.

(g) The ability to expand the Incident Command System.

A fundamental characteristic of any incident command system is that it is initiated as soon as the first response unit arrives. The person in command passes that command along when a responder in higher authority arrives to take control. Pending the arrival of that higher authority, the operational responder will be functioning in command and expanding the operation as needed.

2-3.2 Safety.   The first responder at the operational level shall be capable of the following.

2-3.2.1   Describe the importance of a buddy system in controlling hazardous materials incidents.

One of the purposes of a buddy system is to provide for the safety and accountability of all persons working at an incident. No person involved in operations at an incident is working alone; everyone is either part of a team or has a partner.

2-3.2.2   Identify the advantages and dangers of search and rescue missions at hazardous materials incidents.

This is a broadly stated competency and would require a response in generalities, since search and rescue operations can be so varied in nature and in complexity. If search and rescue is mandated by circumstances, such as indications that people are trapped or missing, then it becomes an obligatory task of responders to take whatever actions are neccesary to locate, rescue, or account for the victims. In performing this task they will also be expanding their size-up of the incident. The hazards associated with the operation are obvious. Rescuers will be exposing themselves to increased hazards and risks.

2-3.2.3   Identify the advantages and hazards associated with the rescue, extrication, and removal of a victim from a hazardous materials incident.

Figure 2.9 Always work in pairs, watching each other closely for signs of giddiness, fatigue, or other forms of abnormal behavior. Even full protective gear and self-contained breathing apparatus may be ineffective against some materials.

Extrication and removal of victims is similarly complex since there are so many complicating variables. There is frequently need for specialized equipment and expertise that would be beyond the normal capability of most operational level responders. If they simply recognize this fact and can identify the situations where special assistance and equipment is needed, then they would appear to be meeting this competency.

2-3.2.4 Describe the precautions to be taken to protect oneself when fighting fire involving hazardous materials.

At this response level, the precautions would depend on the type and severity of the incident, the anticipated actions of the responder, and on proper use of available protective equipment. Safety must be a consideration in every action that is taken at an incident. Proper and early recognition and identification of the material involved is certainly important; assessing the risk is equally vital.

2-3.2.5 Define BLEVE and describe what happens to the container when a BLEVE occurs and how a BLEVE can be prevented.

The acronym BLEVE means a boiling-liquid expanding vapor explosion, and the phenomenon is usually associated with storage of liquefied gases (though it can happen with other stored liquids as well). A BLEVE is a major container failure, into two or more pieces, at a time when the contained liquid is at a temperature well above its boiling point at normal atmospheric pressure. BLEVEs are characterized by rocketing container pieces, rapid formation of a vapor cloud, and a shock wave. A BLEVE can result from impact or from a fire exposure, either of which will weaken the metal in one area.

BLEVEs associated with fire as the cause can be prevented by cooling the container with water. However, this must be done within a few minutes of fire exposure, and there must be adequate water to cover the area of flame impingement so that there is water runoff at that point.

Consult NFPA's Fire Protection Handbook, sixteenth edition, and NFPA'S Liquefied Petroleum Gas Code Handbook for additional detailed information on BLEVE.

2-3.2.6 Describe when it may be prudent to pull back from a hazardous materials incident.

When circumstances at an incident are such that nothing can be done in the way of mitigation, and there is a likelihood that the situation is about to deteriorate, then there is no reason to remain in the immediate vicinity. If flames are impinging on an LPG vessel, for example, and there is no available water to cool the vessel, then it would be prudent to withdraw a safe distance. Many other examples can be used to illustrate this point.

2-3.2.7 Describe the hazards and precautions to be observed when approaching a hazardous materials incident.

Approach to an incident should be from upwind and uphill, whenever possible, and should be calculated and deliberate. Binoculars can assist in identification of the material involved. Monitoring equipment can help to assess the hazard and determine what protective equipment should be used.

2-3.3 Resources and Planning.   The first responder at the operational level shall be capable of the following.

Resources and planning are interdependent. Plans must be formulated in part on the basis of resources that can be brought to bear on the mitigation process. The amount of resources that might be needed for hazardous materials incidents depend in part on plans that have been developed to address potential hazardous incidents.

2-3.3.1 Describe the levels of hazardous materials incidents and levels of hazardous materials incident responder as identified in the local contingency plan.

This will differ from one state to another, and also from one local community to another. Many jurisdictions arrange levels of incidents in three categories (I, II, III) based on degree of severity, with Level III being the highest or most serious. It is frequently usual to arrange levels of response in a somewhat similar fashion, as in fact this standard attempts to do. Whether the response levels are I, II, and III or first responder, technician, and specialist, the same general principles are applicable. The local contingency plan should provide details on these subjects.

2-3.3.2 Describe the need for a hazardous materials response plan and describe the major elements of the plan.

A hazardous materials response plan helps to coordinate the efforts of the various agencies or groups that might become involved in operations. It also serves to establish standard operating procedures. Community planning can also help to prevent or lessen the likelihood of incidents.

There are many different formats used for response plans. One such deals with functions to be performed at operations, and its elements[1] are:

1. Initial notification of response agencies

2. Direction and control

3. Communications among responders

4. Warning systems and emergency public notification

5. Public information and community relations

6. Resource management

7. Health and medical

8. Response personnel safety

9. Personal protection of citizens

10. Fire and rescue

11. Law enforcement

---

[1]From the National Response Team's Hazardous Materials Emergency Planning Guide

12. Ongoing incident assessment

13. Human services

14. Public works

15. Others

2-3.3.3   Describe the importance of coordination between various agencies at the scene of hazardous materials incidents.

Simply stated, if there is a lack of coordination between responding agencies at the scene of an incident, the operation may not succeed, the public will not be well served, and safety of personnel will be jeopardized. The principle purpose of planning is to achieve coordination, and that planning is now a federal mandate.

2-3.3.4   Describe the importance of pre-emergency planning relating to specific sites.

Adequate planning is the most important element in preparing for an emergency at a specific site. It will help assure efficient and safe operations once an incident occurs, and can result in substantial saving of life and property. It requires that responding personnel become familiar with the site and its hazards, and that their operations are based on a sound assessment of the risks involved. It will also familiarize responders with available response routes and with resources at the scene.

2-3.4 Incident Management.   The first responder at the operational level shall be capable of the following.

2-3.4.1   Describe the elements of the Incident Command System to assure coordination of response activities at hazardous materials incidents.

The National Inter-Agency Incident Management System (NIIMS) lists the following components of their incident command system:

- Common terminology

- Modular organization

- Integrated communications

- Unified command structure

- Consolidated action plans

- Manageable span of control

- Predesignated incident facilities

- Comprehensive resource management

2-3.4.2 Given a simulated hazardous materials incident, demonstrate the following skills:

Since simulations can deal with incidents that vary from the simple to the complex, the degree of skill one must demonstrate may vary. Keep in mind, however, that the level of response with which we are dealing will be a major determinant. It is not expected that one would demonstrate knowledge or skill beyond what the response level dictates.

(a) Assume command

If the responder is alone and is first on the scene, command must be assumed. If it is a responding unit, such as a fire company, then the officer in charge of that unit assumes command. Decision-making belongs to the person in charge. One of the first actions taken may be to request additional assistance, if deemed necessary. Until a higher authority arrives, the command operation belongs to the first arriving response officer, and he must begin the process of incident mitigation.

(b) Establish scene control through control zones

The site control process must be put into place quickly if there is a threat of contamination or if contamination exists. Lines of demarcation between the hot zone and warm zone must be established, access to the site must be limited, and a cold zone must be marked off.

(c) Establish a command post.

Retain command responsibilities until duly relieved, or until you are able to terminate the incident with the resources you have on hand. Take effective charge of all activities that are taking place until a superior level of response arrives. This may require that a stationary position be taken so that effective supervision of the ongoing operation is possible.

2-3.4.3 Identify the criteria for determining the location of the control zones for a hazardous materials incident.

Lines of demarcation for the control zones are determined by sampling and monitoring results, an evaluation of the extent of contamination and the path it might take in case of a leak, and the space needed to achieve a successful operation.

2-3.4.4   Describe your organization's standard operating procedures relating to hazardous materials.

This competency is subjective and depends on the responder's local operation.

2-3.5 Recognition of Hazardous Materials.   The first responder at the operational level shall be capable of the following.

2-3.5.1   Given a pesticide label, identify and explain the significance of the following:

EPA is the federal agency that is charged with regulating pesticides. A pesticide is generally defined as any chemical or mixture of chemicals used to control or destroy any living organism that is considered to be a pest, such as some insects.

(a) Name of pesticide

The label will contain the name given the chemical by the manufacturer.

(b) Signal word

There are three signal words that can be used. DANGER is used for the most toxic chemicals, WARNING is less toxic, and CAUTION indicates a relatively minor degree of toxicity.

(c) EPA registration number

The label will contain an EPA registration number that can also be used to establish the identity of the chemical.

(d) Precautionary statement

There will be a caution statement indicating care that must be taken, such as "Keep Out of Reach of Children," "Restricted Use Pesticide," or "Hazard to Humans and Domestic Animals."

(e) Hazard statement

Typical wording might indicate an environmental hazard and would advise avoiding contamination of water supplies.

(f) Active ingredient.

The percentages of each active ingredient will be identified and indicated. Inert ingredients are also shown, but only by percentage.

2-3.6 Classification, Identification, and Verification.   The first responder at the operational level shall be capable of the following.

2-3.6.1 Describe the assistance provided by CHEMTREC, how one is to contact CHEMTREC, and what information the first responder should furnish CHEMTREC.

CHEMTREC stands for Chemical Transportation Emergency Center and is a public service of the Chemical Manufacturers Association. CHEMTREC provides immediate advice by telephone for the on-scene commander at an incident. It also contacts the involved shipper for detailed assistance and response follow-up. It notifies the National Response Center (NRC) of significant incidents, and can also bridge a caller to the NRC for spill reporting. CHEMTREC operates 24 hours a day at a toll-free number: 1-800-424-9300, and can take calls from throughout the United States and Canada. In the District of Columbia, the correct number is 483-7616, and Alaskans must call 0-202-483-7616. Emergency collect calls are accepted.

CHEMTREC can usually provide hazard information warnings and guidance when given the identification number (4 digit), the name of the product, and the nature of the problem. For more detailed information and assistance, or if the product is unknown, attempt to provide as much of the following information as possible:

• Name of caller and a call back number

• Guide number you are using

• Shipper or manufacturer

• Container type

• Rail car or truck number

• Carrier name

• Consignee

• Local conditions

At an incident, try to maintain a phone line connected to CHEMTREC so that they can provide guidance and assistance. CHEMTREC can provide a teleconferencing bridge that allows them to connect experts to your phone line as necessary.

2-3.6.2 Given a Material Safety Data Sheet (MSDS), select and interpret information that is useful in determining the hazards of the chemical.

The major information categories required by OSHA on the MSDS are as follows:

- Manufacturer's name, emergency telephone number, and address; chemical name and synonyms; trade name and synonyms; chemical family and formula

- Hazardous ingredients

- Physical data

- Fire and explosion hazard

- Health hazard data

- Reactivity data

- Spill or leak procedures

- Special protection information

- Special precautions

Your selection and interpretation of useful information would depend on the major hazards presented by the chemical in question.

2-3.7 Chemistry of Hazardous Materials.   The first responder at the operational level shall be capable of the following.

2-3.7.1   Define the following chemical and physical properties and describe their importance in the risk assessment process.

(a) Boiling point

Boiling point is the temperature at which the equilibrium vapor pressure of a liquid equals the total pressure on the surface; the temperature at which a liquid boils when under a total pressure of one atmosphere is the normal boiling point. The lower the boiling point of a flammable liquid, the greater the vapor pressure and, therefore, the greater the fire hazard potential.

(b) Flammable (explosive) limits

The term flammable limits is used interchangeably with explosive limits. The lower flammable limit (LFL) is the minimum concentration of vapor to air below which propagation of a flame will not occur in the presence of an ignition source. The upper flammable limit (UFL) is the maximum vapor to air concentration above which propagation of flame will not occur. If a vapor to air mixture is below the LFL, it is described as being "too lean" to burn; if it is above the UFL, it is "too rich" to burn. When the vapor to air ratio is somewhere between the LFL and UFL, fires and explosions can occur. The mixture is said to be in the flammable or explosive range.

If flammable vapors are suspected or known to be present, it is important to determine the concentration of the vapor in air. Combustible gas instruments are used for this purpose.

(c) Flash point

Flash point of a liquid is the minimum temperature at which vapor is given off in sufficient concentration to form an ignitible mixture with air. Flash point is used as the primary property or characteristic of a liquid to determine the relative degree of flammability. Since it is the vapors of flammable liquids that burn, vapor generation becomes a primary factor in determining the fire hazard.

(d) Ignition (autoignition) temperature

Ignition temperature and autoignition temperature are interchangeable terms. Ignition temperature of a substance, whether solid, liquid, or gas, is the minimum temperature required to initiate or cause self-sustained combustion in the absence of any source of ignition. Assigned ignition temperature should only be looked on as approximations.

Ignition temperatures can be quite high, especially in relation to a liquid's flash point. For example, the flash point of gasoline is −45 °F; its ignition temperature is well over 500 °F.

(e) Specific gravity

Specific gravity is the ratio of the weight of a volume of liquid or solid to the weight of an equal volume of water, with the gravity of water being 1.0. A substance with a specific gravity greater than 1.0 will sink in water; one with a specific gravity less than 1.0 will float in water.

This can be a factor in spill control and in fire fighting. Carbon disulfide is heavier than water, so water can be used to blanket and extinguish a fire in that chemical.

(f) Vapor density

Vapor density measures the weight of a given vapor as compared with an equal volume of air, with air as a value of 1.0. A vapor density greater than 1.0 indicates it is heavier than air; a value less than 1.0 indicates it is lighter.

Vapor density can be important to the responder, since it will determine the behavior of free vapor at the scene of a spill or gas release.

(g) Vapor pressure

Vapor pressure, measured in pounds per square inch (psi) absolute, is the pressure exerted by a volatile substance. It is a measure of a substance's tendency to emit or give off vapors. The higher the vapor pressure, the more volatile the substance, thus the more vapor given off. Vapor pressure increases with temperature.

A flammable substance with a high vapor pressure will be giving off substantial amounts of flammable vapor.

(h) Water solubility.

Water solubility provides information on the degree to which a substance is soluble in water. It can be useful in determining effective extinguishing agents and methods. Alcohol-resistant type foam, for example, is usually recommended for water soluble flammable liquids. Also, water soluble flammable liquids may be extinguished by dilution, although it may be impractical in many cases to do so.

2-3.7.2   Define the following terms:

(a) Alpha radiation

Alpha radiation involves the alpha particle, a positively charged particle that is emitted by some radioactive materials. It is less penetrating than beta and gamma radiation and is not considered to be dangerous unless ingested. If ingested, it will attack internal organs.

(b) Beta radiation

Beta radiation involves the beta particle, which is much smaller but more penetrating than the alpha particle. Beta particles can damage skin tissue, and they can damage internal organs if they enter the body. Full protective clothing, including self-contained breathing apparatus, will protect against most beta radiation.

(c) Gamma radiation.

Gamma radiation is especially harmful since it has great penetrating power. Gamma rays are a form of ionizing radiation with high energy that travels at the speed of light. It can cause skin burns and can severely injure internal body organs. Protective clothing is inadequate in preventing penetration of gamma radiation.

2-3.8 Hazard and Risk Assessment.   Reserved.

2-3.9 Personal Protective Equipment. The first responder at the operational level shall be capable of the following.

NOTE: The terms protective breathing apparatus and respiratory protection are used interchangeably in this document.

The term "respiratory protection" is all-encompassing when dealing with breathing apparatus; it includes protective breathing apparatus and is therefore the term of choice.

2-3.9.1* Identify the respiratory hazards encountered at hazardous materials incidents, and describe the need for proper protective breathing apparatus, as prescribed by OSHA.

A-2-3.9.1 "Employees engaged in emergency response and exposed to hazardous substances shall wear positive pressure self-contained breathing apparatus while engaged in emergency response until such time that the individual in charge of the ICS determines through the use of air monitoring that a decreased level of respiratory protection will not result in hazardous exposures to employees." (OSHA 1910.120, Subpart L, 4iiD.)

Hazardous materials can enter the body through inhalation and, depending on the degree of toxicity, can do extreme harm. The responder can inhale dangerous gases or vapors unless proper respiratory protection is used. Harmful substances that can be inhaled include those that are poisons, irritants, and asphyxiants, among others. Most injuries at incidents occur by inhalation of harmful substances.

2-3.9.2 Identify the physical requirements of the wearer of protective breathing apparatus.

NFPA 1500, *Standard on Fire Department Occupational Safety and Health Program*, requires that all members using SCBA be medically certified by a physician on an annual basis. It also requires that the SCBA users be regularly trained, tested, and certified in the safe and proper use of the equipment. The certifying physician should consult ANSI Z88.6, *Standard for Respiratory Protection — Respirator Use — Physical Qualifications for Personnel*, for guidance on the medical review that is appropriate.

2-3.9.3 Describe the limitations of personnel working with protective breathing apparatus.

There may be reduced visibility with breathing apparatus. SCBA can be somewhat cumbersome and can limit or reduce mobility and access in tight locations. Fatigue will occur more quickly than under normal working conditions. Some people experience varying adverse psychological consequences with respiratory equipment. Facial hair and long hair may interfere with respirator fit and with

wearer vision. Eyeglasses with conventional temple pieces will interfere with the seal of a full facepiece unless a spectacle kit is installed in that responder's facepiece. Gum and tobacco chewing should be prohibited since they may compromise the facepiece fit.

2-3.9.4 List the types of protective breathing apparatus and describe the advantages and limitations of each at hazardous materials incidents.

See Table 2-1. (Reprinted from NIOSH/OSHA/USCG/EPA *Occupational Safety and Health Guidance Manual for Hazardous Waste Site Activities*.)

2-3.9.5 Identify the procedure for cleaning and sanitizing protective breathing apparatus for future use.

Manufacturers and suppliers of the various types of respiratory equipment furnish guidelines and instructions for proper care, maintenance, cleaning, and sanitizing of their equipment. The instructions should be learned and followed.

2-3.9.6 Identify the operational components of the types of protective breathing apparatus provided by the authority having jurisdiction and explain their function.

Positive pressure SCBA consists of: a full facepiece, exhalation valve, breathing tube, regulator, air supply hose, cylinder, cylinder valve, and harness assembly.

Positive pressure supplied-air respirator consists of: a full facepiece, exhalation valve, breathing tube, regulator, belt for regulator, air supply hose, and an air supply.

Air purifying respirators consist of: a full facepiece, exhalation valve, and filter sorbent cartridge. A canister type would have a full facepiece, breathing tube, filter sorbent canister, and harness assembly. Some facepieces are half-mask.

2-3.9.7 Demonstrate the use of positive pressure air-supplied respiratory devices as provided by the authority having jurisdiction.

It is important to point out the advantages and limitations of such equipment during the demonstration. See Table 2-1.

2-3.9.8 Describe the need for specialized protective clothing used at hazardous materials incidents.

The need for and purpose of specialized protective clothing

Table 2-1

Type	Advantages	Limitations
1. Positive pressure self-contained breathing apparatus	• Provides highest level of protection against airborne contaminants and oxygen deficiency.	• Bulky, heavy. • Finite air supply limits work duration. • May impair movement in confined spaces.
2. Positive pressure supplied-air respirators	• Longer work periods than with SCBA. • Less bulky and heavy than SCBA. • Protects against airborne contaminants.	• Impairs mobility. • MSHA/NIOSH limits hose to 300 feet. • As length of hose is increased, minimum approved airflow may not be delivered at the facepiece. • Air line is vulnerable to damage, chemical contamination, and degradation. Decontamination of hoses may be difficult. • Worker must retrace steps to leave work area. • Requires supervision/ monitoring of the air supply line. • Not approved for use in atmospheres immediately dangerous to life or health (IDLH) or in oxygen deficient atmospheres unless equipped with an emergency egress unit such as an escape-only SCBA that can provide immediate emergency respiratory protection in case of air-line failure.
3. Air-purifying respirator [including powered air-purifying respirators (PAPRs)]	• Enhanced mobility. • Lighter in weight than an SCBA.	• Cannot be used in IDLH or oxygen deficient atmosphere (less than 19.5 percent oxygen at sea level). • Limited duration of protection. May be hard to guage safe operating time in field conditions. • Only protects against specific chemicals and up to specific concentrations. • Use requires monitoring of contaminants and oxygen levels. • Can only be used: 1) against gas and vapor contaminants with adequate warning properties, or 2) for specific gases or vapors provided that the service is known and a safety factor is applied or if the unit has an end-of-service-life indicator (ESLI).

(and equipment) is to shield or isolate responders from the chemical, physical, and biological hazards that may be encountered at an incident. Careful selection and use of adequate protective clothing should protect the skin, eyes, face, hands, feet, head, body, and hearing.

2-3.9.9 Describe the application, use, and limitations of the following levels of protective clothing used at hazardous materials incidents:

(a) Structural fire fighting clothing

Structural fire fighting clothing is designed to protect fire fighters against adverse environmental effects during structural fire fighting. Some of these effects could be high temperatures, steam, hot water, and hot particles. This protective clothing is not designed for chemical exposures and may provide only limited protection.

(b) Chemical-protective clothing (nonencapsulating, encapsulating)

Chemical-protective clothing is designed to protect the wearer's skin and eyes from direct chemical contact. It is made from materials that are compatible with specific chemicals and groups of chemicals. Most chemical-protective clothing will not provide thermal protection. No single material will provide protection against all chemicals, and mixtures of chemicals make the prob-

Figure 2.10 Specialized protective clothing should be chosen to suit the hazard encountered.

lem even more complex.

Nonencapsulating clothing is comprised of several components and is designed to provide protection against splashes. It does not offer total body protection from vapors or airborne dusts. Components include head protection, face and eye protection, body protection, hand protection, and foot protection.

Encapsulating clothing is a completely enclosed one piece unit that protects the entire body. The boots and gloves can be separate or an actual part of the garment. There is much stress placed on the wearer of encapsulating clothing. Also, respiratory equipment must be used.

(c) High temperature clothing.

High temperature clothing is designed to protect against short term high temperature exposures. It is of very limited use in chemical incidents.

2-3.9.10 Demonstrate the proper donning, doffing, and usage of all personal protective equipment provided to the first responder by the authority having jurisdiction for use in their normal response activities.

It should be noted that the operational level responder is not expected to use specialized chemical-protective clothing (see 2-3.1).

2-3.9.11 Describe the factors to be considered in selecting the proper respiratory protection at hazardous materials incidents.

Federal regulations require the use of respirator equipment that has been tested and approved by the Mine Safety and Health Administration (MSHA) and NIOSH. Approval numbers are shown on all approved equipment, and only approved equipment should be used.

Positive pressure SCBA can be used: if the atmosphere is or is likely to become IDLH, if the duration of air supply is adequate for completing the assigned task, and if its disadvantages (weight and bulk) will not interfere with performance.

Positive pressure supplied-air respirators can be used provided that: the hose will not impair mobility, the air line is not likely to be damaged or obstructed, air-borne contaminants will not enter the system, and other workers will not be impeded in their operations by the use of an air line.

There is available a combination SCBA/supplied air respirator that allows the user to operate in either mode. It allows for extended work periods in a contaminated area.

Air purifying respirators should not be the choice where the following conditions exist: oxygen deficiency, IDLH concentrations of specific substances, entry into unventilated or confined area where the exposure conditions have not been characterized, presence or potential presence of unidentified contaminants, contaminant concentrations are unknown or exceed designated maximum use concentrations, identified gases or vapors have inadequate warning properties and the sorbent service life is not known and the unit has no end-of-service-life indicator, or high relative humidity that may reduce the protection offered by the sorbent. (Source: Table 8-4, Occupational Safety and Health Guidance Manual for Hazardous Waste Site Activities, NIOSH/OSHA/USCG/EPA.)

2-3.10 Hazardous Materials Control.   The first responder at the operational level shall be capable of the following.

2-3.10.1   Describe the techniques for controlling hazardous materials releases available to the first responder.

Broad general control methods that might be available at this response level could include the following methods, where appropriate: absorption, dilution with water, and dams, dikes, diversions, and dispersion.

2-3.11 Decontamination.   The first responder at the operational level shall be capable of the following.

2-3.11.1   Describe the need for decontamination procedures at hazardous materials incidents.

Decontamination is critical to health and safety at hazardous materials incidents. It serves to protect responders from hazardous substances that may contaminate and eventually permeate the personal protective equipment and other equipment used at the incident. Transmission of harmful substances from one control zone to another is kept at a minimum, and it also serves to protect the environment.

2-3.11.2   Describe the ways that personnel, personal protective equipment, apparatus, tools, and equipment become contaminated and the importance and limitations of decontamination procedures.

Contamination occurs by virtue of contact with hazardous substances at an incident. The importance of decontamination has been pointed out in the commentary for 2-3.11.1. Decontamination methods vary in their effectiveness in removing different substances. The effectiveness must be somehow measured, and procedures should be changed as needed.

Figure 2.11   Some response teams have decontamination units, which are trailers equipped with showers and waste water holding tanks. Units may have positive pressure air, on-demand heated water, and other features.

2-3.11.3   Demonstrate the basic decontamination procedures, as defined by the authority having jurisdiction, for victims, personnel, personal protective equipment, tools, equipment, and apparatus at hazardous materials incidents.

Procedures may vary from one local authority to another. Some decontamination procedures include removal of the contaminant by rinsing, using pressurized air jets, scrubbing and scraping, use of steam jets, evaporation, and extraction. Protective coverings and coatings can be disposed of, as can deeply penetrated articles of clothing. Other decontamination procedures include chemical detoxification or disinfection and sterilization.

The decontamination site should be located in the warm zone at an incident. Factors that have to be considered include the properties of the contaminants, the amount, location, and containment of the contaminants, exposures to personnel, the potential for the substances to permeate or penetrate the equipment, the number and movement of personnel among the control zones, and methods that are available for protecting responders during the decontamination procedures.

2-3.12 Termination Procedures.   The first responder at the operational level shall be capable of the following.

2-3.12.1 Describe the importance of documentation for a hazardous materials incident including training records, exposure records, incident reports, and critique reports.

There are federal regulations requiring training for all responders, so record keeping for this function is not only desirable, but also necessary. Exposure records are equally important for the same reasons. The records will help to establish a medical data base for the responder. Incident reports are normally required in most jurisdictions, and they are important for more than just data gathering reasons. They provide a record of who responded and detail the specifics of the incident. Critiques should be conducted after every incident, and maintaining a record of that procedure will help to assure that the process is performed properly and that lessons will be learned so that future performance will be enhanced.

2-3.12.2 Demonstrate an ability to keep an activity log and exposure records for hazardous materials incidents.

This demonstration will vary from one jurisdiction to another, depending on the format of the activity log and exposure record.

2-3.13 Educational and Medical Requirements. Reserved.

3 Hazardous Materials Technician

3-1 General.

3-1.1 The hazardous materials technician shall meet all of the objectives indicated for the first responder in Chapter 2. In addition, that person shall meet the training and medical surveillance program requirements in accordance with federal Occupational Safety and Health Administration (OSHA) or U.S. Environmental Protection Agency (EPA) regulations.

This response level begins dealing specifically with the individual who is officially designated as a hazardous materials responder. The previous response level dealt with persons who "in the course of their normal duties may be the first on the scene" (awareness) or "whose duties include responding to the scene of emergencies" (operational). This is not to say that the technician's only duty or assignment is hazardous materials response. There could be many other responsibilities, both of a normal and emergency nature. However, when a hazardous materials incident takes place, the technician will most likely be summoned to the scene. It is for this reason that other requirements come into play. Federal regulations require medical surveillance along with specific additional training, the purpose of which is to safeguard and protect the responder from the dangers normally associated with hazardous materials.

3-1.2 Goal. The goal at the hazardous materials technician level shall be to provide the responders with the following competencies to respond safely to hazardous materials incidents:

Note again the emphasis that is put on safety. At each level, the goals state that the purpose is "to respond safely" to incidents.

(a) The ability to implement a safety plan;

Consult NFPA 1501, *Standard for Fire Department Safety Officer*, for information and guidelines. Section 3-7 of that document spells out responsibilities for incident scene safety. A safety plan would become a component of the incident command system and would also concern itself with a critique of the incident.

(b) The ability to classify, identify, and verify known and unknown materials by using basic monitoring equipment;

This is the first mention of monitoring equipment for the responder. It is worth noting that the standard does not specify what type of monitoring equipment should be available to the technician's level. In practice, this will certainly vary from one jurisdiction to another. The Technical Committee gave consideration to the matter of equipment, but decided it was not the intent of the document to mandate different types of equipment for the varying response levels. This is also the first mention of an ability to classify unknown materials by use of monitoring equipment.

(c) The ability to function within an assigned role in the Incident Command System;

Use of the words "an assigned role" implies that there is someone at the scene operating at a higher level of command who is assigning roles and responsibilities. In all likelihood, technician level responders would be assigned to the hazard sector (see definitions), since that assignment deals with mitigation of the incident.

(d) The ability to select and use at least Level B protection in addition to any other specialized personal protective equipment provided to the hazardous materials technician by the authority having jurisdiction;

Level B protection is used when the highest level of respiratory protection is needed, but where a lesser level of skin and eye protection is indicated. Level B is the minimum level recommended on initial site entries until the hazards have been identified and defined by monitoring, sampling, or other dependable methods of analysis. Protective equipment is then dependent on the findings.

Level B personal protective equipment includes the following: positive pressure SCBA, chemical resistant clothing including inner and outer gloves, chemical resistant protective footwear, and intrinsically safe two-way radio communications. There are other items of protective equipment that are optional, such as hard hat and outer boots.

(e) The ability to make hazard and risk assessments;

A hazard and risk assessment at this stage would involve identification of the hazard(s), including the location and quantity, a determination of the extent of the danger posed to people and the environment, and an analysis of the severity of the consequences. Hazard and risk assessments are part of pre-planning required of all local emergency planning groups, but they are also part of the size-up at an actual incident.

(f) The ability to perform advanced hazardous materials control operations within the capabilities of the resources and personal protective equipment available;

The first responder awareness level initiated basic on-scene management. The operation level responder performed control operations within his capabilities. The technician carries the control measures further and becomes involved in hands-on mitigation techniques commensurate with his training, expertise, and equipment.

(g) The ability to select and implement appropriate decontamination procedures;

The operation level is required to have a knowledge of decontamination procedures. The technician performs decontamination procedures. Refer to Supplement VIII for further information on decontamination.

(h) The ability to complete record keeping procedures; and

Record keeping will naturally vary from one jurisdiction to another, dependent on established protocol or standard operating procedures.

There may be standard forms and checklists that assist in the handling of an incident, and there will be forms and reports required in the completion and follow-up period.

Checklists can be used to assist the incident command process and can detail exposures of personnel. Forms can be helpful where an evacuation is necessary, and also in keeping track of responding agencies. There should also be an incident log so that a chronological history of the event can be detailed.

(i) The ability to understand basic chemical, biological, and radiological terms and their behavior.

Section 3-7 lists the terms that this level of response should understand.

3-2 Safety.   The hazardous materials technician shall be capable of the following.

3-2.1*   Describe the components of a site safety plan for a hazardous materials incident.

A-3-2.1   Refer to NIOSH/OSHA/USCG/EPA *Occupational Safety and Health Guidance Manual for Hazardous Waste Site Activities*, October 1985. Also consult 29 CFR 1910.120, *Hazardous Waste Operations and Emergency Response, Final Rule*, for site safety considerations.

The four agency guide points out that a site safety plan should be developed *before* site activities proceed. The plan's purpose is to establish procedures to protect the responder and the public from hazards. It should be pointed out that this guidance manual was developed with a fixed hazardous waste site in mind, and that emergency procedures at a hazardous materials incident may not have the benefit of a site-specific safety plan. Nonetheless, safety plan elements mentioned in the guide can be used to advantage at all incidents. The components of a site safety plan include the following:

1. Planning—this involves having an in-place protocol or standard operating procedure for handling types of incidents, and it should include roles and responsibilities, available equipment, evacuation procedures, and decontamination capabilities, among other things

2. Personnel—this element of the safety plan includes procedures for incident management and designation of a safety sector and a safety officer. It takes into account the potential need for assistance from other agencies and also for sophisticated medical team response

3. Communications—this includes both internal (on-site) and external (off-site)

4. Site mapping

5. Site security and control

6. Evacuation routes and procedures

7. Medical treatment

8. Documentation

The Guidance Manual provides a generic site safety plan developed by the U.S. Coast Guard that lists the following elements as part of its outline:

a. Site description

b. Entry objectives

c. On-site organization and coordination

d. On-site control

e. Hazard evaluation

f. Personal protective equipment

g. On-site work plans

h. Communication procedures

i. Decontamination procedures

j. Site safety and health plan

   1) Designated site safety officer

   2) Emergency medical care

   3) Environmental monitoring

   4) Emergency procedures

   5) Personal monitoring

3-2.2 Given a simulated hazardous materials incident, implement a safety plan.

   The requirement to implement a safety plan at the technician level is an indication and measure of the importance of safety at an incident. A plan cannot be implemented unless and until its elements are known thoroughly.

3-2.3 Identify the criteria to be considered for modifying (increasing or decreasing as conditions warrant) the evacuation areas and/or control zones set up by the first responder.

   There are many variables to be considered in determining safe distances for evacuation and control zones. Those same variables are factors in determining modifications to the originally established distances, and they can include the nature of the incident, the properties of the involved chemical, the size or extent of the release, weather conditions, topography, and the location of the incident. The determining factors present a multiple of combinations to be considered.

3-2.4 Given a specific hazardous material and the reference materials available to the hazardous materials technician, describe the symptoms of exposure to that hazardous material.

   Reference materials available to this level of responder will certainly vary from one jurisdiction to another, but there are some basic references that are commonly used and are helpful in determining symptoms that result from exposures to hazardous materials. These include U.S. DOT's *Emergency Response Guidebook*, which provides information on health hazards; NFPA 49, *Hazardous Chemicals Data*, which lists life hazards; and the NIOSH *Pocket Guide to Chemical Hazards*, which lists symptoms of exposure to specific chemicals.

3-2.5 Describe the symptoms of heat stress.

Heat stress or heat disorders generally fall into three categories: heat cramps, heat exhaustion, and heatstroke.

• Heat cramps – describes a condition that results from physical exertion in high temperatures, characterized by sudden development of severe cramps of the abdominal or skeletal muscles.

• Heat exhaustion – a syndrome resulting from exposure to excessive heat, characterized by prostration and varying degrees of circulatory collapse. The victim may be listless, apprehensive, semicomatose, or in some cases unconscious. The skin is usually ashen, cold, and wet, perspiration is profuse, and blood pressure lowered.

• Heatstroke – a profound disturbance of the heat regulating mechanism characterized by high fever and collapse and sometimes by convulsions, coma, and death. The onset is usually sudden and acute. The victim is flushed, and the skin hot and dry.

3-2.6 Describe the precautions to be observed and followed when responding to incidents for each of the hazardous materials classifications.

Some general precautions may be applied when responding to all hazardous materials incidents, regardless of the classification of the material involved. These would include the following:

• Always employ a cautious approach when a hazardous material is known or suspected to be involved. Do not approach too closely.

• Use every capability available to you to identify the hazard and to size-up the situation — from a safe distance. Make use of both on-site and off-site resources in achieving accurate identification of the involved material.

• Isolate the scene, limit or prohibit close approach or entry as needed, and evacuate if necessary.

• Call for all needed assistance as early as possible.

- Take necessary mitigating action, based on size-up and on the assurance that the operation can be conducted in a safe manner and without unnecessary exposure to responders or to the public.

These general precautions, if properly observed, will lead to a successful operation. The specific hazardous materials classifications represent groups of substances that present precise hazards, and more particular precautions may need to be taken, depending on the circumstances involved. Nonetheless, the general precautions outlined herein are always applicable.

3-3 Resources and Planning.   The hazardous materials technican shall be capable of the following.

3-3.1*   Identify the governmental and private sector agencies that offer assistance during a hazardous materials incident, including their role and the type of assistance or resources available.

A-3-3.1   This would include off-site analytical support in and around the area of the local authority having jurisdiction.

Consult the City of Sacramento Protocol in Supplement IV for some examples of the agencies identified in that plan as being available for providing assistance at incidents. The number of agencies, as well as their availability and resources, will vary from one region of the country to another. The important thing to note, however, is that the technician level should be able to identify those agencies available to his community and the specific resources each agency can provide.

3-3.2   Given a report from a local facility supplied in accordance with a federal, state, or local right-to-know legislative requirement, determine what additional information about the facility may be needed by the authority having jurisdiction and explain how the hazardous materials technician can obtain the information.

The technician can obtain much of the information for pre-planning by a site visit or by an on-site inspection where that approach is appropriate. Given the knowledge of what materials are manufactured, stored, or used, helpful additional information would include the following:

- Exact location of the site

- Physical characteristics such as height and area of structures, type of construction, and nature of occupancy

- Topography

- Site access and building access

**Figure 3.1   The United States Coast Guard can
provide assistance at incidents.**

- Nature and proximity of exposures

- Life hazard, on-site and nearby

- Meteorological conditions, such as prevailing wind

- Storage specifics, e.g., tanks, containers, locations

- Availability of fire protection systems and water

- Runoff routes

- Nature of the materials involved

3-4 Incident Management.   The hazardous materials technician shall be capable of the following.

3-4.1   Given the local hazardous materials contingency plan and standard operating procedures, describe the functions and responsibilities of the hazard sector personnel within the Incident Command System. Functions include:

(a) Safety

The functions and responsibilities of the safety sector at an incident are to evaluate conditions, determine which are unsafe or hazardous, and take steps to assure the safety of personnel. The safety officer has the authority to intervene and halt unsafe operations and acts.

(b) Entry/reconnaissance

This sector would be responsible for assessing the incident, monitoring changes that are occurring, and controlling site-entry.

(c) Information/research

This sector develops information about the incident and releases information to the news media and to other agencies, as determined to be appropriate. Additional duties would include the establishing of liaison with local and federal agencies and private associations, to develop information and to coordinate its dissemination.

(d) Resources

This sector maintains a resource status display and an organization chart and maintains assignment lists. Records must be prepared and maintained for relevant documentation.

Logistical needs must be determined.

(e) Decontamination

This sector is responsible for establishment and control of the entire decontamination process and procedure, in every detail.

(f) Operations.

The operations sector is charged with responsibility for overall management of the incident. The functions include supervising the control operations, which involves assigning personnel in accordance with a pre-incident plan, and determining necessary actions and resource needs.

3-5 Recognition of Hazardous Materials.   The hazardous materials technician shall be capable of the following.

3-5.1   List and define each of the hazardous materials classifications, describe the basic hazards posed by each, and give examples of each.

The classes and classifications are listed in the definitions section under Hazardous Materials. Consult the commentary dealing with that section for examples of each classification.

3-6 Classification, Identification, and Verification.   The hazardous materials technician shall be capable of the following.

3-6.1   Describe the source of, definition of, and circumstances for the use of the following terms:

(a) Hazardous substances

See A-1.3(b)

(b) Hazardous chemicals

See A-1.3(f)

(c) Extremely hazardous substances

See A-1.3(c)

(d) Hazardous wastes

See A-1.3(e)

(e) Hazardous materials

See A-1.3(a)

(f)   Dangerous goods.

Outside the United States, the term Dangerous Goods is used as roughly equivalent to our usage of the term Hazardous Materials, especially as relates to regulations governing transportation of

Figure 3.2   Hazardous materials incidents can vary widely.

materials (goods). The regulations are international and were prepared by the Committee of Experts on the Transport of Dangerous Goods for the United Nations.

3-6.2   Describe the advantages and disadvantages of each of the following information sources:

(a) MSDS

(b) Reference guidebooks

(c) Hazardous materials data bases

(d) Technical information centers (e.g., CHEMTREC, National Response Center)

(e) Technical information specialists

(f) Monitoring equipment.

Advantages of all of the above items are that they serve to provide information and guidance about the substances that are involved, offer advice on remedial action, and help to identify the material. Disadvantages are that the information may not be complete, time is needed to develop the information, and the aids may be either unwieldy to use effectively or expensive to acquire.

3-6.3   Given a specific hazardous material and using the information sources available to the hazardous materials technician, demonstrate extracting appropriate information about the physical characteristics and chemical properties, hazards, and suggested response considerations for that material.

The information sources available to the technician level may vary from one jurisdiction to another. However, certain capabilities and resources are available to all, such as a typical MSDS, reference guidebooks, information centers like CHEMTREC and the National Response Center, and source monitoring equipment. On that basis, the technician should be capable of completing this requirement.

3-6.4   Describe an analysis process for identification of unknown hazardous materials.

Basic recognition and identification measures can often be used to determine the nature of a hazardous material. These include labels, placards, container sizes and shapes, and the like.

An analysis process using monitoring equipment can also be used. For example, the presence of organic vapors can be determined by using a photoionizer or a portable gas chromatograph.

Gross measurements are used at first, then qualitative and quantitative analyses.

Instrumentation is also available to monitor radiation, oxygen deficiency, and combustible gases.

3-6.5   Given a label for a radioactive material, identify the vertical bars, the contents, and the activity and transport index, and define the meaning of each.

Depending on the level of radiation emitted from the package, radioactive materials will bear one of three types of labels, Radioactive White-I, Radioactive Yellow-II, or Radioactive Yellow-III. Radioactive White-I identifies contents that have the lowest level of radiation, and Radioactive Yellow-III the highest level. Radioactive labels are primarily yellow over white. The transport index refers to a number on the yellow label. It indicates the degree of control that is the responsibility of the shipper and determines the number of such packages that are allowed in a single vehicle or storage area. The transport index indicates the dose of radiation measured at three feet from the surface of the package. The label also indicates the contents and activity.

3-7 Chemistry and Toxicology of Hazardous Materials.   The hazardous materials technician shall be capable of the following.

3-7.1   Define the following chemical and physical terms and explain their use in the risk assessment process:

(a) Air reactivity

Reactivity deals with the characteristic of a substance to experience a release of energy or to undergo change. Some materials are self-reactive or can polymerize. Others undergo violent reaction if they contact other materials. Air reactivity describes substances that will ignite or release energy when exposed to air.

(b) Catalysts and inhibitors

A catalyst is a substance that affects the rate of a chemical reaction without being changed itself by the reaction. Catalysts are usually employed to increase the speed of chemical reaction.

An inhibitor is a substance that is used to retard the rate of a chemical reaction. Inhibitors are often added to materials that tend to be self-reactive in order to make those materials more stable.

(c) Concentration

Concentration is a term used to indicate the amount of an active ingredient that is contained in a given solution. Concentrations can be expressed in percent by weight or percent by volume.

(d) Corrosivity

Corrosivity is a measure or tendency of a substance to deteriorate in the presence of another substance or in a particular environment. As an example, steel tanks installed below ground are subject to corrosion. The soil in which they are placed can be measured for corrosivity, and measures can be taken to protect the tank.

(e) Critical temperature and pressure

Critical temperature is the minimum temperature above which a gas cannot be liquefied no matter how much pressure is applied. The critical pressure describes the pressure that must be applied to bring a gas to the liquid state at its critical temperature. A gas cannot be liquefied at a temperature above its critical temperature. It can be liquefied at a temperature below its critical temperature. The lower the temperature, the less the pressure required to bring it to the liquid state.

(f) Instability

Instability is used interchangeably with reactivity. It speaks to the susceptibility of a material to release energy by itself or in combination with other materials.

(g) Oxidation ability

Oxidation (oxidizing) ability is a measure of a substance's propensity to yield oxygen. Most oxidizers contain large amounts of chemically bound oxygen. The oxygen is easily released, especially when heated, and it will accelerate the burning of combustible materials.

(h) pH

The pH of a substance is a numerical measure of its relative acidity or alkalinity. A neutral level is expressed as a pH of 7.0. Designations above that level indicate increasing alkalinity, and designations below indicate increasing acidity. The pH is the accurate determination of the hydrogen ion concentration of a solution.

(i) Polymerization

Polymerization describes what is often a violent reaction, involving the process of forming a compound from several single molecules of the same substance. Polymers are compounds whose large molecules are formed by the repetitious union of many small molecules.

(j) Radioactivity

Radioactivity describes the spontaneous emission of particles or rays from substances that are called radioactive elements.

(k) Self-accelerating decomposition temperature (SADT)

This is the temperature above which the decomposition of an unstable substance continues unimpeded, regardless of the ambient or external temperature.

(l) Strength

Strength is a term used to described the concentration of a solution.

(m) Sublimation

Sublimation is the passing of a substance directly from the solid state to the vapor state, without passing through the liquid state. Solids such as naphthalene (moth balls) are an example. The opposite of sublimation is known as deposition. An increase in temperature increases the rate of sublimation.

(n) Surface tension

Surface tension describes the attractive force between the surface molecules of a liquid. The surface tension of a substance will determine the ability of a liquid to spread on a surface.

(o) Viscosity

Viscosity is a measure of the thickness of a liquid and will determine its ease of flow. Liquids with high viscosity, such as heavy oils, have to be heated to increase their fluidity.

(p) Volatility

Volatility describes the ease with which a liquid or solid can pass into the vapor state. The higher the volatility, the greater the rate of evaporation. Vapor pressure is a measure of a liquid's propensity to evaporate. The higher the vapor pressure, the more volatile the liquid.

(q) Water reactivity.

Water reactivity describes the sensitivity of materials to water, without requiring heat or confinement. Some materials are capable of reacting explosively on exposure to water.

3-7.2 Define the following toxicological terms and explain their use in the risk assessment process:

A knowledge of the meaning and significance of all of the chemical and physical terms listed below is important in determining the nature and the degree of the potential risk that the responder, the public, and the environment face at a given incident. One cannot evaluate the probability of an incident, or the effects thereof, without understanding certain chemical and physical characteristics of materials that might become involved in an incident.

(a) Threshold limit value (TLV-TWA)

This describes an atmospheric concentration of a contaminant to which most workers may be exposed day after day without any ill effects. The TWA indicates the time-weighted average.

(b) Lethal concentration and doses (LD 50/100)

The lethal dosage of a substance is the amount likely to prove fatal if ingested or absorbed. The lethal concentration is the amount of a toxic in air that will, in all likelihood, cause death if inhaled. The LD 50 measures the amount that kills half of the laboratory test animals exposed to the toxic substance. LD 100 is the measured amount that will cause death for all of the exposed laboratory test animals.

(c) Parts per million/billion (ppm/ppb)

This is simply a measurement of the parts of the toxic substance per volume of air, water, or other substance. Regulations are established for allowable amounts of a toxic substance related to parts per million or parts per billion of air, water, or other medium.

(d) Immediately dangerous to life and health (IDLH)

These are exposure concentrations established by NIOSH/OSHA as a guideline for selecting breathing equipment for some chemicals. IDLH concentrations pose an immediate threat of severe exposure to contaminants that are likely to have an adverse effect on health. ANSI defines IDLH as any atmosphere that poses an immediate hazard to life or produces immediate irreversible debilitating effects on health. As a good rule, IDLH concentrations should be an indication that appropriate respiratory protection is essential.

(e) Permissible exposure limit (PEL)

This is the 8 hour time-weighted average or ceiling concentration above which workers may not be exposed. The use of personal protective equipment may be advisable where there is a potential for exposure.

(f) Short-term exposure limit (TLV-STEL)

This is one of the threshold limit values (TLVs) that measures the maximum concentration to which workers can be exposed for a period up to 25 minutes continuously without suffering from irritation, tissue change, or debilitating narcosis. No more than four exposures are permitted per day, with 60 minute intervals required, and the TLV-TWA cannot be exceeded.

(g) Ceiling level (TLV-C).

This is another of the TLVs. It measures the concentration that should not be exceeded even momentarily.

Knowledge of the significance of the above terms is essential for assessing the chemical and physical hazards associated with response to hazardous materials incidents.

3-8 Hazard and Risk Assessment.   The hazardous materials technician shall be capable of the following.

3-8.1   Identify and explain the risk assessment considerations to be made at a hazardous materials incident, including the following:

Items (a) through (f) relate to conditions or circumstances that will allow the responder to assess, characterize, or "size-up" the situation with which he is confronted. They are not all-inclusive. Other unmentioned conditions may be present that will affect the assessment of risk.

(a) Size and type of container and quantity involved;

The size of container establishes quantity parameters. We can establish how much has leaked or may spill. The type of container also helps to assess the risk. For example, steel containers have a different failure mode than do plastic containers when each is exposed to flame; pressure containers offer different risks than atmospheric containers and provide information to the responder about the nature of the product stored.

Figure 3.3 Correct assessment of a situation can be the difference
between successful mitigation of an incident and disaster.

(b) Nature of the container stress;

Containers may be rusted, leaking, bulging, dented, or exposed
to flame or heat. Shutoff valves, if present, may be open, damaged,
or inoperable. If flame is impinging on the vapor space of a
container, there may be sudden and violent failure.

(c) Potential behavior of the container and its contents;

The potential behavior of the container will depend on the
variables that the responder discovers on arrival at the scene. They
include, among other things, the conditions mentioned in (a) and (b)
and the commentary provided for each. The behavior of the contents
of the container will depend on its chemical and physical characteristics and the conditions to which the contents are subjected.

(d) Level of resources available (e.g., personal protective equipment, training, etc.);

The level of resources available will differ vastly from one jurisdiction to another, and from one incident to another even in the
same jurisdiction. What is important about this item is that the
responder must consider the available resources as part of the risk
assessment. There may be incidents where the resources will be
capable of effecting quick and complete resolution and others
where the appropriate action may be temporary inaction. Note

that training is considered as part of the resources brought to the scene. Its importance cannot be overemphasized.

(e) Exposure potential to people, property, environment, and systems; and

These are standard "size-up" items that all responders should be accustomed to evaluating at every incident.

(f) Weather conditions and terrain.

Weather conditions can have a profound effect on the outcome of an incident, either favorably or otherwise. In many incidents, the wind direction becomes a major determining factor and can make a spill or release relatively harmful or totally devastating. Terrain can also be extremely important. Failure of a tank at an elevation, for example, poses a danger to all nearby exposures located at lower levels.

3-8.2 Identify the various monitoring equipment used to monitor and detect the following hazards:

When monitoring for the various types of hazards listed in (a) through (f), it is necessary to establish priorities based on all information that is available as part of your initial size-up and risk assessment. This will help in determining which monitoring equipment needs to be used first and also what personal protective measures must be taken. Initial concerns would be directed toward those conditions that are expected to be immediately dangerous to life and health. Air sampling or collecting would first be conducted downwind and at the perimeter or along the axis of the wind direction by personnel in Level B protection. The results of the monitoring will determine what personal protection is required for additional sampling.

(a) Toxicity

Toxicity can be monitored by use of various instruments, including a portable infrared spectrophotometer, ultraviolet photo-ionization detector, direct reading colorimetric indicator tube, and a flame ionization detector with gas chromatography option.

(b) Flammability

Flammability is monitored by a combustible gas indicator.

(c) Reactivity

Some comments submitted during the Technical Committee Documentation (TCD) period indicated that reactivity cannot be detected by monitoring equipment. The Committee responded to the effect that a thermometer or pyrometer can monitor a polymerizing chemical's reaction.

(d) Radioactivity

Radioactivity can be monitored by use of a gamma radiation survey instrument, but it does not measure alpha or beta radiation.

(e) Corrosivity

Corrosivity can be measured by use of pH papers and strips.

(f) Oxygen deficiency.

Oxygen deficiency is measured by use of an oxygen meter.

3-8.3 Demonstrate the purpose, operation (including interpretation of results and operational and calibration checks), and limitations for the following basic monitoring equipment, in addition to any other monitoring equipment provided to the hazardous materials technician by the authority having jurisdiction:

(a) Combustible gas detector

This instrument, which detects combustible gases and vapors, does not provide a valid reading where there are oxygen deficient conditions. In order to achieve accuracy, the operator must know how to calibrate it properly and must perform this function just prior to use. The accuracy also depends to some degree on the difference between the calibration and sampling temperatures.

(b) Oxygen meter

This device measures the percentage of oxygen in the sampled air. It must be calibrated prior to use.

(c) Colorimetric tubes

These devices monitor specific gases and vapors and measure their concentrations. The operator must judge the stain's end point, and the tube has a limited degree of accuracy. It can also be affected by high humidity.

(d) pH papers and strips

These devices are indicators that will change color to indicate whether the substance being tested is an acid or a base. A fair estimate can be made using these monitoring devices.

(e) CO meter

This measures the level of CO and is specific for that gas.

(f) Radiation detection instruments.

These devices include alpha-beta detectors and gamma radiation survey instruments. Both types are needed, since neither device measures the other type of radiation hazard.

3-8.4 Demonstrate the field maintenance and testing procedures, including operational and calibration checks, for the basic monitoring equipment provided to the hazardous materials technician by the authority having jurisdiction.

3-9 Personal Protective Equipment.   The hazardous materials technician shall be capable of the following.

3-9.1   Given a hazardous materials incident, select the appropriate personal protective equipment to be used in that incident.

The use of personal protective equipment is a federal mandate, by virtue of both OSHA 29 CFR 1910 and EPA 40 CFR 300. There is no combination of equipment that can protect the responder against every hazard. In addition, the equipment can itself present discomfort to the wearer. It is important to select the most appropriate level of protection for the situation, since overprotection can cause harm.

Selection of appropriate respiratory protective equipment and of chemical-protective clothing is a complicated task that requires a great deal of training and experience.

See Table 3-1 for information on respiratory equipment. (Reprinted from NIOSH/OSHA/USCG/EPA *Occupational Safety and Health Guidance Manual for Hazardous Waste Site Activity.*)

3-9.2   Define the following terms as associated with chemical-protective clothing:

(a) Degradation

(b) Penetration

(c) Permeation.

See definitions in 1-3.

3-9.3 Demonstrate how to interpret a chemical compatibility chart for chemical-protective clothing and the limitations and definitions of compatibility charts.

Table 3-1   Relative Advantages and Disadvantages of Respiratory Protective Equipment

Type of Respirator	Advantages	Disadvantages
ATMOSPHERE-SUPPLYING Self-Contained Breathing Apparatus (SCBA)	• Provides the highest available level of protection against airborne contaminants and oxygen deficiency. • Provides the highest available level of protection under strenuous work conditions.	• Bulky, heavy (up to 35 pounds). • Finite air supply limits work duration. • May impair movement in confined spaces.
Positive-Pressure Supplied-Air Respirator (SAR) (also called air-line respirator)	• Enables longer work periods than an SCBA. • Less bulky and heavy than an SCBA. SAR equipment weighs less than 5 pounds (or around 15 pounds if escape SCBA protection is included). • Protects against most airborne contaminants.	• Not approved for use in atmospheres immediately dangerous to life or health (IDLH) or in oxygen-deficient atmospheres unless equipped with an emergency egress unit such as an escape-only SCBA that can provide immediate emergency respiratory protection in case of air-line failure. • Impairs mobility. • MSHA/NIOSH certification limits hose length to 300 feet (90 meters). • As the length of the hose is increased, the minimum approved air flow may not be delivered at the facepiece. • Air line is vulnerable to damage, chemical contamination, and degradation. Decontamination of hoses may be difficult. • Worker must retrace steps to leave work area. • Requires supervision/monitoring of the air supply line.
AIR-PURIFYING Air-Purifying Respirator (including powered air-purifying respirators [PAPRs])	• Enhanced mobility. • Lighter in weight than an SCBA. Generally weighs 2 pounds (1 kg) or less (except for PAPRs).	• Cannot be used in IDLH or oxygen-deficient atmospheres (less than 19.5 percent oxygen at sea level). • Limited duration of protection. May be hard to gauge safe operating time in field conditions. • Only protects against specific chemicals and up to specific concentrations. • Use requires monitoring of contaminant and oxygen levels. • Can only be used (1) against gas and vapor contaminants with adequate warning properties, or (2) for specific gases or vapors provided that the service is known and a safety factor is applied or if the unit has ESLI (end-of-service-life indicator).

See Supplement VII, *Chemical Compatibility Chart for Protective Clothing.*

3-9.4 Demonstrate the maintenance, inspection, and storage procedures for Level B and C chemical-protective clothing, in addition to any other specialized protective equipment provided by the authority having jurisdiction.

Manufacturers will almost always provide instructions for appropriate maintenance, inspection, and storage of the equipment they provide. It is essential that these instructions be followed. Maintenance should only be performed by persons who have been instructed and trained in that skill. Inspection of equipment should take place after each use, including training. In addition, there should be an established inspection frequency and procedure, and records of inspections should be maintained. Proper storage is essential in preventing unnecessary wear and damage to equipment. Failure to store properly can be a leading cause of malfunctioning equipment.

3-9.5 Demonstrate the proper donning, doffing, and usage of Level B and C chemical-protective clothing, in addition to any other specialized protective equipment provided by the authority having jurisdiction.

It is important to have an established written donning and doffing procedure, and training in the procedures is essential toward achieving proficiency for emergency operations.

3-9.6 Describe the four levels of protection as found in the EPA/OSHA publications, list the equipment required for each level, and describe the conditions under which each level is used.

See Table 3-2. (Reprinted from the NIOSH/OSHA/USCG/EPA Guidance Manual.)

3-9.7 Describe at least four conditions that indicate material degradation of chemical-protective clothing after chemical contact.

Degradation can be evidenced by material discoloration, swelling, stiffening, and softening. Chemical permeation, however, can take place with no visible signs to indicate a condition. Checks should also be made for closure failures, seam breaks, tears, and punctures.

The following text is reprinted from NIOSH/OSHA/USCG/EPA *Occupational Safety and Health Guidance Manual for Hazardous Waste Site Activities.*

Table 3-2 EPA/OSHA Protection Levels*

Level of Protection	Equipment	Protection Provided	Should Be Used When:	Limiting Criteria
A	RECOMMENDED: • Pressure-demand, full-facepiece SCBA or pressure-demand supplied-air respirator with escape SCBA. • Fully-encapsulating, chemical-resistant suit. • Inner chemical-resistant gloves. • Chemical-resistant safety boots/shoes. • Two-way radio communications. OPTIONAL: • Cooling unit. • Coveralls. • Long cotton underwear. • Hard hat. • Disposable gloves and boot covers.	The highest available level of respiratory, skin, and eye protection.	• The chemical substance has been identified and requires the highest level of protection for skin, eyes, and the respiratory system based on either: —measured (or potential for) high concentration of atmospheric vapors, gases, or particulates or —site operations and work functions involving a high potential for splash, immersion, or exposure to unexpected vapors, gases, or particulates of materials that are harmful to skin or capable of being absorbed through the intact skin. • Substances with a high degree of hazard to the skin are known or suspected to be present, and skin contact is possible. • Operations must be conducted in confined, poorly ventilated areas until the absence of conditions requiring Level A protection is determined.	Fully-encapsulating suit material must be compatible with the substances involved.

*Based on EPA protective ensembles.

Table 3-2 EPA/OSHA Protection Levels (cont.)

Level of Protection	Equipment	Protection Provided	Should Be Used When:	Limiting Criteria
B	RECOMMENDED: • Pressure-demand, full-facepiece SCBA or pressure-demand supplied-air respirator with escape SCBA. • Chemical-resistant clothing (overalls and long-sleeved jacket; hooded, one- or two-piece chemical splash suit; disposable chemical-resistant one-piece suit). • Inner and outer chemical-resistant gloves. • Chemical-resistant safety boots/shoes. • Hard hat. • Two-way radio communications. OPTIONAL: • Coveralls. • Disposable boot covers. • Face shield. • Long cotton underwear.	The same level of respiratory protection but less skin protection than Level A. It is the minimum level recommended for initial site entries until the hazards have been further identified.	• The type and atmospheric concentration of substances have been identified and require a high level of respiratory protection, but less skin protection. This involves atmospheres: — with IDLH concentrations of specific substances that do not represent a severe skin hazard; or — that do not meet the criteria for use of air-purifying respirators. • Atmosphere contains less than 19.5 percent oxygen. • Presence of incompletely identified vapors or gases is indicated by direct-reading organic vapor detection instrument, but vapors and gases are not suspected of containing high levels of chemicals harmful to skin or capable of being absorbed through the intact skin.	• Use only when the vapor or gases present are not suspected of containing high concentrations of chemicals that are harmful to skin or capable of being absorbed through the intact skin. • Use only when it is highly unlikely that the work being done will generate either high concentrations of vapors, gases or particulates or splashes of material that will affect exposed skin.

Table 3-2   EPA/OSHA Protection Levels (cont.)

Level of Protection	Equipment	Protection Provided	Should Be Used When:	Limiting Criteria
C	RECOMMENDED: • Full-facepiece, air-purifying, canister-equipped respirator. • Chemical-resistant clothing (overalls and long-sleeved jacket; hooded, one- or two-piece chemical splash suit; disposable chemical-resistant one-piece suit). • Inner and outer chemical-resistant gloves. • Chemical-resistant safety boots/shoes. • Hard hat. • Two-way radio communications. OPTIONAL: • Coveralls. • Disposable boot covers. • Face shield. • Escape mask. • Long cotton underwear.	The same level of skin protection as Level B, but a lower level of respiratory protection.	• The atmospheric contaminants, liquid splashes, or other direct contact will not adversely affect any exposed skin. • The types of air contaminants have been identified, concentrations measured, and a canister is available that can remove the contaminant. • All criteria for the use of air-purifying respirators are met.	• Atmospheric concentration of chemicals must not exceed IDLH levels. • The atmosphere must contain at least 19.5 percent oxygen.
D	RECOMMENDED: • Coveralls. • Safety boots/shoes. • Safety glasses or chemical splash goggles. • Hard hat. OPTIONAL: • Gloves. • Escape mask. • Face shield.	No respiratory protection. Minimal skin protection.	• The atmosphere contains no known hazard. • Work functions preclude splashes, immersion, or the potential for unexpected inhalation of or contact with hazardous levels of any chemicals.	• This level should not be worn in the Exclusion Zone. • The atmosphere must contain at least 19.5 percent oxygen.

Permeation and Degradation

The selection of chemical-protective clothing (CPC) depends greatly upon the type and physical state of the contaminants. This information is determined during site characterization (Chapter 6). Once the chemicals have been identified, available information sources should be consulted to identify materials that are resistant to permeation and degradation by the known chemcials. One excellent reference, *Guidelines for the Selection of Chemical-Protective Clothing*, provides a matrix of clothing material recommendations for approximately 300 chemicals based on an evaluation of permeation and degradation data from independent tests, vendor literature, and raw material suppliers. Charts indicating the resistance of various clothing materials to permeation and degradation are also available from manufacturers and other sources. It is important to note, however, that no material protects against all chemicals and combinations of chemcals, and that no currently available material is an effective barrier to any prolonged chemical exposure.

In reviewing vendor literature, it is important to be aware that the data provided are of limited value. For example, the quality of vendor test methods is inconsistent; vendors often rely on the raw material manufacturers for data rather than conducting their own tests, and the data may not be updated. In addition, vendor data cannot address the wide variety of uses and challenges to which CPC may be subjected. Most vendors strongly emphasize this point in the descriptive text that accompanies their data.

Another factor to bear in mind when selecting CPC is that the rate of permeation is a function of several factors, including clothing material type and thickness, manufacturing method, the concentration(s) of the hazardous substance(s), temperature, pressure, humidity, the solubility of the chemical in the clothing material, and the diffusion coefficient of the permeating chemical in the clothing material. Thus permeation rates and breakthrough times (the time from initial exposure until hazardous material is detectable on the inside of the CPC) may vary depending on these conditions.

Most hazardous wastes are mixtures, for which specific data with which to make a good CPC selection are not available. Due to lack of testing, only limited permeation data for multicomponent liquids are currently available.

Mixtures of chemicals can be significantly more aggressive towards CPC materials than can any single component alone. Even small amounts of a rapidly permeating chemical may provide a pathway that accelerates the permeation of other chemicals. Formal research is being conducted on these effects. NIOSH is currently developing methods for evaluating CPC materials against mixtures of chemicals and unknowns in the field. For hazardous waste site operations, CPC should be selected that offers the widest range of protection against the chemicals expected on site. Vendors are now providing CPC material—composed of two or even three different materials laminated together—that is capable of providing the best features of each material.

3-9.8 Demonstrate the use of the following types of respiratory protection when provided by the authority having jurisdiction:

(a) Air purifying respirator

(b) Supplied air respirator (air line respirator).

3-9.9 Describe the factors to be considered in selecting the proper respiratory protection at a hazardous materials incident.

Self-contained breathing apparatus should be worn when the atmosphere is, or is likely to become, immediately dangerous to life and health. Table 3-3 (reprinted from the NIOSHA/OSHA/USCG/EPA Manual) offers advantages and disadvantages of SCBA.

3-10 Hazardous Materials Control. The hazardous materials technician shall be capable of the following.

3-10.1 Describe the basic design and construction features of containers and bulk and nonbulk packaging used to store, process, or transport hazardous materials, which would include, but not be limited to:

(a) Bags

(b) Bottles

(c) Boxes

(d) Cans

(e) Carboys

(f) Cylinders

(g) Drums

(h) Fixed tanks

(i) Intermodal portable tanks

(j) Piping

Table 3-3   Relative Advantages and Disadvantages of SCBA

Type	Description	Advantages	Disadvantages	Comments
ENTRY-AND-ESCAPE SCBA Open-Circuit SCBA	Supplies clean air to the wearer from a cylinder. Wearer exhales air directly to the atmosphere.	Operated in a positive-pressure mode, open-circuit SCBAs provide the highest respiratory protection currently available. A warning alarm signals when only 20 to 25 percent of the air supply remains.	Shorter operating time (30 to 60 minutes) and heavier weight (up to 35 lbs [13.6 kg]) than a closed-circuit SCBA.	The 30- to 60-minute operating time may vary depending on the size of the air tank and the work rate of the individual.
Closed-Circuit SCBA (Rebreather)	These devices recycle exhaled gases (CO_2, O_2, and nitrogen) by removing CO_2 with an alkaline scrubber and replenishing the consumed oxygen with oxygen from a liquid or gaseous source.	Longer operating time (up to 4 hours), and lighter weight (21 to 30 lbs [9.5 to 13.6 kg]) than open-circuit apparatus. A warning alarm signals when only 20 to 25 percent of the oxygen supply remains. Oxygen supply is depleted before the CO_2 sorbent scrubber supply, thereby protecting the wearer from CO_2 breakthrough.	At very cold temperatures, scrubber efficiency may be reduced and CO_2 breakthrough may occur. Units retain the heat normally exchanged in exhalation and generate heat in the CO_2 scrubbing operations, adding to the danger of heat stress. Auxiliary cooling devices may be required. When worn outside an encapsulating suit, the breathing bag may be permeated by chemicals, contaminating the breathing apparatus and the respirable air. Decontamination of the breathing bag may be difficult.	Positive-pressure closed-circuit SCBAs offer substantially more protection than negative-pressure units, which are not recommended on hazardous waste sites. While these devices may be certified as closed-circuit SCBAs, NIOSH cannot certify closed-circuit SCBAs as positive-pressure devices due to limitations in certification procedures currently defined in 30 CFR Part 11.

Table 3-3 Relative Advantages and Disadvantages of SCBA (cont.)

Type	Description	Advantages	Disadvantages	Comments
ESCAPE-ONLY SCBA	Supplies clean air to the wearer from either an air cylinder or from an oxygen-generating chemical. Approved for escape purposes only.	Lightweight (10 pounds [4.5 kg] or less), low bulk, easy to carry. Available in pressure-demand and continuous-flow modes.	Cannot be used for entry.	Provides only 5 to 15 minutes of respiratory protection, depending on the model and wearer breathing rate.

(k) Tank cars

(l) Tank trucks and trailers.

All of the above entries, with the exception of (h) Fixed tanks, are regulated by the U.S. Department of Transportation (DOT) in the *Code of Federal Regulations* (CFR) 49. The design and construction of fixed tanks are governed by a number of codes and standards, including those published by the American Society of Mechanical Engineers (ASME), the American Petroleum Institute (API), and NFPA, among others.

There are excellent articles on intermodal tank containers written by Charles J. Wright of the Union Pacific Railroad Company in NFPA's *Fire Command*, the June and July 1988 issues.

U.S. DOT, in 49 CFR Chapter 1, 173.24, lists standard regulations that are applicable to all packages. Among their requirements are the following, in paraphrased fashion.

• All packages shall be designed and constructed, and the contents limited, so that there is no release, the packaging remains intact, and the contents are not damaging to the packaging.

• Packaging should be strong and tight.

• Specification containers have to be marked as prescribed.

• Use of steel is specified as to type and gauge.

• Lumber is specified by descriptive qualifications.

• Packaging and contents should be compatible to the extent that reactivity is avoided.

• Closures have to be adequate to avoid leaks.

• Nails and similar devices shall not protrude into the interior of the outer packaging.

• Friction during transport should not affect the contents.

There are also regulations for overpacks.

3-10.2* Describe the considerations to be evaluated in implementing hazardous materials control procedures.

A-3-10.2 This should include, but not be limited to, the following control methods: diking, damming, diversion, retention, absorption, directing, dispersing, dilution, plugging, and patching. (*See the Haz Mat Response Team Spill and Leak Guide from Fire Protection Publications, Oklahoma State University.*)

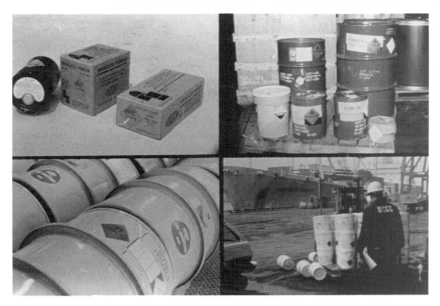

Figure 3.4   Hazardous materials are transported by truck, railroad car, ship, and airplane in a variety of containers, including wooden boxes, metal, plastic, or fiber board drums, and plastic pails (top) and glass carboys in plywood boxes or polystyrene cases, shipping tubes, and multiwall paper bags (bottom).

Consult NFPA 471 and its commentary for a detailed description of control procedures.

3-10.3 Based on the properties of the hazardous materials classes, describe the methods and precautions for controlling releases from nonbulk and bulk packaging and containers of hazardous materials.

NFPA 471 contains recommended procedures for mitigation of incidents, and the handbook commentary elaborates on those procedures.

3-10.4 Based on the hazardous materials classifications, demonstrate the proper selection and use of tools, equipment, and materials available to the hazardous materials technician by the authority having jurisdiction for the control of hazardous materials releases from nonbulk and bulk packaging and containers.

The demonstration capability required by this section will be determined and limited by the availability of mitigating equipment provided to the technician by the local authority having jurisdiction. Although not addressed by this section, improvising can be a valuable asset. Even when tools and equipment are limited or unavailable, it may still be possible to take some action that will help remediate an incident.

Consult NFPA 471 and the handbook commentary for specific information on this requirement.

3-10.5 Describe the maintenance and testing procedures for the tools, equipment, and materials provided to the hazardous materials technician by the authority having jurisdiction for the control of hazardous materials releases.

Fulfilling this requirement will differ from one location to another, since it is dependent on the tools and equipment that are provided to the technician. However, it is vital to successful operations at any emergency incident that personnel assigned to a unit be totally familiar with all of the tools and equipment carried. They should know how to use each properly and efficiently, and also how to maintain them.

3-11 Decontamination. The hazardous materials technician shall be capable of the following.

3-11.1 Identify and describe the advantages and limitations of each of the following methods of decontamination:

(a) Absorption

Absorption can be described as a process in which one substance combines with another by moving into it or entering the interior of that other substance. Absorbents can be used in the decontamination process, but the combined material must be considered and handled as contaminated.

(b) Adsorption

Adsorption is a process whereby the material that is taken up is spread over the surface of the adsorbing material. It is in effect a surface coating process. The adsorbent must be handled as being contaminated.

(c) Chemical degradation

Chemical degradation refers to the molecular breakdown of a material. In the case of a contaminating substance, it can render the material less hazardous. The process can be used to advantage in decontamination, but the extent of degradation must be monitored and verified.

(d) Dilution

Dilution involves the thinning out or weakening of a substance by adding another material, called a diluent. Dilution is a valid method of achieving decontamination, but the end material may still have to be disposed of in accordance with provisions regulating contaminated material.

(e) Disposal

Disposal can be considered as a final step in a process whereby a substance is handed off, in accordance with existing regulations and procedures, to an authorized receiving party, or to a final resting place. All contaminated material must undergo appropriate disposal.

(f) Isolation

Isolation of a contaminated material or substance suggests the establishment of a boundary or perimeter where a material is located or placed and where entry is restricted to authorized personnel. Isolation is not always achievable over a long term, but it does provide a temporary method of dealing with a contaminated substance.

(g) Neutralization

Neutralization refers to the process that occurs when acids and bases neutralize each other, i.e., the characteristic properties of both acid and base disappear or have an equal value. More generally, neutralization is referred to as making a contaminated substance harmless. Where it can be achieved, neutralization is an effective technique in decontamination.

(h) Solidification.

Solidification involves the process of stabilizing a hazardous substance into a solid that has greater integrity. The solidified waste is then less likely to leak or leach out. Where it can be achieved, this is a good method of decontamination.

3-11.2 Describe the considerations associated with the placement, location, and setup of the decontamination site.

At a hazardous materials incident, the decontamination facilities should be placed in the warm zone. The setup of the decontamination site must consider the many variables that may be involved, including the following:

• The severity of the incident and the types of hazardous materials involved

• The amount of material released

• The extent of exposure

• The movement of personnel and equipment between zones

• The methods chosen for decontamination and for protection of responders

3-11.3 Identify sources of technical information for performing decontamination operations.

Sources of information should include the local response protocols that have been established for the decontamination process. In addition, information can be obtained from the following:

• NIOSH/OSHA/USCG/EPA *Occupational Safety and Health Guidance Manual for Hazardous Waste Site Activities*.

• EPA *Standard Operating Safety Guides*

• *Guidelines for Decontamination of Firefighters and Their Equipment Following Hazardous Materials Incidents*, Canadian Association of Fire Chiefs, Inc. (See Supplement VIII of this publication)

3-11.4 Given a simulated hazardous materials incident and using available local resources, demonstrate the implementation of the decontamination procedure.

This requirement to demonstrate an implementation procedure will depend greatly on the availability of local resources.

3-12 Termination Procedures.   The hazardous materials technician shall be capable of the following.

3-12.1   Describe the activities required in terminating the emergency phase of a hazardous materials incident.

Note that the definition (Section 1-3) of termination procedures seems to cover two separate and distinct concepts. The first deals with documenting several specifics, including safety, operations, exposure of personnel, and a critique. Documentation is the important operative word and clearly indicates the need for written reports. The second concept specifies debriefing, post-incident analysis, and the critique.

In Alan Brunacini's *Fire Command*, his treatment of terminating procedures includes returning companies to service and completing command operations. In a complicated operation such as may well occur at a hazardous materials incident, returning companies to service can be a complicated procedure that requires a great deal of coordination.

In *Hazardous Materials — Managing the Incident*, debriefing, post-incident analysis, and critique are covered as distinct and separate activities. Debriefing includes determining exposures of personnel and equipment to contaminants, assigning duties for an effective analysis and critique, and summarizing what each sector did. The post-incident analysis reconstructs the incident. The critique indicates the need to provide written reports to management that suggest ways to improve future operations.

The important point to remember is that documentation is essential. Written reports provide records of what happened, and they are essential to effective operations.

3-12.2   Given a simulated hazardous materials incident, describe the preparation of the locally required report with supporting documentation as necessary.

Note the clear implication that there must be a "locally required report." It has to be more comprehensive than the typical fire report or a "fill in the blanks" type of document. Documentation suggests the need for some detailed narrative. The technician's description of the simulated incident will vary, depending on both the severity of the incident and the locally required report.

3-12.3   Identify the considerations associated with conducting a critique of a hazardous materials incident.

The principal purpose of a critique is to determine and understand the "lessons learned" (a hallmark for decades of the type of article printed in the New York City Fire Department's WNYF magazine in which major or unusual incidents have been described). A critique must have a single person in charge who serves as leader and facilitator. The process must be done in a positive fashion and should not be used as a vehicle to affix blame. There should also be a written report.

3-13 Educational and Medical Requirements. Reserved.

4

Hazardous Materials Specialist

4-1 General.

The specialist is the person in the highest category of response capability. That level is achieved based principally on the amount and type of training received, a demonstrated ability to absorb and use that training, and an ability to use more technically sophisticated equipment at a hazardous materials incident. However, the distinction between specialist and technician is not totally precise as developed in this document. Local jurisdictions may, if they see a compelling need to do so, draw a clear and distinct line of demarcation between these two levels of response.

4-1.1 The hazardous materials specialist shall meet all of the objectives indicated for the first responder and hazardous materials technician. In addition, that person shall meet the training and medical surveillance program requirements in accordance with federal Occupational Safety and Health Administration (OSHA) or U.S. Environmental Protection Agency (EPA) regulations.

Note the requirement that the specialist have all of the competencies acquired by the two lower levels of responder.

The training and medical surveillance requirements can be found specifically in OSHA 1910.120, and the EPA regulations are required by congressional mandate to be identical in this regard with those of OSHA. The regulations are clearly intended to apply to all responders, as well as to those who work at or are involved in cleanup at fixed hazardous waste sites. The mandate for combined and identical regulations from both OSHA and EPA is intended to close a loophole that existed with OSHA requirements. The latter rules have generally not been applied to federal, state, and local employees. EPA regulations will cover that void.

The medical surveillance provisions will require a medical examination upon assignment to a hazardous materials response team and at least annually thereafter. In addition, medical examinations will be required upon termination of the assignment and upon the reporting of symptoms of exposure.

4-1.2 Goal. The goal at the hazardous materials specialist level shall be to provide those persons, whose duties involve response to specialized hazardous materials problems, with the following competencies to respond safely to hazardous materials incidents:

> The standard does not define "specialized hazardous materials problems," so the clear implication is that the specialist level may not be expected to respond to every hazardous materials incident. Policies will vary from one jurisdiction to another. Some protocols call for a tiered response, which indicates that less severe incidents are handled by first responders or technicians. Where dedicated hazardous materials teams operate, there may be a policy that they be called to every hazardous materials incident. Keep in mind that some of the differences may involve semantics, in that designation of a "hazardous materials incident" is bound to differ from one authority to another.

(a) The ability to develop a site safety plan;

> A site safety plan would include among its elements a site map or drawing, establishment of control zones, directions for use of standard operating procedures and deviations therefrom where indicated, creation of a safety sector with a safety officer in command, decontamination procedures, security measures, and communications protocol.

(b) The ability to classify, identify, and verify known and unknown materials by using advanced monitoring equipment provided by the authority having jurisdiction;

> A limiting factor in this requirement is the monitoring equipment that is available. Where additional such equipment is needed, the specialist should know how to procure it, where such a circumstance is feasible. In any event, it is clearly expected that the specialist be capable of using the monitoring equipment that is provided.

(c) The ability to function within an assigned role in the Incident Command System;

> A specialist would operate in control of the hazard sector at most "specialized" hazardous materials problems or incidents. The hazard sector would be charged with actual mitigation of the incident and requires a commander who is capable of dealing with the technical aspects of the incident. There would be subsectors within the overall command of the hazard sector operation at major incidents. These would certainly include safety, communications, and decontamination, among others.

(d) The ability to select and use Level A protection in addition to any other specialized personal protective equipment provided to the hazardous materials specialist by the authority having jurisdiction;

See commentary on 3-9.6.

(e) The ability to perform hazard and risk assessments involving multiple hazards;

See Supplement III, *Beginning the Hazard Analysis Process*, for information on hazard and risk assessments. In addition, the following excerpt, 1-4 Hazards Analysis from *Technical Guidance for Hazards Analysis* by U.S. EPA, FEMA, and DOT should prove helpful. For more information, the complete document should be consulted.

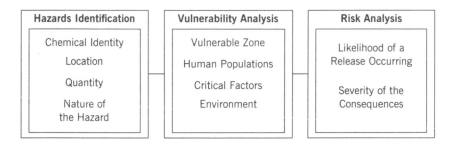

Hazards Identification	Vulnerability Analysis	Risk Analysis
Chemical Identity	Vulnerable Zone	Likelihood of a Release Occurring
Location	Human Populations	
Quantity	Critical Factors	Severity of the Consequences
Nature of the Hazard	Environment	

A hazards analysis[1] is a necessary step in comprehensive emergency planning for a community. Comprehensive planning depends upon a clear understanding of what hazards exist and what risk they pose for various members of the community. This guide follows the definition of "hazards analysis" used in NRT-1 and focuses principally on hazards analysis for airborne releases of EHSs (extremely hazardous substances).

The hazards analysis described in this guide is a 3-step decision-making process to identify the potential hazards facing a community with respect to accidental releases of EHSs. All three steps should be followed even though the level of detail will vary from

---

[1]Hazards analysis as presented in this guidance is intended for use in emergency response planning for EHSs. Its purpose and the meaning of its terminology are different from the purpose and terms used in "risk assessment" as defined by NAS (National Academy of Sciences). Because local communities will be conducting hazards analyses (as described in this guide) and risk assessments under different sections of SARA, a discussion of risk assessment can be found in NAS Press, 1983, *Risk Assessment in the Federal Government: Managing the Process*. Washington, D.C. 191 pp.

site to site. The hazards analysis is designed to consider all potential acute health hazards within the planning district and to identify which hazards are of high priority and should be addressed in the emergency response planning process. The Title III emergency response plan must address all EHSs that are reported to the State Emergency Response Commission (SERC), but other substances including EHSs below their Threshold Planning Quantities may also be included. Depending upon the size and nature of a planning district, the hazards analysis may be complex or relatively easy. Local Emergency Planning Committees that have access to the necessary experts might want to conduct a detailed quantitative hazards analysis. Such a complete analysis of all hazards may not always be feasible or practical, however, given resource and time constraints in individual planning districts.

General information and an approach to understanding the three components of hazards analysis as it is applied to the EHSs are discussed in Chapter 2. A brief overview is presented below.

A. *Hazards identification* typically provides specific information on situations that have the potential for causing injury to life or damage to property and the environment due to a hazardous materials spill or release. A hazards identification includes information about:

• Chemical identities;

• The location of facilities that use, produce, process, or store hazardous materials;

• The type and design of chemical container or vessel;

• The quantity of material that could be involved in an airborne release; and

• The nature of the hazard (e.g., airborne toxic vapors or mists which are the primary focus of this guide; also other hazards such as fire, explosion, large quantities stored or processed, handling conditions) most likely to accompany hazardous materials spills or releases.

B. *Vulnerability analysis* identifies areas in the community that may be affected or exposed, individuals in the community who may be subject to injury or death from certain specific hazardous materials, and what facilities, property, or environment may be susceptible to damage should a hazardous materials release occur. A comprehensive vulnerability analysis provides information on:

• The extent of the vulnerable zones (i.e., an estimation of the area that may be affected in a significant way as a result of a spill or release of a known quantity of a specific chemical under defined conditions);

• The population, in terms of numbers, density, and types of individuals (e.g., facility employees; neighborhood residents; people in hospitals, schools, nursing homes, prisons, day care centers) that could be within a vulnerable zone;

• The private and public property (e.g., critical facilities, homes, schools, hospitals, businesses, offices) that may be damaged, including essential support systems (e.g., water, food, power, communication, medical) and transportation facilities and corridors; and

• The environment that may be affected, and the impact of a release on sensitive natural areas and endangered species.

Chapter 2 discusses vulnerability analysis with a special emphasis on human populations.

C. *Risk analysis* is an assessment by the community of the likelihood (probability) of an accidental release of a hazardous material and the actual consequences that might occur, based on the estimated vulnerable zones. The risk analysis is a judgement of probability and severity of consequences based on the history of previous incidents, local experience, and the best available current technological information. It provides an estimation of:

• The likelihood (probability) of an accidental release based on the history of current conditions and controls at the facility, consideration of any unusual environmental conditions (e.g., areas in flood plains), or the possibility of simultaneous emergency incidents (e.g., flooding or fire hazards resulting in the release of hazardous materials);

• Severity of consequences of human injury that may occur (acute, delayed, and/or chronic health effects), the number of possible injuries and deaths, and the associated high-risk groups;

• Severity of consequences on critical facilities (e.g., hospitals, fire stations, police departments, communication centers);

• Severity of consequences of damage to property (temporary, repairable, permanent); and

• Severity of consequences of damage to the environment (recoverable, permanent).

To have an accurate view of the potential problems in a district, the LEPC would need to address all of the steps in hazards analysis outlined above. Each of the three steps should be followed even if extensive information is not available for each site. The process anticipates that local judgement will be necessary.

(f) The ability to perform specialized hazardous materials control operations within the capabilities of the resources and personal protective equipment available;

See commentary on (c) above, and also consult NFPA 471 and its commentary.

(g) The ability to develop and implement specialized decontamination procedures;

Consult Supplement VIII on decontamination.

(h) The ability to implement record keeping and perform termination procedures; and

See commentary on Section 3-12 and its subsections. The specialist may well be designated the critique lender for some hazardous materials incidents.

Figure 4.1   Adequate hazard analysis and proper risk assessment are important steps in hazardous materials incident mitigation.

(i) An understanding of the chemical, toxicological, biological, and radiological terms and their behavior.

These terms are contained in Section 4-7 and its subsections. Consult the commentary for those subsections.

4-2 Safety. The hazardous materials specialist shall be capable of the following.

4-2.1 Given changes during a hazardous materials incident, determine the adjustment of control zone boundaries and the area of evacuation.

A number of factors determine where the control zone boundaries are going to be located. This starts with a visual survey of the immediate scene of the incident to determine what hazardous materials are involved, together with evaluation of data provided by monitoring equipment. If fire is a factor, or explosion a potential factor, special consideration must be given to the establishment of safe distances. Topography and weather are important pieces of data, together with contamination, travel distances for personnel and equipment, and so on. If any of the factors, data, or information change, then thought must be given to the need for adjusting the control zone boundaries. The same factors come into consideration for determining the need for evacuation and the evacuation distances that are required.

4-2.2 Given a simulated hazardous materials incident, develop a site safety plan.

See commentary on 3-2.1, 3-2.2, and 3-2.3. All of the elements needed for implementation of a site safety plan are contained therein.

4-2.3 Given a simulated hazardous materials incident, function as the safety officer.

Earlier commentary indicated that a specialist could likely function as a hazard sector commander. This provision indicates that a specialist could also be expected to function as a safety officer at an incident. This speaks to the importance of safety at all hazardous materials incidents, as well as to the danger associated with such incidents. Safety must be a pervasive attitude that reaches every responder. To some degree, it can be said that every responder must be his or her own safety officer, concerned with personal well-being, the welfare of fellow responders, and the safety of the community served.

4-3 Resources and Planning. The hazardous materials specialist shall be capable of the following.

4-3.1   Identify the primary federal, state, regional, and local government agencies and describe their regulatory authority and regulations pertaining to the production, transportation, storage, and use of hazardous materials, and the disposal of hazardous wastes.

The primary federal agencies certainly include U.S. EPA, U.S. DOT, and U.S. DOL's OSHA, among others. Each of these agencies generally has what might be described as a state counterpart; also counties, towns, and cities may have some regulatory powers in the same subject areas.

4-3.2   Given a facility that uses, stores, or manufactures hazardous materials, or a transportation corridor, perform a hazard analysis and develop a plan to handle hazardous materials incidents, and recommend what may be needed to reduce the risk to a level within the capabilities of available resources.

In order to begin a valid hazard analysis, one must begin with identification of hazardous materials that may become involved and the quantities that may be present. This is not an impossible task where fixed facilities are located in one's administrative district, but it is a difficult problem to achieve such an identification for transportation, especially of the pass-through nature. Effective preplanning will be concerned with a number of factors, including population density, the environment, and the likelihood and severity of an incident. Evaluation of available resources must be factored in to the study. Recommendations will then vary from one incident to another, based on a total evaluation of the incident, its consequences, and what you have available to deal with it.

4-4 Incident Management.   The hazardous materials specialist shall be capable of the following.

4-4.1   Given a simulated hazardous materials incident, the hazardous materials specialist shall:

Actions will vary according to the type of simulated incident that is presented and the actual resources you have available or that can be summoned.

(a) Perform the duties as the hazard sector officer and coordinate all activities of that sector; and

The hazard sector officer is responsible for the actual mitigation of the incident, namely control of the hands-on remediation, using whatever is deemed necessary. Given a simulated incident, let your imagination have free rein, but base your actions on the actual resources that are available to you.

Figure 4.2   There are over 50,000 chemical compounds
classified as hazardous. 500 new compounds enter the
commercial market each year.

(b) Evaluate information received from on-site and off-site sources and implement appropriate action in response to the information received.

In a simulation, some information items provided to you will clearly be more important than others. You must be able to distinguish what is relevant from that which is extraneous to successful control of an incident, and you must also establish priorities with the relevant items of information. Place special emphasis on safety and caution in all cases.

4-5 Recognition of Hazardous Materials.   Reserved.

Refer to the recognition requirements for first level responders (2-2.5 and 2-3.5).

4-6 Classification, Identification, and Verification.   The hazardous materials specialist shall be capable of the following.

4-6.1   Demonstrate the ability to select and interpret pertinent data on each hazardous material, using information sources provided to the hazardous materials specialist.

This capability depends on what information sources are provided to the specialist. What is noteworthy is the fact that the requirement is to select the appropriate source and to be able to

interpret the data it provides. This requirement assures task-oriented training.

4-6.2   Given at least three unknown chemicals, one of which is a solid, one a liquid, and one a gas, demonstrate the ability to identify or classify, using resources available to the hazardous materials specialist.

Even basic resources can prove adequate to accomplishing this demonstration if the circumstances are favorable. For example, if labeling or placarding is present and visible, the DOT *Emergency Response Guidebook* will prove adequate. However, use of the word "unknown" in the section suggests that the task is more complicated and that some ingenuity may have to be coupled with rather sophisticated monitoring equipment to be successful. Appropriate completion of the competency will depend on circumstances.

4-6.3   Identify at least three references available that indicate the effects of mixing various chemicals.

There are several references that can be used, some of which are clearly more informative than others. Among NFPA publications that can fulfill this requirement are:

• NFPA 49, *Hazardous Chemicals Data*

• NFPA 325M, *Fire Hazard Properties of Flammable Liquids, Gases, and Volatile Solids*

• NFPA 491M, *Manual of Hazardous Chemical Reactions*

• NFPA 704, *Standard System for the Identification of the Fire Hazards of Materials*

Other sources include the following:

• NIOSH *Pocket Guide to Chemical Hazards*

• *Handling Chemicals Safely*, Dutch Association of Safety Experts *et al.* (1980)

• *The Condensed Chemical Dictionary*, Eighth Edition, Van Nostrant Reinhold Co., NY (1971)

• *Classification of Chemical Reactivity Hazards for the Advisory Committee to the U.S. Coast Guard*, National Academy of Sciences. Flynn and Rostow, the Dow Chemical Company, Midland, MI (1970).

For an exhaustive list of references, consult the reference section of NFPA 491M.

4-7 Chemistry and Toxicology of Hazardous Materials.   The hazardous materials specialist shall be capable of the following.

4-7.1   Define the following chemical and physical terms and explain their importance in the risk assessment process:

(a) Compound, mixture

Compounds are substances of definite composition that can be decomposed by simple chemical change into two or more different substances. Common salt, for example, can be decomposed into an active metal (sodium) and a poisonous gas (chlorine). The properties of the substances obtained by the decomposition of a compound are obviously unrelated to the properties of the compound. The new substance that results when two or more elements combine by sharing or transferring electrons are called compounds.

A mixture has no unique set of properties, but rather possesses the properties of the substances of which it is composed. Air is an example of a gaseous mixture. Each of its various components (nitrogen, oxygen, argon, water vapor, and carbon dioxide) displays its own unique properties in the mixture.

(b) Halogenated hydrocarbon

Halogenated hydrocarbon describes a vaporizing liquid that can extinguish a fire by its inhibiting effect on the combustion reaction. Examples include methyl bromide, trifluoromethyl bromide, and chlorobromomethane, among others.

(c) Ionic bond, covalent bond

The attraction that binds unlike ions together is termed an ionic bond, or an electrovalent bond. A shared pair of electrons is called a covalent bond. Compounds whose atoms are joined by covalent bonds are called covalent compounds.

(d) Salt, nonsalt

A salt is a compound comprised of the negatively charged ion from an acid and the positively charged ion from a metal of alkali base. Ionic compounds are commonly called salts. Examples are:

sodium chloride	NaCl or Na^+, Cl^-
potassium chloride	KCl or K^+, Cl^-

(e) Saturated, unsaturated hydrocarbon, aromatics

Hydrocarbons with one or more pairs of carbon atoms joined by multiple bonds are called unsaturated hydrocarbons. Carbon atoms joined by single bonds only are called saturated hydrocarbons. In saturated hydrocarbons, all the carbon atoms are saturated with hydrogen atoms.

Aromatics include any compounds that have the benzene ring in them. The term aromatics is often used to refer to the BTXs, namely benzene, toluene, and xylene. The name aromatics is derived from the characteristic odor of most aromatic compounds.

(f) Solution, slurry

A solution is a mixture in which all of the ingredients or parts are completely dissolved. It is a homogeneous mixture of the molecules, atoms, or ions of two or more different substances.

A slurry is a thin mixture of liquid and fine particles.

(g) Water miscible, immiscible.

Water miscible describes the ability of a substance to mix with water. Alcohol is water miscible. Immiscible indicates an opposite characteristic. Carbon bisulfide is immiscible with water.

4-7.2   Define the following toxicological terms and explain their importance in the risk assessment process:

Toxicology is the science that deals with the harmful effects of substances on living organisms. It is important to be familiar with common toxicological terms in order to be able to perform a proper risk assessment. Forensic toxicology concerns itself with the degree of damage to human beings caused by exposure to specific quantities of a toxic substance. Environmental toxicology is concerned with toxics that affect the health and safety in the workplace, the atmosphere, and in nutrients ingested through food and drink.

(a) Chemical interactions

Chemical interactions describe the effects of chemicals on biologic systems, emphasizing the mechanisms of harmful effects of chemcials and the conditions under which they occur.

(b) Dose-response relationship

In pharmacological terms, dose or dosage range refers to the amount of a drug to be administered. The range will be adjusted based on many factors, including body weight, pathological condition, metabolism, and so on. The response relationship would refer to the anticipated result, in cause-effect fashion.

(c) Effects: local, systemic

Local effects involve a specific and restricted part of an organism. Systemic effects refer to involvement of functions or systems of an organism. Toxics may cause localized reactions, or may initiate conditions that interfere with the proper operation of a body system. Similarly, drugs that are administered for therapeutic reasons can have local or systemic effects.

(d) Exposure: acute, subacute, chronic

An acute exposure is one that involves an intense dose of short duration. Subacute indicates or refers to toxicity after repeated exposures over a long term. Chronic exposure refers to the enduring, long-term period.

(e) Routes of entry: ingestion, absorption, inhalation.

Routes of entry refer to the access routes that toxic substances can take. Ingestion refers to access via the gastrointestinal tract. Absorption indicates percutaneous access, or through the skin. Inhalation indicates access through the respiratory tract.

4-7.3 Define and explain the following radiological terms:

(a) Half-life

Half-life is a measure of the rate of decay of a radioactive material. It indicates the period of time needed for one half of a given amount of a radioactive substance to change to another nuclear form or element.

(b) Time, distance, shielding.

Time, distance, and shielding describe methods of protecting oneself from harmful exposures to radiation. The less time of exposure, the less the dosage. Distance from the exposure serves also to decrease the amount of radiation, inversely to the square of the distance. Shielding refers to the ability to block radiation by use of varying thicknesses of different materials.

4-8 Hazard and Risk Assessment.   The hazardous materials specialist shall be capable of the following.

Figure 4.3   Proper hazard and risk assessment provides the basis for decision making at an incident.

See Supplement III for information on risk analysis.

4-8.1   Given a specific hazardous materials classification, describe the properties and expected behavior emphasizing safety, decision making, and size-up considerations.

> **This competency would require, at the outset, that the specialist level be familiar with the properties of all classifications of hazardous materials. Given that knowledge, coupled with an exhibited ability to perform an effective size-up, to operate as a sector safety officer [4-4.1(a)], and to be able to work in the incident command system, other information would still be needed to perform risk analysis. Properties or characteristics of the hazardous material would not provide enough information, although "size-up considerations" does imply that other items would be considered. The responder would need to know more about the location with all of the ramifications that accompany that aspect, the amount involved, exposure to population and environment, the likelihood of something going amiss, and the consequences of an accident.**

4-8.2   Demonstrate the purpose, operation (including interpretation of results and operational and calibration checks), and limitations of the following specialized monitoring equipment, in addition to that provided to the hazardous materials specialist by the authority having jurisdiction:

Since "demonstrate" requires that the specialist show by actual use, it implies that the items (a) through (d) should be part of the equipment associated with response of hazardous materials specialists. However, the definition of "demonstrate" also indicates that simulation, explanation, illustration, or a combination of these can also be used.

Naturally, the specialist should be intimately familiar with all other monitoring equipment that is provided.

(a) Organic vapor analyzer

A description of an organic vapor analyzer, and a comparison of an OVA with a specific photoionization detector, follows. It is taken from U.S. EPA's *Standard Operating Safety Guides*.

INTRODUCTION

The HNU Photoionizer and the Foxboro Organic Vapor Analyzer (OVA) are used in the field to detect a variety of compounds in air. The two instruments differ in their modes of operation and in the number and types of compounds they detect (Table I-1). Both instruments can be used to detect leaks of volatile substances from drums and tanks, determine the presence of volatile compounds in soil and water, make ambient air surveys, and collect continuous air monitoring data. If personnel are thoroughly trained to operate the instruments and to interpret the data, these instruments can be valuable tools for helping to decide the levels of protection to be worn, assist in determining other safety procedures, and determine subsequent monitoring or sampling locations.

OVA

The OVA operates in two different modes. In the survey mode, it can determine approximate total concentration of all detectable species in the air. With the gas chromatograph (GC) option, individual components can be detected and measured independently, with some detection limits as low as a few parts per million (ppm).

In the GC mode, a small sample of ambient air is injected into a chromatographic column and carried through the column by a stream of hydrogen gas. Contaminants with different chemical structures are retained on the column for different lengths of time (known as retention times) and hence are detected separately by the flame ionization detector. A strip chart recorder can be used to record the retention times, which are then compared to the retention times of a standard with known chemical constituents.

The sample can either be injected into the column from the air sampling hose or injected directly with a gas-tight syringe.

In the survey mode, the OVA is internally calibrated to methane by the manufacturer. When the instrument is adjusted to manufacturer's instructions it indicates the true concentration of methane in air. In response to all other detectable compounds, however, the instrument reading may be higher or lower than the true concentration. Relative response ratios for substances other than methane are available. To correctly interpret the readout, it is necessary to either make calibration charts relating the instrument readings to the true concentration or to adjust the instrument so that it reads correctly. The OVA has an inherent limitation in that it can detect only organic molecules. Also, it should not be used at temperatures lower than about 40 degrees Fahrenheit because gases condense in the pump and column. It has no column temperature control (although temperature control kits are available), and since retention times vary with ambient temperatures for a given column, determinations of contaminants are difficult. Despite these limitations, the GC mode can often provide tentative information on the identity of the contaminants in air without relying on costly, time-consuming laboratory analysis.

HNU

The HNU portable photoionizer detects the concentration of organic gases as well as a few inorganic gases. The basis for detection is the ionization of gaseous species. Every molecule has a characteristic ionization potential (I.P.) which is the energy required to remove an electron from the molecule, yielding a positively charged ion and the free electron. The incoming gas molecules are subjected to ultraviolet (UV) radition, which is energetic enough to ionize many gaseous compounds. Each molecule is transformed into charged ion pairs, creating a current between two electrodes.

Three probes, each containing a different UV light source, are available for use with the HNU. Ionizing energies of the probe are 9.5, 10.2, and 11.7 electron volts (eV). All three detect many aromatic and large molecule hydrocarbons. The 10.2 eV and 11.7 eV probes, in addition, detect some smaller organic molecules and some halogenated hydrocarbons. The 10.2 eV probe is the most useful for environmental response work, as it is more durable than the 11.7 eV probe and detects more compounds than the 9.5 eV probe.

The HNU factory calibration gas is benzene. The span potentiometer (calibration) knob is turned to 9.8 for benzene calibration.

A knob setting of zero increases the response to benzene approximately tenfold. As with the OVA, the instrument's response can be adjusted to give more accurate readings for specific gases and eliminate the necessity for calibration charts.

While the primary use of the HNU is as a quantitative instrument, it can also be used to detect certain contaminants, or at least to narrow the range of possibilities. Noting instrument response to a contaminant source with different probes can eliminate some contaminants from consideration. For instance, a compound's ionization potential may be such that the 9.5 eV probe produces no response, but the 10.2 eV and 11.7 eV probes do elicit a response. The HNU does not detect methane.

The HNU is easier to use than the OVA. Its lower detection limit is also in the low ppm range. The response time is rapid; the meter needle reaches 90% of the indicated concentration in 3 seconds for benzene. It can be zeroed in a contaminated athosphere and does not detect methane.

GENERAL CONSIDERATIONS

Both of these instruments can monitor only certain vapors and gases in air. Many nonvolatile liquids, toxic solids, particulates, and other toxic gases and vapors cannot be detected. Because the types of compounds that the HNU and OVA can potentially detect are only a fraction of the chemicals possibly present at an incident, a zero reading on either instrument does not necessarily signify the absence of air contaminants.

The instruments are non-specific, and their response to different compounds is relative to the calibration setting. Instrument readings may be higher or lower than the true concentration. This can be an especially serious problem when monitoring for total contaminant concentrations if several different compounds are being detected at once. In addition, the response of these instruments is not linear over the entire detection range. Care must therefore be taken when interpreting the data. All identifications should be reported as tentative until they can be confirmed by more precise analysis. Concentrations should be reported in terms of the calibration gas and span potentiometer or gas-select-knob setting.

Since the OVA and HNU are small, portable instruments, they cannot be expected to yield results as accurate as laboratory instruments. They were originally designed for specific industrial applications. They are relatively easy to use and interpret when detecting total concentrations of individually known contaminants in air, but interpretation becomes extremely difficult when trying to quantify the components of a mixture. Neither instrument can be used as an indicator for combustible gases or oxygen deficiency.

The OVA (Model 128) is certified by Factory Mutual to be used in Class I, Division 1, Groups A, B, C, and D environments. The HNU is certified by Factory Mutual for use in Class I, Division 2, Groups A, B, C, and D.

(b) Passive dosimeter

Passive dosimeters are used for monitoring of gases and vapors, and primarily for surveillance of personal exposure. They can, however, be used to survey or monitor an area or site. These

Table I-1   Comparison of the OVA and HNU

	OVA	HNU
Response	Responds to many organic gases and vapors.	Responds to many organic and some inorganic gases and vapors.
Application	In survey mode, detects total concentrations of gases and vapors. In GC mode, identifies and measures specific compounds.	In survey mode, detects total concentrations of gases and vapors. Some identification of compounds possible, if more than one probe is used.
Detector	Flame ionization detector (FID).	Photoionization detector (PID).
Limitations	Does not respond to inorganic gases and vapors. Kit available for temperature control.	Does not respond to methane. Does not detect a compound if probe has a lower energy than compound's ionization potential.
Calibration Gas	Methane.	Benzene.
Ease of Operation	Requires experience to interpret correctly, especially in GC mode.	Fairly easy to use and interpret.
Detection Limits	0.1 ppm (methane).	0.1 ppm (benzene).
Response Time	2–3 seconds (survey mode) for CH_4.	3 seconds for 90% of total concentration of benzene.
Maintenance	Periodically clean and inspect particle filters, valve rings, and burner chamber. Check calibration and pumping system for leaks. Recharge battery after each use.	Clean UV lamp frequently. Check calibration regularly. Recharge battery after each use.
Useful Range	0–1000 ppm.	0–2000 ppm.
Service Life	8 hours; 3 hours with strip chart recorder.	10 hours; 5 hours with strip chart recorder.

devices are of two types, the diffusion sampler and the permeation device. With the diffusion sampler, molecules cross a concentration gradient between the contaminated atmosphere and the indicator material. Permeation devices depend on the natural permeation of a contaminant through a membrane. These devices can be used to pinpoint a single contaminant from a mixture. A membrane that is permeable to a specific contaminant, but impermeable to others, is selected to achieve this end.

Some passive dosimeters are read directly, while others require laboratory analysis.

(c) Personal air monitoring equipment

There are active and passive devices that can be used for personal air monitoring or air sampling. Responders can be equipped with these samplers to indicate the presence of contaminants at specific locations. The devices are attached to the personnel, usually in close proximity to the mouth and nose. The monitoring equipment indicates the potential for the person wearing the device for inhaling the contaminant.

(d) Photoionization detectors.

See commentary on (a).

4-8.3 Demonstrate the field maintenance and testing procedures for the monitoring equipment provided to the hazardous materials specialist by the authority having jurisdiction.

Manufacturer's instructions would be essential to successful completion of this task.

4-9 Personal Protective Equipment. The hazardous materials specialist shall be capable of the following.

4-9.1 Identify the three types of Level A chemical-protective clothing and describe the advantages and disadvantages of each type. (Reference: Type 1, SCBA worn inside the suit; Type 2, SCBA worn outside the suit; and Type 3, breathing air supplied by air line.)

The Type 1 configuration provides a positive pressure inside the suit, which lessens the likelihood of contamination, at least for a short time, even if small leaks occur in the suit. In the event of problems with the SCBA, there will be some air in the suit for a few minutes, during which escape may be achieved. On the negative side, the heavier suit will hasten fatigue and slow down the wearer, it is difficult to replace air cylinders considering that decontamination will be necessary first, the low air pressure alarm may not be heard, and SCBA control valves are not accessible for operating.

The Type 2 arrangement provides greater comfort and access to the SCBA. However, it exposes the SCBA to the contaminated environment, which may cause damage or may render the SCBA expendable. Achieving a seal where the facepiece and hood meet can pose some difficulty. It should be noted that disposable protective hoods and air cylinder covers can help alleviate contamination to the SCBA.

The Type 3 suit allows for prolonged work time since there is ample air supply that provides positive pressure and some cooling effects inside the suit. Disadvantages include the longer work time for the wearer, the air hose length and mobility limitation, and exposure of the air hose to spilled material.

4-9.2 Describe the physical and psychological stresses that can affect users of specialized protective clothing.

There are many physical stresses present for wearers of specialized protective clothing, and their presence will induce a variety of psychological stresses. The physical stresses include fatigue from operating in cumbersome clothing and equipment for the long periods of time needed at hazardous materials incidents, heat stress induced by the working environment, limited mobility, and limited visibility. These physical stresses would certainly bring on anxiety, disorientation, and even fear on the part of the responder.

4-9.3 Describe the testing and maintenance and storage procedures for the type(s) of Level A chemical-protective clothing furnished to the hazardous materials specialist by the authority having jurisdiction.

Consult manufacturer's instructions for prescribed testing, maintenance, and storage procedures.

4-9.4 Demonstrate the correct method of donning and doffing and usage of the type(s) of Level A chemical-protective clothing, including communication equipment, furnished to the hazardous materials specialist by the authority having jurisdiction.

Demonstration would require a hands-on performance. There should be an established written procedure for the donning and doffing of all chemical-protective clothing. Manufacturers of the clothing should furnish prescribed procedures for this activity, in addition to instructions for cleaning, testing, monitoring, and stowing the equipment.

4-9.5 Describe safety procedures for personnel wearing Level A chemical-protective clothing.

There are general safety procedures that should be part of a program that applies to all responders who will wear chemical-protective clothing. These include medical and physical fitness programs that incorporate medical surveillance, periodic scheduled medical examinations, and maintenance of medical records for all personnel.

Different jurisdictions will have varying response protocols. When Level A chemical-protective clothing is used, however, there are some procedures that should prevail everywhere. Medical monitoring of personnel should be a part of the operation. Some protocols require this even on initial entry to the site. The buddy system is essential, and there should be an adequate backup capability that can act quickly and appropriately in the event of an emergency. Contamination should be avoided to the ultimate extent possible. There must be adequate and continuous communication capability, visual contact where possible, and an appropriate escape route for personnel. Operating time should be monitored and limited.

It is essential that safety considerations be given major emphasis in the standard operating procedures.

Figure 4.4   It is important that all crews operating in a hostile environment have both a primary and a secondary form of communications. Do not overlook the need for effective hand signals when operating in contaminated atmospheres.

4-9.6 Given a hazardous materials incident, determine the selection of chemical-protective clothing.

> See commentary on 3-9.6.

4-9.7 Demonstrate the ability to log completely the use, repair, and testing of chemical-protective clothing.

> **This demonstration will depend on standard operating procedures, availability of forms for record keeping or the need to maintain a log without use of a form, and manufacturer's instructions for the specific type of chemical-protective clothing.**

4-9.8 Given an incident involving a hazardous material that is flammable and corrosive and has other hazardous properties, describe the considerations to be evaluated for protecting personnel against a flash fire when wearing chemical-protective clothing.

> **Even given the specifics of the situation, namely a substance that is flammable, corrosive, and otherwise harmful or unfriendly, there are still variables at work that permit latitude in a description of what to do and what to avoid. These include the degree of hazard associated with the substance, presence of fire, nearby sources of ignition, location, ambient conditions, exposures, and so on almost without end. A given in the described incident is that personnel will be wearing chemical-protective clothing, so an assumption can be made that a proper decision has been made to take some remedial action. A good rule of thumb for any incident involving the presence of flammable vapor is to avoid entering the vapor cloud. Even close approach may be unwise, since ignition can be explosive or otherwise catastrophic. Given the suggestion in the text that our concern should be to protect against a "flash fire," we have to assume a fairly benign, nondevastating ignition might take place. There are available some chemical-protective ensembles that have flash protective characteristics, but their effectiveness is limited and only for brief exposures. Chemical-protective clothing itself does not offer protection against a flash fire.**

4-9.9 Identify three methods of cooling Level A chemical-protective clothing and describe the advantages and disadvantages of each method.

> **Cooling can be achieved by the use of one of two cooling garments or by use of an air line cooling unit.**
>
> **(a) The air line circulates cool, dry air through the suit, but they are not economical because so much air is needed.**
>
> **(b) A jacket or vest with compartments for ice; these pose storage and recharge problems.**

(c) Circulating cool water through tubes by use of a pump. This system presents ice storage problems and added weight.

4-10 Hazardous Materials Control. The hazardous materials specialist shall be capable of the following.

4-10.1 Describe the purpose of, equipment required, procedures for, and safety precautions used with the following techniques for hazardous materials control:

(a) Transferring liquids and gases

The purpose of transferring liquids and gases is often associated with removal from a damaged or suspect containment vessel to one that is suitable and undamaged. The task is generally performed by people with particular expertise and responsibility for the operation, under the supervision and control of responders. The equipment will vary with the type of vessel involved and the hazardous material being transferred. However, the equipment must be appropriate and suitable. Transfer procedures are generally established in advance by good industry practice. Only essential personnel should be involved in the process, and the operation is in the overall command, in most cases, of the hazard sector officer.

(b) Flaring liquids and gases

Flaring liquids and gases is a procedure that can be used to deplete the amount of hazardous material involved or to depressurize a vessel. It should be performed only by persons who are trained in the technique. In some cases, where the burning is already underway at a vent or pressure relief valve when responders arrive, a prescribed procedure will be to allow the flaring process to continue. Determination must be made that the flaring does not present a source of ignition to other spilled flammable substances.

(c) Hot tapping

Hot tapping describes a situation in which welding or cutting is done on a container or piping while it contains a flammable liquid or gas. It is an emergency procedure where circumstances necessitate the work and where the contents cannot be removed prior to performance of the work. The work must be done only by those who have been trained to do such procedures. There are established procedures for hot work. Consult API 2201 of the American Petroleum Institute entitled *Welding or Hot Tapping on Equipment Containing Flammables*.

(d) Vent and burn.

Vent and burn describes a highly specialized and seldom practically used technique that involves the use of shaped charges that are designed to create an opening in a container and set fire to its contents. It was used successfully at the Livingston, Louisiana train derailment in 1982, but under the control and direction of experts. The purpose is to deplete the supply of contained hazardous flammable substances when other methods are unavailable.

4-10.2   Describe the maintenance and testing procedures for the tools and equipment provided to the hazardous materials specialist by the authority having jurisdiction for the control of hazardous materials releases.

The specialist's description will depend on what equipment is available, on established maintenance and testing procedures, and, in most cases, on directions received from the manufacturer.

4-11 Decontamination.   The hazardous materials specialist shall be capable of the following.

4-11.1   Given a simulated hazardous materials incident, and using available local resources, obtain the necessary technical information to develop a decontamination procedure.

See Supplement VIII for information on a decontamination procedure.

4-12 Termination Procedures.   The hazardous materials specialist shall be capable of the following.

4-12.1   Given a simulated incident, prepare any locally required incident reports with supporting documentation.

The specialist's preparation will depend on local reports, which may differ substantially from one jurisdiction to another.

4-12.2   Given a simulated incident, conduct a critique.

The very nature of emergency incidents, where decisions are made quickly, under adverse conditions, and without benefit of complete information, often result in less than perfect operations. That is one of the principal justifications for the post-incident critique. If conducted properly, the critique will serve to improve future operations. It should never be done haphazardly, nor should it be simply a description of what each person or unit did. Those participating in the critique must look on the process as one that allows them to recount what went wrong. The decision-makers at the incident must be similarly disposed. Mutual respect must be present.

Figure 4.5   When the nature of the chemical remains unknown
and it is necessary to commit emergency forces to the hazard
sector, complete decontamination procedures are implemented. As
knowledge about the hazard is gained, the procedures can be
downgraded as necessary.

If the critique can uncover shortcomings in preplanning or in standard operating procedures, then effective changes can be instituted to improve future operations.

Critiques of complex incidents can be simplified by adequate preparation on the part of the critique leader, and the process can be enhanced by the use of visual aids.

Every incident should be subjected to the critique process. Complex incidents should be concluded with a written report on the operation, and the report should offer recommendations based on the critical analysis. The lessons learned from one incident can serve to overcome the shortcomings at future operations.

4-13 Educational and Medical Requirements.   Reserved.

5 Referenced Publications

5-1 The following documents or portions thereof are referenced within this standard and shall be considered part of the requirements of this document. The edition indicated for each reference is the current edition as of the date of the NFPA issuance of this document.

5-1.1 NFPA Publication. National Fire Protection Association, Battery-march Park, Quincy, MA 02269.

NFPA 704-1985, *Standard System for the Identification of Fire Hazards of Materials*.

5-1.2 Other Publications.

5-1.2.1 US Government Publications. US Government Printing Office, Superintendent of Documents, Washington, DC 20402.

Title 29 CFR Part 1910.120

Title 29 CFR Part 1910.1200

Title 40 CFR Part 261.33

Title 40 CFR Part 302

Title 40 CFR Part 355

Title 49 CFR Parts 170-179

Emergency Response Guidebook-1987, US Department of Transportation DOT P 5800.4

Appendix A

The material contained in Appendix A is included in the text within this *Handbook* and therefore is not repeated here.

Appendix B

Hazardous Materials Management

This Appendix is not a part of the requirements of this NFPA document, but is included for information purposes only.

B-1 Incident Commander.

B-1.1 Role.

(a) Responsible for the direction and coordination of all aspects of the incident, from initial response through stabilization.

(b) Operates within the scope of an integrated emergency management plan.

(c) Operates under clear procedures for notification and utilization of nonlocal resources (including private, and state and federal government sector personnel).

(d) Directs resources (private, governmental, and others) with expected task assignments and on-scene activities, based on their capabilities.

(e) Provides management overview, technical review, and logistical support to private and government sector personnel.

(f) Provides focal point for information transfer to media and local elected officials.

(g) Provides subsequent documentation of the hazardous materials incident.

(h) Advises on the reporting requirements of federal, state, and local agencies.

(i) Conducts a critique of the incident.

B-1.2 Training. The incident commander has completed at least first responder qualifications and maintains retraining requirements.

B-2 Private Sector Personnel. The private sector is likely to respond with personnel who fit into two distinct categories:

Private Sector Manager — Has overall responsibility for the direction and coordination of all activities of his/her own organization in conjunction with activities of the rest of the response community; will generally remain outside the hot zone.

Private Sector Technician — Functions as technical resource to private sector manager; is qualified to work within perimeter of hot zone.

B-2.1 Private Sector Manager.

B-2.1.1 Role.

(a) Provides management review of all company and contractor personnel employed.

(b) Reports to command post upon arrival.

(c) Provides detailed written work proposal and summaries to incident commander as requested.

(d) Obtains consensus support from incident commander on all phases of activity before proceeding with those activities.

(e) Provides safety and health program for company personnel.

(f) Ensures that company and contractor personnel have the appropriate personal protective equipment necessary for activities within the hot zone.

(g) Assesses the potential of both short- and long-term effects of a hazardous materials incident, including environmental protection considerations.

(h) Remains on the scene until incident is stabilized.

(i) Remains as contact person until incident is terminated.

(j) Operates under detailed company contingency plan.

(k) Has authority to commit company resources and money to support response effort.

B-2.1.2 Training. The private sector manager has completed the first responder level of training and maintains retraining requirements.

B-2.2 Private Sector Technician.

B-2.2.1 Role.   If the private sector technician responds without the private sector manager, the private sector technician will also assume role of and perform as the private sector manager.

(a) Reports to private sector manager upon arrival who reports same to incident commander.

(b) Provides technical support as requested from incident commander through private sector manager.

(c) Works in appropriate protective equipment within the perimeter of the hot zone, as requested by incident commander through private sector manager.

(d) Operates under detailed safety and health program and company contingency plan.

(e) Provides detailed work proposal and summaries to private sector manager, who relays that information to incident commander.

(f) Obtains consensus support from incident commander through private sector manager on all phases of activities before proceeding with those activities.

B-2.2.2 Training.   The private sector technician has completed the first responder through hazardous material technician levels of training and maintains retraining requirements. This training may be specifically oriented towards those products or containers handled.

B-3 Government Sector.   The government sector (state and federal) is likely to respond with personnel who fit into two distinct categories:

Government Sector Manager — Has overall responsibility for the direction and coordination of all activities of own organization in conjunction with activities of rest of response community; will generally remain outside the hot zone perimeter.

Government Sector Technician — Functions as technical resource to government sector manager; will be qualified to work within perimeter of hot zone.

B-3.1 Government Sector Manager.

B-3.1.1 Role.

(a) Provides management review of all government and contractor personnel employed.

(b) Reports to command post upon arrival.

(c) Provides detailed written work proposal and summaries to incident commander, as requested.

(d) Obtains consensus support from incident commander on all phases of activity before proceeding with those activities.

(e) Provides safety and health program for governmental personnel.

(f) Ensures that governmental personnel have the appropriate personal protective equipment necessary for activities within hot zone.

(g) Assesses the potential of both short- and long-term effects of a hazardous materials incident, including environmental protection considerations.

(h) Remains on the scene until incident is stabilized.

(i) Remains the contact person until incident is terminated.

(j) Operates under detailed government contingency plan.

(k) Has authority to commit government resources and money to support response effort.

B-3.1.2 Training.   The government sector manager has completed the first responder level of training and maintains retraining requirements.

B-3.2 Government Sector Technician.

B-3.2.1 Role.   If the government sector technician responds without the government sector manager, the government sector technician will also assume role of and perform as the government sector manager.

(a) Reports to government sector manager upon arrival, who reports same to incident commander.

(b) Provides technical support as requested from incident commander through government sector manager.

(c) Works in appropriate protective equipment within the perimeter of the hot zone, as requested by incident commander through government sector manager.

(d) Operates under detailed safety and health program and government contingency plan.

(e) Provides detailed work proposal and summaries to government sector manager, who relays that information to incident commander.

(f) Obtains consensus support from incident commander through government sector manager on all phases of activities before proceeding with those activities.

B-3.2.2 Training.   The government sector technician has completed the first responder through hazardous material technician levels of training.

Appendix C

Referenced Publications

C-1 The following documents or portions thereof are referenced within this standard for informational purposes only and thus are not considered part of the requirements of this document. The edition indicated for each reference is the current edition as of the date of the NFPA issuance of this document.

C-1.1 National Fire Academy Publication. National Fire Academy, Federal Emergency Management Agency, Emmitsburg, MD.

Hazardous Materials Incident Analysis.

C-1.2 Oklahoma State University, Fire Protection Publications, Stillwater, OK 74078-0008.

Haz Mat Response Team Spill and Leak Guide, 1984.

C-1.3 U.S. Government Publication. U.S. Government Printing Office, Superintendent of Documents, Washington, DC 20402.

Occupational Safety and Health Guidance Manual for Hazardous Waste Site Activities.

NFPA 471

Recommended Practice for

Responding to Hazardous Materials Incidents

1989 Edition

NOTICE: An asterisk (*) following the number or letter designating a paragraph indicates explanatory material on that paragraph in Appendix A. Material from Appendix A is integrated with the text and is identified by the letter A preceding the subdivision number to which it relates.

Information on referenced publications can be found in Chapter 8.

1 Administration

1-1* Scope. This practice applies to all organizations that have responsibilities when responding to hazardous materials incidents and recommends standard operating guidelines for responding to such incidents. It specifically covers planning procedures, policies, and application of procedures for incident levels, personal protective equipment, decontamination, safety, and communications.

A-1-1 Many of the recommendations in this document are based on United States federal laws and regulations that were in effect at the time of adoption. Users should carefully review laws and regulations that may have been added or amended or that may be required by other authorities. Users outside the jurisdiction of the United States should determine what requirements may be in force at the time of application of this document.

The NFPA definition of recommended practice is as follows: a document containing only advisory provisions (using the word "should" to indicate recommendations) in the body of the text.

Note that this document does not limit its application to fire departments. It is the intent of the committee that the practices outlined in this recommended practice be suitable for use by, and have application to, all persons who might be called on to respond to a hazardous materials incident. Fire service personnel come immediately to mind as front line emergency responders, as, on frequent occasions, do police department responders. Other affected entities might include government and industrial responders, as well as workers engaged in operations that involve everyday dealing with hazardous materials, in the broadest sense of the definition.

1-2 Purpose. The purpose of this document is to outline the minimum requirements that should be considered when dealing with responses to hazardous materials incidents and to specify operating guidelines for responding to hazardous materials incidents. It is not the intent of this recommended practice to restrict any jurisdiction from using more stringent guidelines.

Hazardous materials incidents will differ widely in terms of intensity, duration, and significance, as will the capabilities and

143

preparedness of responders. Every incident brings a new set of variables, and subjective judgements must be made on appropriate courses of action. Nonetheless, there are helpful guidelines that are suitable for all incidents, and that is the underlying purpose of this recommended practice.

1-3 Application. The recommendations contained in this document should be followed by organizations that respond to hazardous materials incidents and by incident commanders responsible for managing hazardous materials incidents.

The federally mandated establishment of local emergency planning committees (LEPCs) will serve to identify those organizations that have any likelihood of being summoned for assistance at hazardous materials incidents. Incident commanders are (or should be) aware of their potential for confronting such incidents, and they should prepare themselves for the eventualities that may occur.

1-4 Definitions.

Authority Having Jurisdiction. The "authority having jurisdiction" is the organization, office or individual responsible for "approving" equipment, an installation or a procedure.

NOTE: The phrase "authority having jurisdiction" is used in NFPA documents in a broad manner since jurisdictions and "approval" agencies vary as do their responsibilities. Where public safety is primary, the "authority having jurisdiction" may be a federal, state, local or other regional department or individual such as a fire chief, fire marshal, chief of a fire prevention bureau, labor department, health department, building official, electrical inspector, or others having statutory authority. For insurance purposes, an insurance inspection department, rating bureau, or other insurance company representative may be the "authority having jurisdiction." In many circumstances the property owner or his designated agent assumes the role of the "authority having jurisdiction"; at government installations, the commanding officer or departmental official may be the "authority having jurisdiction."

Confinement. Those procedures taken to keep a material in a defined or local area.

Containment. Those procedures taken to keep a material in its container.

For purposes of clarity, a distinction is made between containment and confinement. A large fuel oil tank is designed to contain its stored product. The dike surrounding the tank is intended to confine the fuel oil in the event of a spill. The use of plugging or patching techniques falls under containment, while improvising a dike to direct or limit a spill is confinement.

Contaminant/Contamination. A substance or process that poses a threat to life, health, or the environment.

Figure 1.1   Containment involves keeping a hazardous material in a
suitable container.

A contaminant possesses some property or characteristic that
renders it threatening. Water can cause flooding, which is life
threatening, but it is not a contaminant.

Control.   The procedures, techniques, and methods used in the mitigation
of a hazardous materials incident, including containment, extinguishment,
and confinement.

This includes whatever measure is taken to address an incident.
It could include containment and confinement, among many
other available procedures or actions.

Control Zones.   The designation of areas at a hazardous materials incident
based upon safety and the degree of hazard. Many terms are used to describe
the zones involved in a hazardous materials incident. For purposes of this
standard, these zones shall be defined as the hot, warm, and cold zones.

See commentary under the definition in NFPA 472.

Decontamination (Contamination Reduction).   The physical and/or chemical process of reducing and preventing the spread of contamination from
persons and equipment used at a hazardous materials incident.

In the broad sense, decontamination includes the preventive measures that are taken that will protect against contamination. To the practical extent possible, contamination is to be avoided. The narrow meaning involves the safe and effective physical removal of contaminants or the use of a chemical treatment that will render the contaminant less harmful. Intelligent planning, use of proper protective equipment, and deployment of control zones are major factors to be considered in dealing with the problem. The extent of decontamination needed is based largely on the degree of harm associated with the contaminant.

Degradation.   A chemical action involving the molecular breakdown of a protective clothing material due to contact with a chemical. The term degradation may also refer to the molecular breakdown of the spilled or released material to render it less hazardous.

The degradation process can also occur if protective clothing is not properly maintained and stored. Physical circumstances can accelerate degradation. When used in association with mitigation efforts, degradation can refer to the breakdown of complex chemicals into simpler forms. We speak of products as being biodegradable, meaning that they can be broken down by bacteria into basic elements. Bacteria have been used successfully in breaking down underground spills of petroleum products.

Emergency.   A sudden and unexpected event calling for immediate action.

Implicit in this definition is the notion that the event is an untoward one that requires urgent action for control or remediation.

Environmental Hazard.   A condition capable of posing an unreasonable risk to air, water, or soil quality and to plants or wildlife.

The usual occurrence involves a release or potential release of a hazardous material that will endanger the environment. The significance of an environmental hazard should not be given less concern than more immediate visible occurrences like fire or explosion, since people are a part of the environment, and since public health and welfare are severely affected by environmental hazards. The Valdez oil spill in early 1989 is an example of an environmental hazardous incident that will have long lasting and widespread adverse effects.

Hazard/Hazardous.   Capable of posing an unreasonable risk to health and safety (Department of Transportation). Capable of doing harm.

Hazard Sector.   That sector of an overall incident command system that deals with the actual mitigation of an incident. It is directed by a sector officer and principally deals with the technical aspects of the incident.

Figure 1.2 An emergency can happen at fixed facilities or on any transportation corridor that passes through the community.

Figure 1.3 On April 21, 1980, fire erupted at a chemical disposal facility in Elizabeth, New Jersey. Fire fighters encountered exploding barrels of unknown chemicals, toxic fumes, and searing heat.

There may be several sectors at an incident apart from the hazard sector. These might deal with safety, public information, medical needs, logistics, and so on. Each sector has someone designated as the person in charge, and that person is under the command of, and reports to, the incident commander.

Hazard Sector Officer.   The person responsible for the management of the hazard sector.

Hazardous Material.*   A substance (gas, liquid, or solid) capable of creating harm to people, property, and the environment. See specific regulatory definitions in Appendix A.

Class.   The general grouping of hazardous materials into nine categories identified by the United Nations Hazard Class Number System, including:

Explosives

Gases (compressed, liquefied, dissolved)

Flammable Liquids

Flammable Solids

Oxidizers

Poisonous Materials

Radioactive Materials

Corrosive Materials

Other Regulated Materials

Classification.   The individual divisions of hazardous materials called "hazard classes" in the United States and "divisions" in the United Nations system, including:

Explosives A	Flammable Solids
Explosives B	Oxidizers
Explosives C	Organic Peroxides
Blasting Agents	Poison Gases
Nonflammable Gases	Poison B
Flammable Gases	Irritating Materials
Flammable Liquids	Etiologic Agents
Combustible Liquids	Radioactive Materials

Corrosive Materials	ORM C
ORM A	ORM D
ORM B	ORM E

A-1-4 Hazardous Material. There are many definitions and descriptive names being used for the term hazardous material, each of which depends on the nature of the problem being addressed.

Unfortunately, there is no one list or definition that covers everything. The United States agencies involved, as well as state and local governments, have different purposes for regulating hazardous materials that, under certain circumstances, pose a risk to the public or the environment.

(a) *Hazardous Materials.* The United States Department of Transportation (DOT) uses the term *hazardous materials*, which covers eight hazard classes, some of which have subcategories called classifications, and a ninth class covering other regulated materials (ORM). DOT includes in its regulations hazardous substances and hazardous wastes as an ORM-E, both of which are regulated by the Environmental Protection Agency (EPA), if their inherent properties would not otherwise be covered.

(b) *Hazardous Substances.* EPA uses the term *hazardous substance* for the chemicals which, if released into the environment above a certain amount, must be reported and, depending on the threat to the environment, federal involvement in handling the incident can be authorized. A list of the hazardous substances is published in 40 CFR Part 302, Table 302.4.

(c) *Extremely Hazardous Substances.* EPA uses the term *extremely hazardous substance* for the chemicals which must be reported to the appropriate authorities if released above the threshold reporting quantity. Each substance has a threshold reporting quantity. The list of extremely hazardous substances is identified in Title III of Superfund Amendments and Reauthorization Act (SARA) of 1986 (40 CFR Part 355).

(d) *Toxic Chemicals.* EPA uses the term *toxic chemical* for chemicals whose total emissions or releases must be reported annually by owners and operators of certain facilities that manufacture, process, or otherwise use a listed toxic chemical. The list of toxic chemicals is identified in Title III of SARA.

(e) *Hazardous Wastes.* EPA uses the term *hazardous wastes* for chemicals that are regulated under the Resource, Conservation and Recovery Act (40 CFR Part 261.33). Hazardous wastes in transportation are regulated by DOT (49 CFR Parts 170-179).

(f) *Hazardous Chemicals.* The United States Occupational Safety and Health Administration (OSHA) uses the term *hazardous chemical* to denote any chemical that would be a risk to employees if exposed in the work place. Hazardous chemicals cover a broader group of chemicals than the other chemical lists.

(g) *Hazardous Substances*. OSHA uses the term *hazardous substance* in 29 CFR Part 1910.120, which resulted from Title I of SARA and covers emergency response. OSHA uses the term differently than EPA. Hazardous substances, as used by OSHA, cover every chemical regulated by both DOT and EPA.

Incident.   The release or potential release of a hazardous material into the environment.

This is an occurrence that requires some order of response from the emergency response community.

Incident Command System.   An organized system of roles, responsibilities, and standard operating procedures used to manage and direct emergency operations.

Incident Commander.   The person responsible for all decisions relating to the management of the incident. The incident commander is in charge at the incident.

While one person is (and should be) in overall responsible charge of an incident, his decisions will be based on the flow of information that is made available to him. At a major incident of

Figure 1.4   Some materials are multiple hazards, falling in a number of classifications.

prolonged duration, it would not be unusual to have dozens of agencies, departments, administrations, and institutions playing some role in advising the person in charge. However, the decision-making falls to that one person in charge – the incident commander.

Mitigation.   Actions taken to prevent or reduce product loss, property damage, human injury or death, and environmental damage due to the release or potential release of hazardous materials.

All actions taken to address a hazardous materials incident fall under the broad meaning of the term mitigation. The mitigation process serves to resolve or remediate the situation. Final mitigation can be an enduring, long-lasting procedure.

Monitoring Equipment.   Instruments and devices used to identify and quantify contaminants.

In addition to determining the level of protective equipment needed, monitoring equipment can be useful in establishing control zone boundaries and also in evaluating the ill effects of exposure to the contaminants. Identification of the class to which a contaminant belongs usually comes first. Determination of the amount or concentration of the contaminant is not as easily achieved.

Figure 1.5   Potential releases of hazardous materials must also be dealt with during the mitigation process.

National Contingency Plan.* Policies and procedures of the federal agency members of the National Oil and Hazardous Materials Response Team. This document provides guidance for responses, remedial action, enforcement, and funding mechanisms for hazardous materials incident responses.

A-1-4 National Contingency Plan. See Code of Federal Regulations: 40 CFR, Part 300, Subchapters A through J.

The National Contingency Plan came about as a result of Presidential Executive Order 12580 of January 23, 1987. It reads, in part, as follows:

Superfund Implementation

By the authority vested in me as President of the United States of America by Section 115 of the Comprehensive Environmental Response, Compensation, and Liability Act of 1980, as amended (42 U.S.C. 9615 *et seq.*) ("the Act"), and by Section 301 of Title 3 of the United States Code, it is hereby ordered as follows:

Section 1. National Contingency Plan.
(a) (1) The National Contingency Plan ("the NCP") shall provide for a National Response Team ("the NRT") composed of representatives of appropriate Federal departments and agencies for national planning and coordination of preparedness and response actions, and regional response teams as the regional counterpart to the NRT for planning and coordination of regional preparedness and response actions.

(2) The following agencies (in addition to other appropriate agencies) shall provide representatives to the National and Regional Response Teams to carry out their responsibilities under the NCP: Department of State, Department of Defense, Department of Justice, Department of the Interior, Department of Agriculture, Department of Commerce, Department of Labor, Department of Health and Human Services, Department of Transportation, Department of Energy, Environmental Protection Agency, Federal Emergency Management Agency, United States Coast Guard, and the Nuclear Regulatory Commission.

(3) Except for periods of activation because of a response action, the representative of the Environmental Protection Agency ("EPA") shall be the chairman and the representative of the United States Coast Guard shall be the vice chairman of the NRT and these agencies' representatives shall be co-chairs of the Regional Response Teams ("the RRTs"). When the NRT or an RRT is activated for a response action, the chairman shall be the EPA or United States Coast Guard representative, based on whether the release or threatened release occurs in the inland or coastal zone, unless

otherwise agreed upon by the EPA and United States Coast Guard representatives.

(4) The RRTs may include representatives from State governments, local governments (as agreed upon by the States), and Indian tribal governments. Subject to the functions and authorities delegated to Executive departments and agencies in other sections of this Order, the NRT shall provide policy and program direction to the RRTs.

(b)(1) The responsibility for the revision of the NCP and all of the other functions vested in the President by Sections 105(a), (b), (c), and (g), 125, and 301(f) of the Act is delegated to the Administrator of the Environmental Protection Agency ("the Administrator").

(2) The function vested in the President by Section 118(p) of the Superfund Amendments and Reauthorization Act of 1986 (Public Law 99-499) ("SARA") is delegated to the Administrator.

(c) In accord with Section 107(f)(2)(A) of the Act and Section 311(f)(5) of the Federal Water Pollution Control Act, as amended [33 U.S.C. 1321(f)(5)], the following shall be among those designated in the NCP as Federal trustees for natural resources;

(1) Secretary of Defense;

(2) Secretary of the Interior;

(3) Secretary of Agriculture;

(4) Secretary of Commerce;

(5) Secretary of Energy;

(d) Revisions to the NCP shall be made in consultation with members of the NRT prior to publication for notice and comment. Revisions shall also be made in consultation with the Director of the Federal Emergency Management Agency and the Nuclear Regulatory Commission in order to avoid inconsistent or duplicative requirements in the emergency planning responsibilities of those agencies.

(e) All revisions to the NCP, whether in proposed or final form, shall be subject to review and approval by the Director of the Office of Management and Budget ("OMB").

Additional sections of the Superfund Executive Order cover the following subjects:

Response and related authorities

Cleanup schedules

Enforcement

Liability

Litigation

Financial responsibility

Employee protection

Management of the Hazardous Substance Superfund and Claims

Federal facilities and General provisions

Penetration.   The movement of a material through a suit's closures, such as zippers, buttonholes, seams, flaps, or other design features of chemical-protective clothing, and through punctures, cuts, and tears.

> **Protection against penetration is vital. Proper storage, mainte-nance, and testing of protective clothing and equipment must be part of a response team's standard procedures.**

Permeation.   A chemical action involving the movement of chemicals, on a molecular level, through intact material.

> **See commentary on this subject in the definition section of NFPA 472. Note the difference between penetration and perme-ation. The latter takes place at a molecular level through the intact material itself.**

Protective Clothing.   See Chapter 5 for definitions.

Response.   That portion of incident management in which personnel are involved in controlling a hazardous materials incident.

> **This is a broad all-inclusive term that encompasses every indi-vidual who serves in the mitigation and control of an incident.**

Sampling.   Sampling is the process of collecting a representative amount of gas, liquid, or solid for analytical purposes.

> **Sampling is an essential procedure used to evaluate an incident. It must begin as early as possible in the incident in order to evaluate conditions, and it should continue, periodically, until deemed by the incident commander to be no longer necessary.**

Initial sampling will help determine what courses of emergency action are necessary, especially as relates to response personnel, public safety, and environmental impact.

Stabilization.   The period of an incident where the adverse behavior of the hazardous material is controlled.

This can be considered as an intermediate step in the mitigation process. An analogy may be made to a fire fighting operation where the fire is declared "under control." There may be hours or days of work remaining to be done, but the situation is considered to be "in hand" and will not worsen, yet the response forces may be a long way from termination procedures.

Waste Minimization.   Treatment of hazardous spills by procedures or chemicals designed to reduce the hazardous nature of the material and/or to minimize the quantity of waste produced.

One of the immediate goals of all initial responders and response units should be to minimize the quantity of waste that is produced. Their actions should be directed at controlling, or especially at not increasing, the extent of the hazard. Subsequent

Figure 1.6   Industry personnel may have instruments that can be used to identify the hazard. For example, a portable photoionizing trace gas analyzer is routinely used to detect a wide variety of organic and inorganic vapors. This equipment may be available to responding fire fighters.

Figure 1.7   Ideally, hazardous materials should be handled according
to regulations from the time of their generation through their disposal.
Response to a hazardous materials incident must also be geared to the
ultimate goal of safe disposal, with minimal contamination.

responders with special equipment and expertise can resort to
treatments or procedures that can reduce or neutralize the nature
of the waste.

2

Incident Response Planning

2-1 Planning is an essential part of emergency preparedness. The development of both facility response plans and community emergency plans is required by numerous state and federal laws, including SARA, Title III, "The Emergency Planning and Community Right to Know Act of 1986." Planning guides and reference materials are listed in Appendix B-2.4.

To the extent possible, everyone in the response community who may become involved should also be included in the incident planning process. The purpose of planning is to develop the capability of handling an incident as effectively and efficiently as is possible. The planning should also be a continuing process, and periodic testing of the plan is essential to success.

Figure 2.1 In preparing a hazardous materials emergency response plan, study the plans from other communities.

157

2-2   A planning team is necessary to develop the hazardous materials emergency plan. Local, state, and federal planning guidelines should be reviewed and consulted by the planning team when preparing plans for hazardous materials incidents.

The National Response Team is responsible for coordinating federal planning, preparedness, and response activities that relate to oil and hazardous substance releases. The Superfund Amendments and Reauthorization Act (SARA) of 1986 empowers the National Response Team to develop plans for hazardous substances emergencies. Consult Supplement V for an edited version of the Hazardous Materials Emergency Planning Guide. As the title of the Guide indicates, the publication is concerned principally with community planning for emergencies.

2-3   A planning team is necessary to develop the hazardous material emergency plan.

As the Emergency Planning Guide points out, a team approach encourages participation of the entire community in the planning process. Several communities can combine to form a hazardous materials advisory council (HMAC), which then becomes a resource for the individual planning team. Members of the team should be selected on the basis of certain considerations, according to the NRT. These include ability, commitment, authority, expertise, and compatibility. Members should be representative of all elements of the community.

2-4   As a minimum, an annual review and update of the hazardous material emergency plan is necessary.

It is important that someone be designated as the appropriate person responsible for keeping the plan current. The NRT recommends that the plan be reviewed every six months, but in any event at least annually. Title III of SARA requires an annual review.

2-5   As a minimum, a training exercise should be conducted annually to determine the adequacy and effectiveness of the hazardous material emergency plan.

The training exercise can be a full-scale simulation or a tabletop exercise. It should be followed by a review or critique. Training, however, must be ongoing in order to keep the plan updated. The plan should indicate which community organization has responsible charge of the exercise.

Figure 2.2   Realistic training exercises are good indicators of the
effectiveness of an emergency plan.

3 Response Levels

3-1* Table 3-1 is a planning guide intended to provide the user with assistance in determining incident levels for response and training. Potential applications to a jurisdiction's response activities may include development of standard operating procedures; implementation of a training program using the competency levels of NFPA 472, *Standard for Professional Competence of Responders to Hazardous Materials Incidents*; acquisition of necessary equipment; and development of community emergency response plans. When consulting this table, the user should refer to all of the incident condition criteria to determine the appropriate incident level.

A-3-1 These incidents can be considered as requiring either offensive operations or defensive operations.

Offensive operations include actions taken by a hazardous materials responder, in appropriate chemical-protective clothing, to handle an incident in such a manner that contact with the released material may result. This includes: patching or plugging to slow or stop a leak; containing a material in its own package or container; and cleanup operations that may require overpacking or transfer of a product to another container.

Defensive operations include actions taken during an incident where there is no intentional contact with the material involved. This includes: elimination of ignition sources, vapor suppression, and diking or diverting to keep a release in a confined area. It requires notification and possible evacuation, but does not involve plugging, patching, or cleanup of spilled or leaking materials.

Jurisdictions have the responsibility to develop standard operating procedures that equate levels of response to levels of training indicated in NFPA 472, *Standard for Professional Competence of Responders to Hazardous Materials Incidents*. Depending on the capabilities and training of personnel, first responder operational level may equate to incident level one, the technician level may equate to incident level two, and the specialist may equate to incident level three.

Response personnel should operate only at that incident level that matches their knowledge, training, and equipment. If conditions indicate a need for a higher response level, then additional personnel, appropriate training, and equipment should be summoned.

Several response organizations have established a tiered response capability. As an example, if a reported incident is deemed to constitute a level one condition, the responding unit(s) will have the capability of handling it. A level two incident will require greater response capability, and a level three incident will need even more sophisticated equipment and highly trained personnel.

It is not the intent of Table 3-1 that response capabilities be absolutely determined on the basis of the incident level; nor is it intended that every jurisdiction must have tiered response capability. The table is intended to serve as a guide in the event one needs assistance in determining the relative real or potential seriousness of an incident.

Some of the terms and conditions relating to the Table warrant explanation.

Product Identifications:

Placard not required refers to U.S. Department of Transportation regulations. However, the absence of a placard should not be taken as an assurance that the contents are harmless.

NFPA 0 or 1 in all categories is a reference to NFPA 704, *Standard System for the Identification of the Fire Hazards of Materials*. This standard deals with a labeling system that advises on three hazard conditions: health, flammability, and reactivity. There are five degrees of intensity, ranging from 0 through 4. A zero or 1 in all three hazard conditions would indicate relatively low hazard.

All ORM A,B,C, and D. ORM means Other Regulated Materials. It is U.S. Department of Transportation language, and the A through D categories are defined as follows:

A. A material that has an anesthetic, irritating, noxious, toxic, or other similar property and that can cause extreme annoyance or discomfort to passengers and crew in the event of leakage during transportation.

B. A material (including a solid when wet with water) capable of causing significant damage to a transport vehicle from leakage during transportation.

C. A material that has other inherent characteristics not described as an ORM A or B, but which make it unsuitable for shipment unless properly identified and prepared for transportation.

Table 3-1 Planning Guide for Determining Incident Levels, Response, and Training

Incident Level	One	Two	Three
Incident Conditions Product Identifications	Placard not required, NFPA 0 or 1 all categories, all ORM A, B, C, and D.	DOT placarded, NFPA 2 for any categories, PCBs without fire, EPA regulated waste.	Poison A (gas), explosives A/B, organic peroxide, flammable solid, materials dangerous when wet, chlorine, fluorine, anhydrous ammonia, radioactive materials, NFPA 3 & 4 for any categories including special hazards, PCBs & fire, DOT inhalation hazard, EPA extremely hazardous substances, and cryogenics.
Container Size	Small (e.g., pail, drums, cylinders except one-ton, packages, bags).	Medium (e.g., one-ton cylinder, portable containers, nurse tanks, multiple small packages).	Large (e.g., tank cars, tank trucks, stationary tanks, hopper cars/trucks, multiple medium containers).
Fire/Explosion Potential	Low.	Medium.	High.
Leak Severity	No release or small release contained or confined with readily available resources.	Release may not be controllable without special resources.	Release may not be controllable even with special resources.
Life Safety	No life threatening situation from materials involved.	Localized area, limited evacuation area.	Large area, mass evacuation area.
Environmental Impact (Potential)	Minimal.	Moderate.	Severe.
Container Integrity	Not damaged.	Damaged but able to contain the contents to allow handling or transfer of product.	Damaged to such an extent that catastrophic rupture is possible.

D. A material such as a consumer commodity that, though otherwise subject to the regulations of this subchapter, presents a limited hazard during transportation due to its form, quantity, and packaging.

PCBs without Fire. Polychlorinated biphenyls (PCBs) present serious health threats to skin and the liver. Even without the added hazard of fire, they are sufficiently harmful to responders to warrant a level two condition.

EPA Regulated Waste. These can be found in 40 CFR 261 (Code of Federal Regulations).

Poison A. Examples are arsine, hydrocyanic acid, or phosgene. These are extremely dangerous poisons.

Explosives A/B. Examples of A include dynamite, dry TNT, or black powder. The B category includes propellent explosives, rocket motors, or special fireworks.

Organic Peroxide. Many organic peroxides are highly flammable, and most will decompose readily when heated. In some cases, the decomposition can be violent.

Flammable Solid. Examples are pyroxylin plastics, magnesium, or aluminum powder.

Materials Dangerous when Wet. Included in this category are sodium and potassium metals and calcium carbide.

Chlorine. A greenish yellow gas that is highly toxic and irritating.

Fluorine. A yellow gas that is extremely reactive and intensely poisonous.

Anhydrous Ammonia. A very toxic and corrosive gas.

Radioactive Materials. Materials that spontaneously emit ionizing radiation having a specific activity greater than 0.002 microcuries per gram.

DOT Inhalation Hazard. Inhalation hazards are measured in terms of TLV/TWA (threshold limit value/time weighted average). It is the concentration of a material to which an average healthy person may be repeatedly exposed for 8 hours each day, 40 hours per week, without suffering adverse health effects.

EPA Extremely Hazardous Substances. EPA has published a list of 366 such substances.

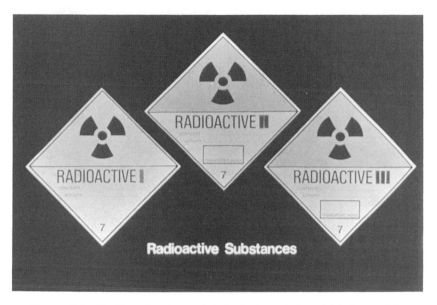

Figure 3.1   Radioactive materials include cobalt, which is used in
medicine, and plutonium and uranium, which are
transported by the nuclear power industry.

Cryogenics. Cryogenics are extremely cold liquefied gases that can have temperatures well below $-200\ °F$. Cryogenic liquids can cause severe damage to skin or other body tissues.

Container Size. The rationale at work is that the larger the container, the greater the potential for risk, and hence the increase in level of incident condition.

Fire/Explosion Potential. The assumption in each case is that the incident is not simply a fire, but that there is some hazardous material involved. Where there is no fire, level one may be appropriate, depending on other prevailing conditions. If a container is involved in fire, then a level three may be more appropriate. It is conceivable that a fire involving a container can be handled safely by a responding fire department without the assistance of any hazardous materials response personnel. Nonetheless, appropriate authorities would have to be notified and alerted to the situation. It is also vitally important to keep in mind that containers that are involved in fire can overpressurize and fail in a catastrophic manner. Every precaution must be taken when approaching containers that are exposed to fire.

Leak Severity. The selection of levels is obviously dependent on the extent of the leak and the likelihood of its being controlled.

Figure 3.2   Hazardous wastes can be found stored in containers
that may be corroded or unsecured. Poorly contained wastes
may mix, forming new compounds more unstable and toxic
than their original components.

Life Safety. The number of people potentially exposed is a major determining factor in selection of the appropriate level.

Environmental Impact. As can be seen, the terms used are general. Judgement is required, and experts should be consulted. Environmental impacts may not be known at the start of an incident, and they are frequently more severe than anticipated.

Container Integrity. Extreme care must be taken with damaged containers prior to allowing transfer. Once again, people who are expert in this field should be consulted. If doubt exists, the incident should be considered as level three.

4 Site Safety

4-1 Emergency Incident Operations.

4-1.1 Emergency incident operations should be conducted in compliance with Chapter 6 of NFPA 1500, *Standard on Fire Department Occupational Safety and Health Program*, or 29 CFR 1910.120 or EPA.

All fire fighters can be considered to be categorized as at least first responders. Some will simply meet first responder awareness level, but most will be qualified as first responder operational level. (*See NFPA 472, Standard for Professional Competence of Responders to Hazardous Materials Incidents.*) The first responder category will also apply to most police officers, public utility workers, emergency medical technicians, employees at hazardous waste facilities, and drivers of trucks carrying hazardous materials. In many of these cases, the first responder awareness level will apply. In any event, OSHA 1910.120 has application, and specified training will be required. Chapter 6 of NFPA 1500 deals with emergency operations and covers organization, safety requirements, and incidents involving special hazards. This last category certainly is intended to apply to hazardous materials incidents.

4-1.2 An incident command system should be implemented at all hazardous materials incidents. Operations should be directed by a designated incident commander and follow established written standard operating procedures.

NOTE: Though in the following text 29 CFR 1910.120 is cited, it should be understood that some states will adopt these regulations under state OSHA plans and others will adopt these regulations through adoption of a similar regulation established by EPA and appropriate state agencies.

Note that a specific incident command system is not cited. For further information, the reader can consult publications on the subject provided by NFPA, the National Fire Academy, FEMA, or Fire Protection Publications of Oklahoma State University.

4-1.3   An emergency response plan describing the general safety procedures that are to be followed at an incident should be prepared in accordance with 29 CFR 1910.120. These procedures should be thoroughly reviewed and tested.

All elements of the response community are required to have an emergency response plan outlining standard operating procedures for hazardous materials incidents. The plan should include a number of subject areas, including an incident command procedure, preplanning, site control, evacuation policies, decontamination procedures, medical response plans, training, and several others.

4-2 Ignition Sources.   Ignition sources should be eliminated whenever possible at incidents involving releases, or probable releases, of ignitable materials. Whenever possible, electrical devices used within the hot zone should be certified as intrinsically safe by recognized organizations.

Most ignition sources are well known and easily recognized, such as matches, smoking materials, or open flames. Pilot lights, incendive static discharges, and electrical sparks often receive less attention. Whenever flammable vapor is present, sources of ignition should be removed.

4-3 Safety Officer.

Consult NFPA 1501, *Standard for Fire Department Safety Officer*, for information on safety officer operations that will be helpful to non-fire service responders also.

4-3.1   A safety officer should be designated by and report to the incident commander.

NOTE: Under this section the safety officer is given specific responsibilities. It should be understood that even though these duties are to be carried out by the safety officer, the incident commander still has overall responsibility for the implementation of these tasks.

The incident commander is in overall charge of a hazardous materials incident and is ultimately responsible for the outcome of all operations undertaken at the scene. Part of that responsibility includes the designation of a safety officer to oversee all operations from the standpoint of assuring that safe practices are carried out and observed.

4-3.2   The safety officer should provide the incident commander with recommendations on the establishment of the control zones at each emergency incident, based on the identification and evaluation of the hazards.

At a typical hazardous materials incident, the safety officer's primary responsibility will concern itself with establishment and safe operation of the control zones. There may well be a need for several assistants at major incidents, and other safety considerations beyond the control zone operation will apply. Since dangers associated with hazardous materials incidents are both multiple and complex, the safety officer's role is vitally important. Training and experience are essential. Keeping the incident commander informed is one of the most important responsibilities of the safety officer.

4-3.3 The safety officer should maintain control and security of entry and exit of all personnel between the various zones.

Personnel moving from one control zone to another must be made aware of conditions and of what is expected of them. This can be done by use of posted instructions, or by safety personnel at each access point. All nonessential personnel must be kept out of the control zone areas.

4-3.4 The safety officer is responsible for implementing the safety plan.

The safety officer must also develop the safety plan and make all responders aware of its requirements. There are certain elements

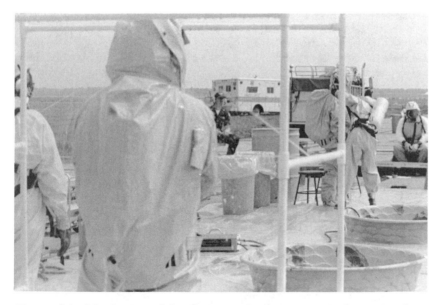

Figure 4.1   The layout of the decontamination sector is determined in part by the extent of the procedure to be followed. But, generally, response personnel will move along a series of stations, moving from the most heavily contaminated area to clean areas.

of a plan that can be applied at all incidents, and responders should be aware of these prior to an occurrence. These include a prohibition against smoking or eating in the hot and warm zones, no matches or lighters in these zones, and the need to check in and out of each zone at an appropriate access point. Another key element of every safety plan is implementation of a buddy system.

4-3.5 The safety officer should make the final decision on entry/no entry, corrective actions, respiratory and personal protective clothing, monitoring and sampling methods, and when personnel should be withdrawn or evacuated.

As can be concluded from the list of responsibilities assigned, the safety officer becomes the key operating agent at a hazardous materials incident. Nonetheless, the safety officer is not in overall command of the incident. Cooperation, coordination, and communication with the incident commander will determine the relative success of the operation.

4-3.6 The safety officer should ensure that the proper decontamination procedures are in place before entry.

At some situations, an emergency entry, with appropriate protective clothing, must be made before decontamination procedures are established. Such would be the case when loss of life or possibility of serious injury to a victim is likely. In such instances, an emergency decontamination procedure can be established. If immediate medical treatment is essential, it may be necessary to delay the decontamination process. On the other hand, if serious contamination is causing the injury, then decontamination must take place at once.

4-3.7 The safety officer should monitor and maintain communications between the entry personnel and him/herself, and with the incident commander.

The primary means of communication are radio and telephone. Good radio procedure at an emergency dictates that nonessential transmissions be eliminated.

4-3.8 The safety officer should ensure that a backup team wearing the appropriate level of personal protective equipment is ready at all times during entry team operations.

Responders in the hot zone should always be in visual or radio contact with safety personnel in the cold zone in the event that immediate assistance is required.

Figure 4.2   Two-way radios link site commanders with entry personnel.

4-3.9   The safety officer should ensure that all other elements of safety are in place and that emergency medical services with transport capabilities are available.

Other elements of safety would include proper signs, lighting, safe refueling areas, inspection of equipment, and training in standard operating procedures.

It is recommended that emergency medical service personnel receive training in hazardous materials response procedures.

4-3.10   The safety officer should ensure that all pertinent information is gathered and recorded. Pertinent documents, manifests, and reports should be collected and safeguarded.

In order to implement an effective medical surveillance program for the protection of response personnel, proper record keeping is essential.

4-3.10.1 Incident Site Log.   A record of the incident should be maintained for future reference, and as a minimum should contain the following:

(a) Location

(b) Date

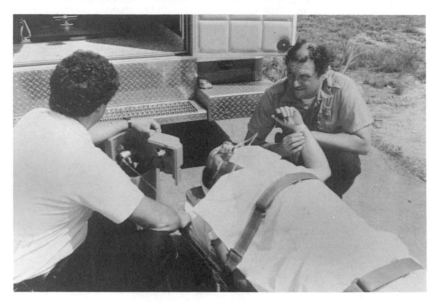

Figure 4.3   All injuries sustained at the scene of a hazardous materials
incident should be recorded.

(c) Name, description, source, quantity, and cause of release

(d) Weather information

(e) Names and job assignments for all personnel involved

(f) Injuries to personnel and public

(g) Corrective action taken

(h) Chronological recording of events

(i) Entry and exit times of the entry personnel

(j) Method of recording exposure of personnel to hazardous materials

(k) Resource personnel data.

It would also be important to list and describe the type of
personal protective equipment that was worn by each person, and
the duration of usage.

4-4 Control Zones.   Control zone names have not been consistently applied
at incidents. The intent of this section is to show areas of responder control.
The various zones or areas at a typical emergency response site are shown in
Figure 4-4.

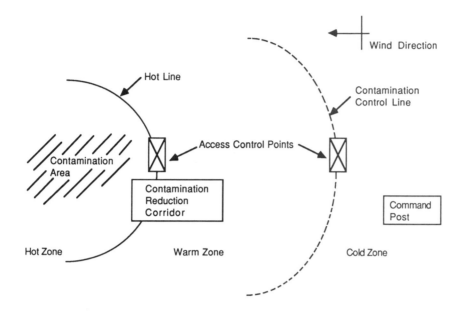

Figure 4-4 **Diagram of control zones.**

Control zones serve to reduce the contamination of personnel and equipment by controlling and directing their operations and movements at the incident. A site map that shows wind direction and topography will prove helpful. Establishing boundaries for the various control zones is based on monitoring of the incident. Personnel should move only through the access control points in order to maintain control of the site and to prevent spread of contamination across zones.

4-4.1 Hot Zone. Area immediately surrounding a hazardous materials incident, which extends far enough to prevent adverse effects from hazardous materials releases to personnel outside the zone. This zone is also referred to as the exclusion zone or restricted zone in other documents.

NOTE: Access into the hot zone is to be limited to those persons necessary to control the incident. A log is to be maintained at the access control point to record entry and exit time of all personnel in the hot zone.

Reasons for entry of personnel into the hot zone, where contamination is likely to occur, include the need for sampling, clean up, or spill control measures. The boundary lines should be clearly delineated by hazard tape, signs, or rope. It may be appropriate to have more than one access control point per zone in order to have separate entrance and exit points. All personnel within the hot zone should wear the level of protective equipment determined by the incident commander. It is possible that differing levels of protection are appropriate in the same area, depending on the specific task being performed.

4-4.2 Warm Zone.   The area where personnel and equipment decontamination and hot zone support takes place. It includes control points for the access corridor and thus assists in reducing the spread of contamination. This is also referred to as the decontamination, contamination reduction, or limited access zone in other documents.

The severity of contamination should decrease as one moves from the hot line toward the cold zone because of the effective decontamination procedures in the warm zone. There will need to be a decontamination line for personnel and another for heavy equipment within the contamination reduction corridor.

Personnel entering the warm zone from the cold zone should be wearing the level of protection required for operating in the warm zone. Personnel leaving the warm zone for the cold zone should remove the protective clothing worn in the warm zone.

The warm zone will have to be large enough to accommodate all of the decontamination procedures that will take place therein.

4-4.3 Cold Zone.   This area contains the command post and such other support functions as are deemed necessary to control the incident. This is also referred to as the clean zone or support zone in other documents.

Personnel in the cold zone may wear normal work clothes. The cold zone should be upwind of the hot zone and as far away as is practical. Support functions might include site security, medical support, reserve equipment, and a field laboratory.

4-5 Communications.

Effective handling of any emergency situation depends on establishment and implementation of a coordinated communications program. Everyone operating at the scene must be involved, in varying degrees as necessary, in the communications loop. In addition, the communications network must extend beyond the immediate scene of operations, since dispatching centers are also part of the system.

Hazardous materials incidents are often more complicated in nature than routine emergencies, and this fact adds more emphasis on an effective communications system. Contact with many other agencies is often essential, and the amount of information available to those operating at the scene may be limited, especially in the early stages of an incident.

Figure 4.4   Maintaining effective communications is a top priority at any hazardous materials incident.

4-5.1 When personal protective clothing or remote operations inhibit communications, an effective means of communications, such as radios, should be established.

Radio equipment has improved substantially in recent years. Every consideration should be given to having workers equipped with two-way radio equipment, particularly where the wearing of protective clothing causes isolation of the individual.

4-5.2 The frequencies employed in these radios should be "dedicated" and not used or shared with other local agencies.

Where multi-channel radios are available to personnel at the scene of operations, the incident commander can designate a particular channel for all on-scene communications.

4-5.3 Communication should be supplemented by a prearranged set of hand signals and hand-light signals to be used when primary communication methods fail. Hand-lights employed for this purpose should be in accordance with NFPA 70, *National Electrical Code*®, for use in hazardous environments.

Where hand signals are to be employed, they should be simple to use and understand, and they should be limited in number to commands that are essential. Since an operation may well involve personnel from different disciplines and agencies, it is important

that everyone involved in the operation understand and be able to use the appropriate hand or hand-light signals.

4-6 Monitoring Equipment.

Also consult the commentary in NFPA 472 for additional information on monitoring equipment.

4-6.1 Monitoring equipment operates on several different principles and measures different aspects of hazardous materials releases. Examples of this equipment are:

(a) Oxygen meters

Air contains 21 percent oxygen by volume under normal conditions. Sixteen percent is needed to support human life. Atmospheres with less than 19.5 percent are generally considered deficient.

There are several instruments that can be used to determine the oxygen level, and many of these are small and easy to use. One such type of meter measures the partial pressure of oxygen in air by use of an electrochemical sensor, and the reading is then converted to the oxygen concentration.

(b) Combustible gas indicator (explosimeter)

This instrument measures the concentration of a combustible gas or vapor. The combustible gas is burned in the instrument, and measurement of the increased heat caused by the burning provides information relating to the actual concentration of the gas or vapor.

(c) Carbon monoxide meter

Carbon monoxide is a colorless, odorless, and tasteless toxic gas that can prove fatal in very small concentrations. There are several instruments that can be used to measure its presence. Some of these use a catalytic combustion process and then measure the heat that is produced to determine how much carbon monoxide is present.

(d) pH meter

(e) Radiation detection instruments

There are four types of radiation: alpha particles, beta particles, gamma rays, and neutron particles. Radiation detection equipment includes survey meters and dosimeters. The dosimeter will measure accumulated exposure to gamma radiation. The survey

Figure 4.5   Monitoring equipment can be used to measure various
types of hazardous materials present at an incident.

meters will detect and measure beta and gamma radiation. Neutron radiation measuring devices are very expensive, but incidents
resulting in neutron emissions are rare.

(f) Colorimetric detector tubes

These devices are widely used to evaluate airborne gases and
vapors, and they are also known as detector tubes or gas indicators. They are easy to operate and can detect many contaminants.

The devices have a pump and a colorimetric indicator tube that
contains a substance that will react when contaminated air is
drawn through the tube. Manufacturer's instructions and conversion
tables help to determine the concentration of the contaminant.

(g) Organic vapor analyzer

These are direct reading devices that are capable of detecting all
organic vapors. They can be certified as suitable for use in Class I,
Division I atmospheres. Manufacturer's instructions must be followed.

(h) Photoionization meter

These types of monitoring devices will indicate the presence of
many organic and some inorganic gases and vapors. Initially, the

total concentration of gases or vapors will be detected. Other probes can then be used to identify some of the compounds. This meter uses ultraviolet radiation to ionize molecules. A current is thereby produced that is proportionate to the number of ions present.

(i) Air sampling devices

One type of air sampling device collects a specific volume of air. The other passes a determined volume of air through some medium that removes the contaminant for sampling. The first type of metering technique is called instantaneous sampling; the second is known as integrated sampling.

(j) Other meters to measure specific products such as chlorine, hydrogen sulfide, or ethylene oxide

Metering instruments can be made selective to a specified chemical or gas. For example, various chemicals absorb infrared energy at specified frequencies, and the instrument can be configured to limit the frequencies it will test to those that are characteristic of a specific gas or chemical. Where a single chemical is present, the device will indicate that presence. Where more than one chemical is present, other methods will also have to be used.

(k) pH paper or strips

These monitor the corrosivity of a substance by measuring whether it is an acid or alkaline material.

(l) Organic vapor badge or film strip

This type of area sampling device involves sampling of the breathing zone of the wearer and measuring the dose to which the badge is exposed. It measures personal exposure over a period of time. Some of these devices are battery operated. Miniature dosimeters have been recently developed so that a worker can wear it on a lapel, so the exposure over a specified sampling period can be measured.

(m) Mercury badge

(n) Formaldehyde badge or strip.

4-6.2 All monitoring equipment should be operationally checked prior to use and periodically calibrated as per manufacturer's specifications.

5

Personal Protective Equipment

5-1 General. It is essential that personal protective equipment meeting appropriate NFPA and OSHA standards be provided, maintained, and used. Protection against physical, chemical, and thermal hazards must be considered when selecting personal protective equipment.

Personal protective equipment includes structural fire fighting clothing, chemical-protective clothing, and high temperature protective clothing. It is not essential that the responder to hazardous materials incidents be equipped with all three types. Many hazardous materials incidents will not involve fire at all, and, in those cases, structural fire fighting clothing will therefore not be needed. Since first responders at the awareness level or operational levels may include police officers, emergency medical service personnel, and fixed site workers, one would not expect them to be equipped with structural fire fighting clothing.

It is most important that whatever personal protective equipment is used, such equipment meet relevant standards.

Personal protective equipment must be used carefully, since there are hazards associated with its use. It can provide a false sense of security and can present other disadvantages, including rapid fatigue, heat stress, limited vision, and impaired communications. Selection of appropriate protection is important, since overprotection can cause needless hardship.

5-1.1 A written personal protective equipment program should be established in accordance with 29 CFR Part 1910.120. Elements of the program should include personal protective equipment selection and use, storage, maintenance, and inspection procedures, and training considerations. The selection of personal protective clothing should be based on the hazardous materials and/or conditions present and be appropriate for the hazards encountered.

Note that providing a complete range of protective equipment is not enough. There must be a knowledge of which items to use, an ability to use them correctly, and a maintenance, storage, and

179

inspection program that will ensure the serviceability of the equipment when it is needed.

The OSHA fire brigade standard [OSHA 1910.156(e)] establishes different criteria for incipient fire fighting versus structural fire fighting. If a fire fighting brigade is trained and equipped only for incipient fires, then it is important that its operations not exceed that capability. This same principle should be applied to the handling of hazardous materials incidents. You should never operate beyond the stage for which you are trained and equipped. To do so presents a danger not only to yourself, but to a much wider community. Safety and self-protection is not the sole concern at work. An inappropriate action may have wide ranging negative consequences.

5-1.2 Protective clothing and equipment used to perform fire suppression operations, beyond the incipient stage, should meet the requirements of Chapter 5 of NFPA 1500, *Standard on Fire Department Occupational Safety and Health Program*. Structural fire fighting protective clothing is not intended to provide chemical protection to the user.

The following excerpts from NFPA 1500 are important and worth citing.

"5-1.1 The fire department shall provide each member with the appropriate protective clothing and protective equipment to provide protection from the hazards of the work environment to which the member is or may be exposed. Such protective clothing and protective equipment shall be suitable for the tasks that the member is expected to perform in that environment.

5-1.2 Protective clothing and protective equipment shall be used whenever the member is exposed or potentially exposed to the hazards for which it is provided.

5-1.3 Members shall be fully trained in the care, use, inspection, maintenance, and limitations of the protective clothing and protective equipment assigned to them or available for their use.

5-1.4 Protective clothing and protective equipment shall be used and maintained in accordance with manufacturers' instructions. A maintenance and inspection program shall be established for protective clothing and protective equipment. Specific responsibilities shall be assigned for inspection and maintenance."

It is important to note that structural fire fighting clothing is not designed to offer chemical protection to the user. Every fire department and every fire fighter is generally considered to have at least first responder awareness level capability and, even more likely, first responder operational capability. Their structural fire fighting equipment is governed by OSHA regulation and by NFPA 1500. Yet that equipment should not be considered as appropriate protection for exposure to chemical spills, even though it may provide protection in some cases. In other words, structural fire fighting equipment and chemical-protective clothing constitute two different categories, even though self-contained breathing apparatus may be common to each category.

It is conceivable that a structural fire in a chemical facility would require the use of chemical-protective clothing rather than structural fire fighting clothing, even though the situation is clearly a structural fire incident. Judgement must be used, and that decision falls to the officer in charge of the incident.

5-2 Respiratory Protective Equipment.

5-2.1 Self-contained breathing apparatus (SCBA) should meet the requirements of NFPA 1981, *Standard on Open-Circuit Self-Contained Breathing Apparatus for Fire Fighters*.

Open-circuit self-contained breathing apparatus (SCBA) provides air from a cylinder to the mask, and the wearer exhales directly to the atmosphere. Closed-circuit SCBA, on the other

Figure 5.1 Respiratory protection must be the primary concern in nonencapsulated protective clothing ensembles. Positive pressure self-contained breathing apparatus will provide the highest level of respiratory protection.

hand, involves rebreathing of the exhaled, recycled gases that have been chemically scrubbed and supplemented with oxygen from a supply source. The latter type of apparatus has a longer operating time.

5-2.2   Personal alert safety systems should meet the requirements of NFPA 1982, *Standard on Personal Alert Safety Systems (PASS) for Fire Fighters.*

NFPA 1500 requires fire service members involved in rescue, fire fighting, or other hazardous duty to use a PASS device. The standard also recommends it be worn on the protective clothing and be used whether or not SCBA is being used. The PASS device sounds an alarm signal to warn of lack of motion on the part of the wearer after 30 (± 5) seconds.

5-2.3 Air Purifying Respirators.   These devices are worn to filter particulates and contaminants from the air. They should only be worn in atmospheres where the type and quantity of the contaminants are known and sufficient oxygen is known to be present.

Air purifying respirators do not have an air source, so their use is very limited. They must rely on a filter element that is designed to purify the ambient air in which they are used. They cannot be used in atmospheres that contain less than a 19.5 percent oxygen level or that contain contaminants that are immediately dangerous to life or health. When they are used, it is essential that there be monitoring of the oxygen level and of the contaminated atmosphere.

5-3 Chemical-Protective Clothing.

5-3.1   Chemical-protective clothing (CPC) is made from special materials and is designed to prevent the contact of chemicals with the body. Chemical-protective clothing is of two types: totally encapsulating and nonencapsulating.

Selection of the appropriate type of chemical-protective clothing is all-important. It is essential that the wearer be adequately trained in the use and limitations of the protection and that manufacturer's instructions be followed. Totally encapsulating suits are generally one-piece suits that are designed and intended to protect the wearer against gases and vapors. Nonencapsulating suits generally come in component parts and will not protect against gases and vapors since they are not gastight.

Factors to consider in acquiring protective clothing include cost, ease of decontaminating, and the anticipated life of the garment.

5-3.2   A variety of materials are used to make the fabric from which clothing is manufactured. Each material will provide protection against certain specified chemicals or mixtures of chemicals. It may afford little or no protection against certain other chemicals. It is most important to note that there is no material that provides satisfactory protection from all chemicals. Protective clothing material must be compatible with the chemical substances involved, consistent with manufacturers' instructions.

Knowledge of the limitations of the particular protective garment is essential to safe operation at a hazardous materials incident. This presupposes an awareness of the identification of the material involved and of the manufacturer's instructions regarding the garment's performance capabilities against specified chemicals for specified exposure times.

5-3.3 Performance requirements must be considered in selecting the appropriate chemical-protective material. These would include chemical resistance, permeation, penetration, flexibility, abrasion, temperature resistance, shelf life, and sizing criteria.

The purpose of chemical-protective clothing is to protect the wearer from exposure. In order to accomplish that end, performance requirements are established. There are many variables at work, since no single material protects against every chemical, and since different types of materials are more effective than others in protecting against specific chemicals. Combinations of chemicals is a complicating factor. In addition, there are some chemicals for which there is no effective enduring protective material.

5-3.3.1 Chemical resistance is the ability of the material, from which the protective garment is made, to prevent or reduce degradation and permeation of the fabric by the attack chemical. Degradation is a chemical action involving the molecular breakdown of the material due to contact with a chemical. The action may cause the fabric to swell, shrink, blister, discolor, become brittle, sticky, or soft, or to deteriorate. These changes permit chemicals to get through the suit more rapidly or to increase the probability of permeation.

Manufacturers of chemical-protective clothing will furnish charts that explain the chemical resistance a product offers against degradation and permeation. However, manufacturer's information can be misleading, since their conclusions may be based on old data, raw material performance, inconsistent test methods, and laboratory type (rather than real world) usage.

Mixtures of chemicals can significantly increase the rate of degradation of protective clothing. A chemical that can permeate the clothing can offer entry to other chemicals in the mixture to further degrade the clothing. Some protective clothing is now layered in order to offer additional protection against mixtures of chemicals.

5-3.3.2   Permeation is a chemical action involving the movement of chemicals, on a molecular level, through intact material. There is usually no indication that this process is occurring. Permeation is defined by two terms, permeation rate and breakthrough time. Permeation rate is the quantity of chemical that will move through an area of protective garment in a given period of time, usually expressed as micrograms of chemical per square centimeter per minute. Breakthrough time is the time required for the chemical to be measured on the inside surface of the fabric. The most desirable protective fabric is one that has the longest breakthrough time and a very low permeation rate. Breakthrough times and permeation rates are not available for all the common suit materials and the variety of chemicals that exist. Manufacturer's data and reference sources should be consulted. Generally if a material degrades rapidly, permeation will occur rapidly.

Additional variables that may affect permeation and breakthrough time are associated with actual use conditions, such as temperature and humidity. Others include clothing thickness and the concentration of the chemical it is intended to resist.

5-3.3.3 Penetration.   Penetration is the movement of material through a suit's closures, such as zippers, buttonholes, seams, flaps, or other design features. Torn or ripped suits will also allow penetration.

Zippers and closures on totally encapsulating suits are designed and intended to be gastight, and the suit will be designed for internal positive pressure operation. There should be no penetration of vapor. Nonencapsulating suits should protect against splashes, but not all body parts are covered. The clothing should protect against splash penetration.

5-4 Thermal Protection.

5-4.1 Proximity Suits.   These suits provide short duration and close proximity protection at radiant heat temperatures as high as 2000 °F and may withstand some exposure to water and steam. Respiratory protection must be provided with proximity suits.

Proximity suits are not intended or designed for fire entry. These suits can be a two-piece or three-piece ensemble made of a heat reflective material with layers of insulating linings. The wearer must also use special heat reflective mittens and boots. This ensemble is often worn over other protective clothing.

5-4.2 Fire Entry Suits.   This type of suit provides protection for brief entry into total flame environment at temperatures as high as 2000 °F. This suit is not effective or meant to be used for rescue operations. Respiratory protection must be provided with fire entry suits.

Figure 5.2   Proximity suits are primarily used for close proximity, short duration exposures to both flame and radiant heat temperatures as high as 2000 °F. They will withstand exposures to steam, liquids, and weak corrosive chemicals.

This suit is comprised of coat, pants, boots, gloves, and hood, each of which is made up of many layers of flame retardant materials. The outer layer is usually aluminized.

5-4.3 Overprotection Garments.

These garments are worn in conjunction with chemical-protective encapsulating suits.

5-4.3.1 Flash Cover Protective Suit.

Flash cover suits are neither proximity nor fire entry suits. They provide limited overprotection against flash-back only. They are worn outside of other protective suits and are used only when the risks require them.

5-4.3.2 Low Temperature Suits.

Low temperature suits provide some degree of protection of the encapsulating chemical-protective clothing from contact with low temperature gases and liquids. They are worn outside of the encapsulating chemical-protective clothing and are used only when the risk requires them.

5-5 Levels of Protection.

Personal protective equipment is divided into four categories based on the degree of protection afforded.

NOTE: An asterisk (*) after the description indicates optional, as applicable.

Figure 5.3   Fire entry suits offer effective protection for short duration entry into total flame environments such as one may find at petrochemical fires. They can withstand prolonged exposures to radiant heat levels as high as 2000 °F.

Those who have not been adequately trained in the selection and use of personal protective equipment should not be permitted to operate in such equipment at a hazardous materials incident. The training should be thorough and frequent so that the responder becomes intimately familiar with the limitations and handicaps associated with the wearing of such equipment. Training on usage should include selection, donning, operating, testing, cleaning, maintaining, and caring for the clothing.

During an incident, it is conceivable that the need to change the level of protection may become apparent. The change can be from a high level of protection to a lower one, or the reverse. Such a decision is made by the safety officer, based on an evaluation of the hazards presented to personnel.

5-5.1 Level A.   To be selected when the greatest level of skin, respiratory, and eye protection is required. The following constitute Level A equipment; it may be used as appropriate.

5-5.1.1   Pressure-demand, full facepiece, self-contained breathing apparatus (SCBA), or pressure-demand supplied air respirator with escape SCBA,

Figure 5.4 Full facepiece, self-contained breathing apparatus (SCBA).

approved by the National Institute of Occupational Safety and Health (NIOSH).

5-5.1.2 Totally encapsulating chemical-protective suit. Totally encapsulating chemical-protective suit (TECP suit) means a full body garment that is constructed of protective clothing materials; covers the wearer's torso, head, arms, and legs; has boots and gloves that may be an integral part of the suit, or separate and tightly attached; and completely encloses the wearer by itself or in combination with the wearer's respiratory equipment, gloves, and boots. All components of a TECP suit, such as relief valves, seams, and closure assemblies, should provide equivalent chemical resistance protection.

5-5.1.3 Coveralls.*

5-5.1.4 Long underwear.*

5-5.1.5 Gloves, outer, chemical-resistant.

5-5.1.6 Gloves, inner, chemical-resistant.

5-5.1.7 Boots, chemical-resistant, steel toe and shank.

5-5.1.8 Hard hat (under suit).*

5-5.1.9 Disposable protective suit, gloves, and boots (depending on suit construction, may be worn over totally encapsulating suit).*

5-5.1.10 Two-way radios (worn inside encapsulating suit).

5-5.2 Level B. The highest level of respiratory protection is necessary but a lesser level of skin protection is needed. The following constitutes Level B equipment; it may be used as appropriate.

5-5.2.1 Pressure-demand, full facepiece, self-contained breathing apparatus (SCBA), or pressure-demand supplied air respirator with escape SCBA, NIOSH approved.

5-5.2.2 Hooded chemical-resistant clothing (overalls and long-sleeved jacket, coveralls, one- or two-piece chemical-splash suit, disposable chemical-resistant overalls).

5-5.2.3 Coveralls.*

5-5.2.4 Gloves, outer, chemical-resistant.

5-5.2.5 Gloves, inner, chemical-resistant.

5-5.2.6 Boots, outer, chemical-resistant, steel toe and shank.

5-5.2.7 Boot-covers, outer, chemical-resistant (disposable).*

5-5.2.8 Hard hat.

5-5.2.9 Two-way radios (worn inside encapsulating suit).

5-5.2.10 Face shield.*

5-5.3* Level C. The concentration(s) and type(s) of airborne substance(s) is known and the criteria for using air purifying respirators are met. The following constitute Level C equipment; it may be used as appropriate.

A-5-5.3 Refer to OSHA 29 CFR 1910.134.

5-5.3.1 Full-face or half-mask, air purifying respirators (NIOSH approved).

5-5.3.2 Hooded chemical-resistant clothing (overalls, two-piece chemical-splash suit, disposable chemical-resistant overalls).

5-5.3.3 Coveralls.*

5-5.3.4 Gloves, outer, chemical-resistant.

5-5.3.5 Gloves, inner, chemical-resistant.

5-5.3.6 Boots, outer, chemical-resistant, steel toe and shank.

5-5.3.7 Boot-covers, outer, chemical-resistant (disposable).*

5-5.3.8 Hard hat.

5-5.3.9   Escape mask.*

5-5.3.10   Two-way radios (worn under outside protective clothing).

5-5.3.11   Face shield.*

5-5.4 Level D.   A work uniform affording minimal protection, used for nuisance contamination only. The following constitute Level D equipment; it may be used as appropriate.

5-5.4.1   Coveralls.

5-5.4.2   Gloves.*

5-5.4.3   Boots/shoes, chemical-resistant, steel toe and shank.

5-5.4.4   Boots, outer, chemical-resistant (disposable).*

5-5.4.5   Safety glasses or chemical-splash goggles.

5-5.4.6   Hard hat.

5-5.4.7   Escape mask.*

5-5.4.8   Face shield.*

5-6   The types of hazards for which levels A, B, C, and D protection are appropriate are described below.

5-6.1   Level A protection should be used when:

5-6.1.1   The hazardous material has been identified and requires the highest level of protection for skin, eyes, and the respiratory system based on either the measured (or potential for) high concentration of atmospheric vapors, gases, or particulates; or the site operations and work functions involve a high potential for splash, immersion, or exposure to unexpected vapors, gases, or particulates of material that are harmful to skin or capable of being absorbed through the intact skin;

5-6.1.2   Substances with a high degree of hazard to the skin are known or suspected to be present, and skin contact is possible; or

5-6.1.3   Operations must be conducted in confined, poorly ventilated areas, and the absence of conditions requiring Level A have not yet been determined.

In any event, the totally encapsulating protective clothing has to be compatible with the particular substance that is involved in the incident.

5-6.2   Level B protection should be used when:

5-6.2.1 The type and atmospheric concentration of substances have been identified and require a high level of respiratory protection, but less skin protection;

NOTE: This involves atmospheres with IDLH (immediately dangerous to life and health) concentrations of specific substances that do not represent a severe skin hazard, or that do not meet the criteria for use of air-purifying respirators.

5-6.2.2 The atmosphere contains less than 19.5 percent oxygen; or

5-6.2.3 The presence of incompletely identified vapors or gases is indicated by a direct-reading organic vapor detection instrument, but the vapors and gases are known not to contain high levels of chemicals harmful to skin or capable of being absorbed through the intact skin.

Level B protection cannot protect against vapors or gases that contain high concentrations of chemicals that are harmful to skin or are capable of being absorbed through the skin. Additionally, Level B would not be used when the work being performed is likely to generate high concentrations of gases or splashes of material that will affect the exposed skin.

5-6.2.4 The presence of liquids or particulates is indicated but they are known not to contain high levels of chemicals harmful to skin or capable of being absorbed through the intact skin.

5-6.3 Level C protection should be used when:

5-6.3.1 The atmospheric contaminants, liquid splashes, or other direct contact will not adversely affect or be absorbed through any exposed skin;

5-6.3.2 The types of air contaminants have been identified, concentrations measured, and an air-purifying respirator is available that can remove the contaminants; and

5-6.3.3 All criteria for the use of air-purifying respirators are met.

Level C protection would not be appropriate where atmospheric concentrations of chemicals exceed IDLH levels, nor where the atmosphere contains less than 19.5 percent oxygen.

5-6.3.4* Atmospheric concentration of chemicals must not exceed IDLH levels. The atmosphere must contain at least 19.5 percent oxygen.

A-5-6.3.4 Refer to OSHA 29 CFR 1910.134.

5-6.4 Level D protection should be used when:

5-6.4.1 The atmosphere contains no known hazard; and

5-6.4.2 Work functions preclude splashes, immersion, or the potential for unexpected inhalation of or contact with hazardous levels of any chemicals.

NOTE: Combinations of personal protective equipment other than those described for Levels A, B, C, and D protection may be more appropriate and may be used to provide the proper level of protection.

Level D protection would not be appropriate for personnel operating in the warm zone, and the atmosphere must contain at least 19.5 percent oxygen.

6 Incident Mitigation

6-1 Control.

6-1.1 This chapter will address those actions necessary to assure confinement and containment (the first line of defense) in a manner that will minimize risk to both life and the environment in the early, critical stages of a spill or leak. Both natural and synthetic methods can be employed to limit the releases of hazardous materials so that effective recovery and treatment can be accomplished with minimum additional risk to the environment or to life.

A popular definition of hazardous materials is one offered by Ludwig Benner in his DECIDE action guide: "Something that jumps out of its container at you when something goes wrong and hurts or harms the thing it touches." An important element of that definition is that the "something" that is harmful is normally controlled or contained. When it is outside of its normal controlling element a hazardous materials incident occurs. It is only reasonable, then, that mitigation of an incident must deal with control of the material that is presenting the problem. The types of control are divided into confinement and containment, and the methods of mitigation are considered to be either physical or chemical. The purposes behind this approach are to present some order to the process and to simplify it for better understanding.

6-2 Types of Hazardous Materials.

All hazardous materials can be generally subdivided into three categories, based on the principal characteristic that makes them harmful or dangerous: chemical, biological, or radioactive. Of course, other safety hazards certainly exist at all emergency occurrences. These include typical dangers associated with the response itself and with the operations that are needed up to the controlling of the incident.

6-2.1 Chemical Materials. Those materials that pose a hazard based upon their chemical and physical properties.

Examination of the U.S. Department of Transportation list of hazard classes indicates that most of the classes would fall under

193

the chemical hazard type of material. The effect of exposure to chemical hazards can be either acute or chronic.

6-2.2 Biological Materials.   Those organisms that have a pathogenic effect to life and the environment and can exist in normal ambient environments.

Examples of biological hazards would include those requiring an Etiologic Agents label on packaging, such as for toxins or microorganisms that cause disease (cholera, tetanus, botulism). Disease-causing organisms might be found in waste from hospitals, laboratories, and research institutions.

6-2.3 Radioactive Materials.   Those materials that emit ionizing radiation.

U.S. DOT lists three classes of radioactive materials, with Class I being the least harmful. Packaging requirements for radioactive materials will vary depending on the varying hazard potentials presented by the material itself. The three types of harmful radiation emitted by radioactive materials are alpha, beta, and gamma.

6-3 Physical States of Hazardous Materials.   Hazardous materials may be classified into three states, namely gases, solids, and liquids. They can be stored or contained at a high or low pressure. All three states may be affected by the environment in which the incident occurs. The emergency responder must take into account conditions such as heat, cold, rain, or wind, which can have a significant effect on the methods used to accomplish a safe operation.

All matter exists in one of the three states listed. There are properties associated with each state that will have a bearing on how a specific material will appear or behave in the environment. For example, a liquid with a boiling point below 100 °F will tend to give off substantial vapor at ambient temperatures. A gas with a vapor density substantially heavier than air may collect at low points and migrate along the ground until it becomes mixed with air.

6-4 Methods of Mitigation.   There are two basic methods for mitigation of hazardous materials incidents, physical and chemical. Table 6-4.1 lists many physical methods and Table 6-4.2 lists many chemical methods that may be acceptable for mitigation of hazardous materials incidents. Recommended practices should be implemented only by personnel appropriately prepared by training, education, or experience.

Some of the methods listed for mitigation of the incident require a high degree of specialized training on the part of the responder and also require the use of sophisticated technical equipment. On the other hand, some of the mitigation efforts might be carried out by first responder operational level personnel. Diking or blanketing of a liquid spill of diesel fuel can be

Table 6-4.1   Physical Methods of Mitigation[5]

Method	Chemical				Biological				Radiological			
	Gases		Liq.	Sol.	Gases		Liq.	Sol.	Gases		Liq.	Sol.
	LVP*	HVP**			LVP	HVP			LVP	HVP		
Absorption	yes	yes	yes	no	no	no	yes[4]	no	no	no	yes	no
Covering	no	no	yes	yes	no	no	yes	yes	no	no	yes[3]	yes[3]
Dikes, Dams, Diversions, & Retention	yes	yes[6]	yes	yes	no	no	yes	yes	no	no	yes	yes
Dilution	yes	yes	yes	yes	no	no	no	no	yes	no	yes	yes
Overpack	yes	no	yes	yes	yes	no	yes	yes	yes	no	yes	yes
Plug/Patch	yes	yes	yes	yes	yes	yes	yes	yes	yes	yes	yes	yes
Transfer	yes	no	yes	yes	yes	no	yes	yes	yes	no	yes	yes
Vapor Suppression (Blanketing)	no	no	yes	no	no	no	yes	yes	no	no	no	no
Venting[1]	yes	yes	yes	no	yes	no	no	no	yes[2]	no	no	no

*Low Vapor Pressure
**High Vapor Pressure

1. Venting of low vapor pressure gases is recommended only when an understanding of the biological system is known. Venting is allowed when the bacteriological system is known to be nonpathogenic, or if methods can be employed to make the environment hostile to pathogenic bacteria.
2. Venting of low vapor pressure radiological gases is allowed when the gas(es) is/are known to be alpha or beta emitters with short half lives. Further, this venting is only to be allowed after careful consultation with a certified health physicist.
3. Covering should be done only after consultation with appropriate experts.
4. Absorption of liquids containing bacteria is permitted where the absorption bacteria or environment is hostile to the bacteria.
5. For substances involving more than one type, the most restrictive control measure should be used.
6. Water dispersion on certain vapors and gases only.

easily accomplished in many cases. Transferring that same product from a damaged tank truck to another tank truck would require more specialized training and equipment beyond what would be expected of the first responder. Vent and burn techniques would only be attempted by highly specialized personnel. In every case, good judgement must be used.

6-4.1 Physical Methods.   Physical methods of control involve any of several processes or procedures to reduce the area of the spill, leak, or other release mechanism. In all cases, methods used should be acceptable to the incident commander. The selection of personal protective clothing should be based on the hazardous materials and/or conditions present and should be appropriate for the hazards encountered. Refer to Table 6-4.1.

NOTE: Procedures described in 6-4.1.1 through 6-4.1.8 should be completed only by personnel trained in those procedures.

6-4.1.1 Absorption.   Absorption is the process in which materials hold liquids through the process of wetting. Absorption is accompanied by an increase in the volume of the sorbate/sorbent system through the process of swelling. Some of the materials typically used as absorbents are sawdust, clays, charcoal, and polyolefin type fibers. These materials can be used for confinement but it should be noted that the sorbed liquid can be desorbed under mechanical or thermal stress. When absorbents become contaminated, they retain the properties of the absorbed hazardous liquid, and they are, therefore, considered to be hazardous materials and must be treated and disposed of accordingly.

There are many commercially available products that are suitable for use as absorbents. Different types are intended and designed for specific types of spilled materials. The use of absorbents can help to reduce vapor generation, and it also serves to facilitate cleanup procedures. For further information on absorbents, consult ASTM-F-716-82, *Testing of Sorbent Performance.*

6-4.1.2 Covering.   Refers to a temporary form of mitigation for radioactive, biological, and some chemical substances such as magnesium. It should be done after consultation with a certified health physicist (in the case of radioactive materials) or other appropriate experts.

6-4.1.3 Dilution.   Refers to the application of water to water miscible hazardous materials. The goal is to reduce the hazard to safe levels.

Water should not be used indiscriminately or without knowledge of what the effect will be. The addition of water to a liquid spill can add to confinement problems. Many materials will react with water and thereby increase the intensity of the incident. Even with water soluble materials, the amount of water necessary to achieve a safe level could well render dilution an impractical approach. Nonetheless, it is a viable option in many instances.

6-4.1.4 Dikes, Dams, Diversions, and Retention.   These refer to the use of physical barriers to prevent or reduce the quantity of liquid flowing into the environment. Dikes or dams usually refer to concrete, earth, and other barriers temporarily or permanently constructed to hold back the spill or leak. Diversion refers to the methods used to physically change the direction of flow of the liquid. Vapors from certain materials, such as liquefied petroleum gas (LPG), can be dispersed using a water spray.

These techniques are the most commonly employed methods of control of releases, because improvisation is available and simple methods of confinement can be devised with a little ingenuity. On the other hand, substantial liquid spills of hazardous materials can pose insurmountable problems. In addition to the techniques listed, trenches can be used to collect spilled liquids, and pumps can serve to transfer the materials to containers or to a containment system. Earthen dikes or dams can be erected quickly under favorable conditions, and even sand bags can be used to assist in the damming effort. Commercial boom devices are available and are widely used in controlling spills, especially spills on waterways.

6-4.1.5 Vapor Dispersion.   Vapors from certain materials can be dispersed or moved using a water spray. With other products, such as liquefied petroleum gas (LPG), the gas concentration can be reduced below the lower

Figure 6.1   A fine water spray can be used to disperse some hazardous vapors.

flammable limit through rapid mixing of the gas with air, using the turbulence created by a fine water spray. Reducing the concentration of the material through the use of water spray may bring the material into its flammable range.

6-4.1.6 Overpacking.
The most common form of overpacking is accomplished by the use of an oversized container. Overpack containers should be compatible with the hazards of the materials involved. If the material is to be shipped, DOT specification overpack containers must be used. (The spilled materials still should be treated or properly disposed of.)

If it is possible to do so, a leaking drum or container should be temporarily repaired to reduce spillage prior to its being placed in an overpack container. Reducing a leak in a container can sometimes be accomplished simply by repositioning the container. Holes can be covered and temporary patches applied. Overpack containers typically have a formfitting gasket in a lid that can be tightly secured with a ring type closure.

Placing a leaking container into an overpack drum or container can be accomplished by placing the overpack on its side and sliding the smaller container in, by lowering the overpack over the leaking container and then tipping it upright, or by use of mechanical equipment to raise and lower a leaking container into the overpack container.

It is important that the overpack container be labeled in accordance with U.S. DOT regulations for the particular product it is carrying.

6-4.1.7 Plug and Patch.
Plugging and patching refers to the use of compatible plugs and patches to reduce or temporarily stop the flow of materials from small holes, rips, tears, or gashes in containers. The repaired container may not be reused without proper inspection and certification.

There are many plugging and patching techniques available to the responder to assist in stemming the flow of material from a container. Limiting or restricting a leak is an important condition in the mitigation process, so it is an essential skill to be mastered. At all times, however, the safety of the responder must be paramount.

Plugs can be of the inflating type using foam, water, or air or of the formed tapered type. Patches are also available in an assortment that includes magnetic steel, fabrics, adhesives, or epoxies. Piping can be patched with devices that are similar to hose clamps.

6-4.1.8 Transfer.
Transfer refers to the process of moving a liquid, gas, or some forms of solids, either manually, by pump, or pressure transfer, from a

Figure 6.2   One method of overpacking involves lowering the
overpack over the leaking container, then tipping it upright.

leaking or damaged container or tank. Care must be taken to ensure the
pump, transfer hoses and fittings, and container selected are compatible with
the hazardous material. When flammable liquids are transferred, proper concern for electrical continuity (such as bonding/grounding) must be observed.

**When transfer of materials must be made from one tank truck to
another, it should be done by personnel who are skilled and
practiced in the procedure. The incident commander will be in
charge of the operation and is responsible for assuring that proper
precautions are taken. However, reliance must be placed on the
experience of industry personnel who are appropriately trained
and equipped to perform the transfer operation.**

**In addition to the concern for grounding or bonding, it is
essential that all electrical equipment used for the transfer of
flammable liquids be of an approved type for such usage. NFPA 70,
The National Electrical Code (NEC), should be consulted for
additional information on this subject.**

6-4.1.9* Vapor Suppression (Blanketing).   Vapor suppression refers to the
reduction or elimination of vapors emanating from a spilled or released
material through the most efficient method or application of specially designed agents. The recommended vapor suppression agent is a polar solvent
aqueous film forming foam, proportioned at the same concentration regardless of the nature of the spill.

NOTE: Vapor suppression can also be considered a chemical method of mitigation.

A-6-4.1.9   One technique available for handling a spill of a hazardous liquid is the application of foams to suppress the vapor emanating from the liquid. This technique is ideally suited for liquid spills that are contained, i.e. diked. It can also be used where the spill is not confined. In all cases this technique should only be undertaken by personnel who have been trained in the use of foam agents for vapor suppression. Training in the use of foam as a fire extinguishing agent is not sufficient to qualify an individual for foam application as a vapor suppressing agent.

Vapor suppressing foam agents vary in their effectiveness depending on a number of factors. These factors can include the type of foam, the expansion ratio of the foam, the 25 percent drainage time of the foam, the rate of application of the foam, and the application (gpm/sq ft) density (gallons/sq ft) of the foam. These variables serve to emphasize the need for training of the person selecting this technique for applying foam as a vapor suppressing medium.

There are two general categories of foam agents used as fire extinguishing agents, namely chemical foam and mechanical foam. Chemical foam is produced by a chemical reaction between salts of weak acids and weak bases. It is becoming obsolete in the United States and has never been tested as a vapor suppressing foam agent. Mechanical foam agents are produced by mechanically mixing a dilute solution of the foam concentrate and water with air, producing an expanded foam. Mechanical foam agents have been tested both as fire extinguishing agents and as vapor suppressing agents.

Several types of mechanical foam will be briefly described. For more detailed information, consult NFPA 11, *Standard for Low Expansion Foam and Combined Agent Systems*, or the foam agent or equipment manufacturer's literature.

Mechanical foam agents can be broken down into two broad categories that reflect the chemical composition of the surfactants used to produce the foam concentrate. Protein foam agents are derived from a hydrolyzed protein substance such as hoof and horn meal or feather meal. These are naturally occurring substances. The second category of mechanical foam agents are synthetically derived. The two subdivisions within the synthetic foam area reflect the surfactant composition. Synthetic foams may contain surfactants of the hydrocarbon surfactant type, i.e. detergents; or more commonly, they may contain surfactants that are fluorinated and are known generically as AFFF. AFFF is an acronym for the terminology Aqueous Film Forming Foam, indicating that these foam agents form an aqueous film on a hydrocarbon liquid surface in spite of the fact that the hydrocarbon is less dense than water.

A more practical way of differentiating synthetic foams is based on their expansion ratios. Most synthetic foams that are composed strictly of hydrocarbon surfactants are used either as medium expansion or high expansion foam agents. Low expansion agents are those foam agents that when mechanically mixed with air will produce expansion ratios of 20 to 1 or less. Medium expansion refers to expansion ratios between 200 to 1 and 20 to 1. High expansion foam agents refer to expansion ratios in excess of 200 to 1 and typically are in the vicinity of 750 to 1. Synthetic foam agents derived from hydrocarbon surfactants can be used as low expansion foams but they generally are not used in that fashion.

Within the two types of mechanical foams, several variations exist. Low expansion foams consist of six categories that are differentiated by chemical composition. The first and oldest of the low expansion foam agents are those derived from hydrolyzed proteins, to which are added stabilizers, freezing point depressants, and fungicide agents. Protein foams are characterized by a consistent mass of foam bubbles that have relatively poor mobility but good burnback resistance and good vapor suppression characteristics when used on fuels of moderate volatility.

A second type of low expansion foam agent is derived from protein foam to which a fluorochemical surfactant has been added. They are called fluoroprotein foams and have improved mobility on fuel surfaces, slightly improved burnback resistance, and the same vapor suppressing characteristics as the protein foams.

A third type of low expansion foams are the synthetic hydrocarbon type surfactants or detergent foams used in the expansion ratio categories of medium and high expansion. They have, however, been used as low expansion foams. Their mobility on a fuel surface and their vapor suppressing characteristics are not as good as the other foam agents in this category.

A fourth category of low expansion foams are those referred to as AFFF. AFFF are foams that are formulated from synthetic fluorochemical and hydrocarbon surfactants in conjunction with solvents. They provide an aqueous film from the draining foam bubbles whose surface characteristics allow it to spread on a hydrocarbon fuel surface in spite of the lower density of a hydrocarbon fuel such as gasoline. The mobility of the AFFF on a hydrocarbon fuel surface is superior to that of the other low expansion foam concentrates. However, the burnback resistance of this foam is not as good as that of the protein or fluoroprotein foams. Its vapor suppressing characteristics would be as good as most of the other foam agents due to the aqueous film.

A fifth category involves a modification of the fluoroprotein foam. This type of low expansion foam is referred to by the acronym FFFP, standing for Film Forming Fluoroprotein Foam agent. These agents are simply fluoroprotein foam agents to which enough fluorochemical surfactant has been added to provide the ability to form the aqueous film that gives mobility comparable to the AFFF products. However, adding this fluorochemical weakens the burnback characteristics to the point that they are comparable to the AFFF agent. They are also comparable to the AFFF products as vapor suppressing agents.

The sixth and final type of low expansion foam has an ability to extinguish not only hydrocarbon fires but those fires involving certain types of hydrocarbons that are referred to as polar solvents. These are materials that have appreciable water miscibility or solubility and are typified by materials such as alcohols, ketones, and ethers. Some common examples of these types of materials would be methanol, ethanol, isopropanol, acetone, methyl ethyl ketone, dimethyl ether, diethyl ether, isopropyl ether, etc. The five low expansion foam agents previously discussed are effective only on the hydrocarbon type fuels such as gasoline. If those products are applied to a fire involving a fuel such as methanol, the foam will dissolve and the fuel is said to be foam destructive. It should be clearly understood that the five previous compositions for low expansion foams are effective only on water insoluble fuels such as gasoline.

There are two basic compositions that allow for the use of a low expansion foam extinguishing agent on a polar solvent. The most widely used composition is based on a conventional AFFF type low expansion foam to which a material known as a polysaccharide has been added. The polysaccharide forms an insoluble membrane when it comes in contact with a water miscible fuel such as methanol or ethanol. This membrane acts in much the same way that the aqueous film acts when that same product is used on a conventional hydrocarbon fuel such as gasoline. Until recently, this type of product was available in a configuration that required that it be proportioned as a 3 percent product, i.e. 3 parts of concentrate to 97 parts of water, when used on hydrocarbon fuels such as gasoline; and as a 6 percent product, i.e. 6 parts of concentrate to 94 parts of water, when used on a polar solvent such as ethanol. This problem has been resolved and there are now products available that will proportion at 3 percent on either type of fuel substance. The polysaccharide also stabilizes the AFFF product so the burnback resistance is equivalent to protein and fluoroprotein foam. On this basis, the agent to use as a vapor suppressant would be a product that is capable of being used regardless of the type of fuel substrate, that can be proportioned at the same concentration regardless of the fuel substrate, and that has excellent mobility as well as burnback resistance, thereby providing vapor suppressing characteristics.

A second composition of this type has been derived from the film forming fluoroprotein products that allows them to be used on a polar solvent fuel as well as on a normal hydrocarbon fuel. They have to be proportioned at two different concentrations and therefore are not as attractive, although they will perform adequately in a vapor suppressing application.

The use of fire fighting foam as a vapor suppressant involves some considerations that are different than those required for fire extinguishing agents. One of the obvious issues that is of concern is the stability of the foam blanket as a function of time. This can be generally approached by looking at what is referred to as the foam quality. Foam quality is generally measured in terms of three parameters: foam expansion ratio, foam 25 percent drainage time, and foam viscosity. Foam expansion ratio refers to the volume of foam obtained from a unit volume of foam liquid dilute solution. The expansion ratio for low expansion foams is less than 20 to 1. Typical values for expansion ratio for some of the low expansion foam agents are in the range of 6 to 1 and 12 to 1 when used in aspirating equipment. AFFF, however, can be used in nonaspirating equipment such as water fog nozzles, and the expansion ratios obtained there can be as low as 3 to 1 or 2 to 1.

The 25 percent drainage time is the time that is required for 25 percent of the foam liquid to drain from the foam. This is the property that is generally used to measure the stability of the foam, and it, in combination with the expansion ratio, will determine how the thickness of the foam blanket will vary with time. For example, a typical application of foam could involve an expansion ratio of 10 to 1 with a 25 percent drain time of 10 minutes. This means that whatever thickness is attained on the fuel surface with a 10 to 1 expanded foam, 25 percent of it or 2.5 units of it will disappear within 10 minutes. For purposes of simplicity, we can assume that foam drainage is a linear function and that in the course of draining, approximately 40 minutes will be required to deplete the foam from the fuel surface. This makes several assumptions regarding the environmental conditions such as wind velocity, wind direction, ambient temperature, etc.

An important factor is the vapor pressure of the fuel that is being used as the substrate by the vapor suppressant. Fuel vapor pressure can vary widely depending on the type of fuel. For example, the vapor pressure of unleaded regular gasoline is certainly considerably higher than the vapor pressure associated with a fuel oil. The key to using foam as a vapor suppressing agent is to have a good thick foam blanket continuously on the fuel surface. This requires that common sense, good judgement, and the use of manufacturer's recommendations for the foam agent be followed at all times.

There is another category of foam agents that are used as vapor suppressants. These particular foam agents are not fire extinguishing agents and should never be used for that purpose. They are specifically formulated to be used as vapor suppressants in spills involving materials that are acidic or caustic. These materials are generally synthetic in nature and are designed to be used in a certain way. At present, there exists no standard that addresses materials of this type nor are there any approval or testing procedures. These products should be used in the equipment specified by the manufacturer and with the application conditions that are specified by the manufacturer.

It is important to recognize that there are some limitations in the use of foam fire extinguishing agents and of foam vapor suppressing agents that must be observed. In the case of the former, the foam blanket must be maintained in such a fashion that there are no openings for vaporization to occur that could result in fire or in exposure to human life. There are a number of low expansion foam agents that can be used for vapor suppression. There are disadvantages and advantages associated with all of them. Recognize that both medium expansion and high expansion foams can be used for vapor suppression but both are technique dependent, less mobile on the fuel surface, and in general result in foams that are more impaired by climatic conditions such as wind velocity, wind direction, precipitation, etc. Foam agents that are designed specifically to be used as vapor suppressants for spills involving acidic or caustic materials are not fire extinguishing agents. They can never be used when a fire is involved with the acidic or basic material. They are vapor suppressants and the manufacturer's recommendations and specifications regarding how they are used, and in what type of equipment, must be followed.

While vapor suppression or blanketing does not change the nature of the hazardous material, it is an important procedure in mitigation in that it can greatly reduce the immediate hazard and danger associated with the presence of uncontrolled vapor. In addition, it buys additional time to provide measures that will bring about complete control and resolution of the incident, under safer circumstances.

6-4.1.10 Venting.   Venting is a process that is used to deal with liquids or liquefied compressed gases where a danger, such as an explosion or mechanical rupture of the container or vessel, is considered likely. The method of venting will depend on the nature of the hazardous material. In general, it involves the controlled release of the material to reduce and contain the pressure and diminish the probability of an explosion.

6-4.2 Chemical Methods.   Chemical methods of control involve the application of chemicals to treat spills of hazardous materials. Chemical methods may involve any one of several actions to reduce the involved area affected by the release of a hazardous material. In all cases, methods used should be acceptable to the incident commander. The selection of personal protective clothing should be based on the hazardous materials and/or conditions present and appropriate for the hazards encountered. Refer to Table 6-4.2.

NOTE: The procedures described in 6-4.2.1 through 6-4.2.10 should only be used by personnel trained in those procedures.

6-4.2.1 Adsorption.   Adsorption is the process in which a sorbate (hazardous liquid) interacts with a solid sorbent surface. The principal characteristics of this interaction are:

(a) The sorbent surface is rigid and no volume increase occurs as is the case with absorbents.

(b) The adsorption process is accompanied by heat of adsorption whereas absorption is not.

(c) Adsorption occurs only with activated surfaces, i.e., activated carbon, alumina, etc.

NOTE: Spontaneous ignition can occur through the heat of adsorption of flammable materials, and caution should be exercised.

Table 6-4.2   Chemical Methods of Mitigation

Method	Chemical				Biological				Radiological			
	Gases		Liq.	Sol.	Gases		Liq.	Sol.	Gases		Liq.	Sol.
	LVP*	HVP**			LVP	HVP			LVP	HVP		
Adsorption	yes	yes	yes	no	yes³	yes	yes³	no	no	no	no	no
Burn	yes	yes	yes	yes	yes	yes	yes	yes	no	no	no	no
Dispersion/Emulsification	no	no	yes	yes	no	no	yes³	no	no	no	no	no
Flare	yes	yes	yes	no	yes³	yes	yes	no	no	no	no	no
Gelation	yes	no	yes	yes	yes³	no	yes³	yes³	no	no	no	no
Neutralization	yes¹	yes⁴	yes	yes²	no	no	yes³	no	no	no	no	no
Polymerization	yes	no	yes	yes	no	no	no	no	no	no	no	no
Solidification	no	no	yes	no	no	no	yes³	no	yes	no	yes	no
Vapor Suppression	yes	yes	yes	yes	yes	yes	yes	yes	yes	yes	yes	yes
Vent/Burn	yes	yes	yes	no	yes	yes	yes	no	no	no	no	no

*Low Vapor Pressure
**High Vapor Pressure

1. Technique may be possible as a liquid or solid neutralizing agent and water can be applied.
2. When solid neutralizing agents are used, they must be used simultaneously with water.
3. Technique is permitted only if resulting material is hostile to the bacteria.
4. The use of this procedure requires special expertise and technique.

Figure 6.3   Chemical methods of hazardous materials mitigation
include the use of neutralizing agents.

The word "sorbents" generally includes both absorbents and
adsorbents. Adsorbents act in such a way that the internal structure of the material is not penetrated. Adsorbents can be natural
or synthetic materials and can be used on liquid spills on land and
also to some degree on water. They should be nonreactive to the
spilled material. Porous clay and sand are examples of adsorbents.

6-4.2.2 Controlled Burning.   For purposes of this practice, controlled combustion is considered a chemical method of control. However, it should only
be used by qualified personnel trained specifically in this procedure.

In some emergency situations where extinguishing a fire will result in large,
uncontained volumes of contaminated water, or threaten the safety of responders or the public, controlled burning is used as a technique. It is advised
that consultation be made with the appropriate environmental authorities
when this method is used.

There are occasions where extinguishment of a fire, with accom-
panying runoff of large amounts of contaminated waste, is the
improper approach. Such was the case with the Sandoz fire in
Basel, Switzerland that resulted in major contamination of the
Rhine River. At the Sherwin Williams fire in Dayton, Ohio, a
decision was made not to use heavy fire streams because the
building was essentially lost, and because contaminated runoff

might pollute a major water supply for the city. The latter decision was praised by environmentalists and fire fighting experts alike.

Another aspect of the controlled burning approach is actually to incinerate the spilled hazardous material. There are transportable incinerators that are designed to promote combustion of spilled materials, especially for oil spill situations.

6-4.2.3 Dispersion, Surface Active Agents, and Biological Additives. Certain chemical and biological agents can be used to disperse or break up the materials involved in liquid spills. The use of these agents results in a lack of containment and generally results in spreading the liquid over a much larger area. Dispersants are most often applied to spills of liquids on water. The dispersant breaks down a liquid spill into many fine droplets, thereby diluting the material to acceptable levels. Use of this method may require the prior approval of the appropriate environmental authority.

Dispersants result generally in oil in water emulsions, since the chemicals used will reduce the surface tension between oil and water. Chemical dispersants would not be used where increased biological damage would result, so it is essential that environmental authorities be consulted. A surface active agent also results in increased emulsification and dispersion of a spill. Surface cleaning equipment is also available for soil surface cleaning. It involves agitation of the soil surface with water to form a slurry. The contamination is then removed through a separation process that takes place in a specially designed sand separator.

Biological additives can be used to degrade certain hazardous materials spilled on land or in water by biochemical oxidation and biochemical accelerators.

6-4.2.4 Flaring. Flaring is a process that is used with high vapor pressure liquids or liquefied compressed gases for the safe disposal of the product. Flaring is the controlled burning of material in order to reduce or control pressure and/or dispose of a product.

6-4.2.5 Gelation. Gelation is the process of forming a gel. A gel is a colloidal system consisting of two phases, a solid and a liquid. The resulting gel is considered to be a hazardous material and must be disposed of properly.

When gelling agents are used on hazardous chemicals, the resulting gel is more easily cleaned up by either mechanical or physical methods. Gels can be used on spilled liquids in water, and to a lesser degree on land.

6-4.2.6 Neutralization. Neutralization is the process of applying acids or bases to a spill to form a neutral salt. The application of solids for neutralizing can often result in confinement of the spilled material. Special formulations are available that do not result in violent reactions or local heat generation during the neutralization process. In cases where special neutralizing formulations are not available, special considerations should be given to protecting persons applying the neutralizing agent, as heat is generated and violent reactions may occur. One of the advantages of neutralization is that a hazardous material may be rendered nonhazardous.

The pH scale is used to categorize compounds as acids or bases. A value of 7 on the scale is neutral, with descending values denoting increasing acidity. Levels above 7 up to 14 denote bases, with the higher values indicating increasingly stronger bases. It is possible to neutralize acidic or basic spills by mixing with a neutralizing fluid.

6-4.2.7 Polymerization.   A process in which a hazardous material is reacted in the presence of a catalyst, of heat or light, or with itself or another material to form a polymeric system.

6-4.2.8 Solidification.   Solidification is the process whereby a hazardous liquid is treated chemically so that a solid material results. Adsorbents can be considered an example of a solidification process. There are other materials that can be used to convert hazardous liquids into nonhazardous solids. Examples are applications of special formulations designed to form a neutral salt in the case of spills of acids or caustics. The advantage of the solidification process is that a spill of small scale can be confined relatively quickly and treatment effected immediately.

There are commercially available adsorbents that can be used to solidify oily wastes that are water insoluble. The spilled liquids are adsorbed to granules to form a solid, nonflowing mixture. This makes the resulting product safer and more easily transported to an appropriate disposal facility.

6-4.2.9 Vapor Suppression.   The use of solid activated materials to treat hazardous materials so as to effect suppression of the vapor off-gasing from the materials. This process results in the formation of a solid that affords easier handling but that may result in a hazardous solid that must be disposed of properly.

6-4.2.10 Venting and Burning.   This involves the use of shaped charges to vent the high vapor pressure at the top of the container and then with additional charges to release and burn the remaining liquid in the container in a controlled fashion.

In the Livingston, Louisiana derailment that occured in 1982, shaped explosive charges were used to "vent and burn" where the tank cars were too badly damaged to attempt transfer of product. This was done some eight days into the incident. Vent and burn is

Figure 6.4   Venting and burning techniques have been used in at least
one train derailment.

a highly sophisticated technique that should be attempted only by
adequately trained specialists and under very controlled circumstances.

7 Decontamination

7-1 Decontamination Plan. At every incident involving hazardous materials there is a possibility that response personnel and their equipment will become contaminated. The contaminant poses a threat, not only to the persons contaminated, but to other personnel who may subsequently have contact with them or the equipment.

Decontamination can be achieved by removing or neutralizing the contaminants that have accumulated on clothing and equipment. It requires an organized and well ordered procedure, hence the need for a plan in order to carry it out successfully. The plan must take into account measures that will minimize contamination to begin with, and ultimate responsibility for implementation of the plan falls to the incident commander.

7-1.1 Incident responders should have an established procedure to minimize contamination or contact, to limit migration of contaminants, and to properly dispose of contaminated materials. Decontamination procedures should begin upon arrival at the scene, should provide for an adequate number of decontamination personnel, and should continue until the incident commander determines that decontamination procedures are no longer required. Decontamination of victims may be required.

During the course of an incident, the decontamination plan will have to be revised to address changing conditions.

Good operating practices can reduce the extent of contamination to clothing and equipment. For example, sampling and monitoring equipment can be protected by bagging, with a limited number of openings as needed for operation. Disposable outer garments also help, and covering the source of the contamination, where possible, can greatly reduce the need for extensive decontamination procedures.

7-2 Personal Protective Equipment. Before personal protective equipment is removed it should be decontaminated. During doffing of personal protective equipment, the clothing should be removed in a manner such that the outside surfaces do not touch or make contact with the wearer. A log of

211

Figure 7.1   Dilution is one method of decontamination. It reduces
the concentration of the contaminant to a level at which it is
no longer harmful. Water is used, except when the potential
exists for a chemical reaction.

personal protective equipment used during the incident should be maintained. Personnel wearing disposable protective equipment should go through the decontamination process and the disposable protective equipment should be disposed of in accordance with established procedures.

Some heavy contamination can be removed physically by wiping off, wetting down, and allowing to dry. Chemical means can also be used to deactivate the contaminant. A combination of physical and chemical means can also be used.

The log of personal protective equipment used during an incident should record the type of equipment used, duration of use, decontamination procedures used, and type of chemicals to which the equipment was exposed. It should also record the name of the person using the equipment.

Disposable protective equipment should be placed in plastic bags pending final disposal.

7-3 Decontamination.   Decontamination consists of removing the contaminants by chemical or physical processes. The conservative action is always to assume contamination has occurred and to implement a thorough, technically sound decontamination procedure until it is determined or judged to be unnecessary.

Figure 7.2   For some contaminants, protective clothing can never be adequately decontaminated and must be disposed of. Yet sufficient decontamination must take place to allow workers to safely remove the protective clothing.

There are established methods for testing the effectiveness of decontamination procedures, and they should be used. If the decontamination procedure is not working effectively, revisions in the process may have to be made.

Test methods include visual examination in both natural and ultraviolet light, laboratory analysis of a wipe sample, analysis of the used cleaning solution, and laboratory analysis of a segment of protective clothing.

7-3.1   Procedures for all phases of decontamination must be developed and implemented to reduce the possibility of contamination to personnel and equipment. Reference guides for the development of decontamination procedures can be found in Appendix B-2.4(d) and B-2.4(f). Assuming protective equipment is grossly contaminated, use appropriate decontamination methods for the chemicals encountered.

Some decontamination procedures may actually present additional hazards. It is possible that the decontamination solution may react with the chemical to which the clothing was exposed, that it may itself permeate or degrade some protective clothing, or that it may emit vapors that are harmful. Compatibility should be determined prior to use.

7-3.2 Outer clothing should be decontaminated prior to removal. The outer articles of clothing, after removal, should be placed in plastic bags for later additional decontamination, cleaning, and/or inspection. In some cases, they may have to be overpacked into containers for proper disposal. Water or other solutions used for washing or rinsing may have to be contained, collected, containerized, and analyzed prior to disposal.

The use of plastic drop cloths on which contaminated outer clothing can be placed is a common and useful procedure. It is also helpful to have lined containers on hand for packing the contaminated clothing and equipment. Metal or plastic drums are effective for storage of washing or rinsing solutions.

7-3.2.1 Initial procedures should be upgraded or downgraded as additional information is obtained concerning the type of hazardous materials involved, the degree of hazard, and the probability of exposure of response personnel.

If the decontamination method being used is not effectively removing contaminants, then different and additional measures have to be taken in order to prevent further contamination. If necessary, specialists should be consulted.

The decontamination plan may have to be changed whenever the type of protective clothing and equipment changes, when site conditions change, or when new information is received.

7-3.3 Decontaminating Solutions.   Using solutions containing chemicals to alter or change contaminants to less hazardous materials should only be done after consultation with persons experienced and familiar with the hazards involved. The use of detergent-water washing solutions is more prevalent, but its effectiveness against certain contaminants may be low. It is less risky however than using chemical solutions.

If time is not an overriding factor, and if specialists are not on site or immediately available, then the detergent-water solution method of preliminary decontamination is the safest and most appropriate approach. See commentary on 7-3.1 regarding hazards associated with use of chemical solutions.

7-3.4 Decontamination of Equipment.   Many types of equipment are very difficult to decontaminate and may have to be discarded as hazardous wastes. Whenever possible, other pieces of small equipment should be disposable or made of nonporous material. Monitoring instruments and some types of sampling equipment can be placed in plastic bags (with only the detecting element exposed) to minimize potential contamination problems.

Some equipment and tools can be covered with a strippable coating that can be removed during decontamination.

7-3.5 Large items of equipment, such as vehicles and trucks, should be subjected to decontamination by high pressure water washes, steam, or special solutions. Water or other solutions used for washing or rinsing may have to be contained, collected, containerized, and analyzed prior to disposal. Consultation with appropriate sources should be utilized to determine proper decontamination procedures.

Decontamination of large items of equipment will require different supplies on hand. These might include adequately sized storage tanks or collection systems, pumps or drains, long handled brushes, and wash booths or containers.

7-3.6 Decontamination Team. Personnel assigned to the decontamination team should wear an appropriate level of personal protective equipment and may require decontamination themselves.

The decontamination team members closest to the hot zone may require a higher level of protective clothing than team members closest to the cold zone. The level of protection required varies with the decontamination equipment that is in use. Selection of the protective clothing should be made by a qualified person.

Figure 7.3   In some cases, fire fighting equipment and trucks may require decontamination as well.

8 Referenced Publications

8-1 The following documents or portions thereof are referenced within this recommended practice and should be considered part of the recommendations of this document. The edition indicated for each reference is the current edition as of the date of the NFPA issuance of this document.

8-1.1 NFPA Publications. National Fire Protection Association, Batterymarch Park, Quincy, MA 02269.

NFPA 70-1990, *National Electrical Code*

NFPA 472-1989, *Standard for Professional Competence of Responders to Hazardous Materials Incidents*

NFPA 1500-1987, *Standard on Fire Department Occupational Safety and Health Program*

NFPA 1981-1987, *Standard on Open-Circuit Self-Contained Breathing Apparatus for Fire Fighters*

NFPA 1982-1988, *Standard on Personal Alert Safety Systems (PASS) for Fire Fighters*.

8-1.2 US Government Publication. US Government Printing Office, Superintendent of Documents, Washington, DC 20402.

Title 29 CFR Part 1910.120.

Appendix A

The material contained in Appendix A is included in the text within this *Handbook* and therefore is not repeated here.

Appendix B

Suggested Reading List

This Appendix is not a part of the recommendations of this NFPA document, but is included for information purposes only.

B-1 Introduction.

This list provides the titles of references and organizations that may be of value to those responding to hazardous materials incidents. This list can be expanded based on personal preferences and requirements.

The references are categorized by subject. The title, author, publisher, and place of publication are given for each. The year of publication is not always given because many are revised annually. The user should attempt to obtain the most recent edition.

The last section lists sources of these references as well as other information that might be useful. Usually, these agencies or associations will provide a catalog on request. Where available, phone numbers are also listed.

B-2 References.

B-2.1 Industrial Hygiene (Air Sampling and Monitoring, Respiratory Protection, Toxicology).

(a) *Air Sampling Instruments for Evaluation of Atmospheric Contaminants*, American Conference of Governmental Industrial Hygienists, Cincinnati, OH.

(b) *Direct Reading Colorimetric Indicator Tubes Manual*, American Industrial Hygiene Association, Akron, OH.

(c) *Fundamentals of Industrial Hygiene*, National Safety Council, Chicago, IL.

(d) *Industrial Hygiene and Toxicology*, Frank A. Patty, John Wiley and Sons, Inc., New York, NY.

(e) *Manual of Recommended Practice for Combustible Gas Indicators and Portable, Direct Reading Hydrocarbon Detectors*, American Industrial Hygiene Association, Akron, OH.

(f) *NIOSH/OSHA Pocket Guide to Chemical Hazards*, DHHS No. 85-114, NIOSH, Department of Health and Human Services, Cincinnati, OH.

(g) *Occupational Health Guidelines for Chemical Hazards*, DHHS No. 81-123, NIOSH, Department of Health and Human Services, Cincinnati, OH.

(h) *Occupational Safety and Health Standards*, Title 29, Code of Federal Regulations, Part 1910.120. "Hazardous Waste Operations and Emergency Response Final Rule."

(i) *TLVs Threshold Limit Values and Biological Exposure Indices (Threshold Limit Values for Chemical Substances and Physical Agents in the Workroom Environment)*, American Conference of Governmental Industrial Hygienists, Cincinnati, OH.

B-2.2 Chemical Data.

(a) *Chemical Hazard Response Information System (CHRIS)*, U.S. Coast Guard, Washington, DC, Commandant Instruction M.16565.12A.

(b) *CHRIS–A Condensed Guide to Chemical Hazards*, U.S. Coast Guard, Commandant Instruction M16565.11a.

(c) *The Condensed Chemical Dictionary*, G. Hawley, Van Nostrand Reinhold Co., New York, NY.

(d) *CRC Handbook of Chemistry and Physics*, CRC Press, Boca Raton, FL.

(e) *Dangerous Properties of Industrial Materials*, N. Irving Sax, Van Nostrand Reinhold Co., New York, NY.

(f) *Effects of Exposure to Toxic Gases*, Matheson.

(g) *Emergency Handling of Hazardous Materials in Surface Transportation*, Association of American Railroads.

(h) *Farm Chemicals Handbook*, Farm Chemicals Magazine, Willoughby, OH.

(i) *Firefighter's Handbook of Hazardous Materials*, Baker, Charles J., Maltese Enterprises, Indianapolis, IN.

(j) *Fire Protection Guide to Hazardous Materials*, National Fire Protection Association, Quincy, MA.

(k) *Hazardous Materials Handbook*, Meidl, J.H., Glencoe Press.

(l) *The Merck Index*, Merck and Co., Inc., Rahway, NJ.

(m) *Emergency Action Guides*, Association of American Railroads.

B-2.3 Safety and Personnel Protection.

(a) *A Guide to the Safe Handling of Hazardous Materials Accidents*, ASTM STP 825. American Society for Testing and Materials, Philadelphia, PA.

(b) *Fire Protection Handbook*, National Fire Protection Association, Quincy, MA.

(c) *Guidelines for Decontamination of Firefighters and their Equipment Following Hazardous Materials Incidents*, Canadian Association of Fire Chiefs (May 1987).

(d) *Guidelines for the Selection of Chemical Protective Clothing.* Volume 1: Field Guide, A. D. Schwope, P.P. Costas, J. O. Jackson, D. J. Weitzman. Arthur D. Little, Inc., Cambridge, MA (March 1983).

(e) *Guidelines for the Selection of Chemical Protective Clothing.* Volume 2: Technical and Reference Manual, A.D. Schwope, P.P. Bostas, J.O. Jackson, D.J. Weiteman, J.O. Stull; Arthur D. Little, Inc., Cambridge, MA. 3rd Edition (February 1987).

(f) *Hazardous Materials*, Warren Isman and Gene Carlson, Glencoe Press, 1981.

(g) *Hazardous Materials Emergencies Response and Control*, John R. Cashman, Technomic Publishing Company, Lancaster, PA (June 1983).

(h) *Hazardous Materials for the First Responser*, International Fire Service Training Association, Stillwater, OK (1988).

(i) *Hazardous Materials: Managing the Incident*, Gregory Noll, Michael Hildebrand and James Yvorra, Fire Service Publications, Stillwater, OK (1988).

(j) *Handling Radiation Emergencies*, Purington and Patterson, NFPA, Batterymarch Park, Quincy, MA.

(k) *Hazardous Materials Injuries, A Handbook for Pre-Hospital Care*, Douglas R. Stutz, Robert C. Ricks, Michael F. Olsen, Bradford Communications Corp., Greenbelt, MD.

(l) *National Safety Council Safety Sheets*, National Safety Council, Chicago, IL.

(m) *Radiological Health–Preparedness and Response in Radiation Accidents*, U.S. Department of Health and Human Services, Washington, DC.

(n) *SCBA–A Fire Service Guide to the Selection, Use, Care, and Maintenance of Self-Contained Breathing Apparatus*, NFPA, Batterymarch Park, Quincy, MA.

(o) *Standard First Aid and Personal Safety*, American Red Cross.

B-2.4 Planning Guides.

(a) *A Fire Department's Guide to Implementing Title III and the OSHA Hazardous Materials Standard* (August 1987). William H. Stringfield, International Society of Fire Service Instructors, Ashland, MA.

(b) *Federal Motor Carrier Safety Regulations Pocketbook*, U.S. Department of Transportation, J. J. Keller and Associates, Inc.

(c) *Hazardous Chemical Spill Cleanup*, Noyes Data Corporation, Ridge Park, New Jersey.

(d) *Occupational Safety and Health Guidance Manual for Hazardous Waste Site Activities*, NIOSH/OSHA/USCG/EPA, U.S. Department of Health and Human Services, NIOSH.

(e) *Hazardous Materials Emergency Planning Guide*, (March 1987). National Response Team.

(f) *Standard Operating Safety Guides*, Environmental Response Branch, Office of Emergency and Remedial Response, U.S. Environmental Protection Agency.

(g) *1987 Emergency Response Guidebook–Guidebook for Hazardous Materials Incidents*, DO P 5800.3 USDOT, Materials Transportation Bureau, Attn: DMT-11. Washington, DC 20590.

B-3 Agencies and Associations.

Agency for Toxic Substances Disease Registry
Shamlee 28 S., Room 9
Centers for Disease Control
Atlanta, GA 30333
404/452-4100

American Conference of Governmental Industrial Hygienists
6500 Glenway Avenue–Building D-5
Cincinnati, OH 45211
513/661-7881

American Industrial Hygiene Association
475 Wolf Ledges Parkway
Akron, OH 44311-1087
216/762-7294

American National Standards Institute, Inc.
1430 Broadway
New York, NY 10018
212/354-3300

American Petroleum Institute (API)
1220 L St. N.W. 9th Floor
Washington, DC 20005
202/682-8000

Association of American Railroads
50 F St. N.W.
Washington, DC 20001
202/639-2100

Chemical Manufacturer's Association
2501 M St. N.W.
Washington, DC 20037
202/877-1100

CHEMTREC
800/424-9300

The Chlorine Institute
2001 L. Street N.W.
Washington, DC 20001
202/639-2100

Compressed Gas Association
1235 Jefferson Davis Highway
Arlington, VA 22202
703/979-0900

The Fertilizer Institute (TFI)
1015 18th St. N.W.
Washington, DC 20036
202/861-4900

International Society of Fire Service Instructors
30 Main Street
Ashland, MA 01721
617/881-5800

National Fire Protection Association
Batterymarch Park
Quincy, MA 02269
617/770-3000

Spill Control Association of America
Suite 1575
100 Renaissance Center
Detroit, MI 48243-1075

U.S. Department of Transportation
Materials Transportation Bureau
Office of Hazardous Materials Operations
400 7th St. S.W.
Washington, DC 20590
202/366-4555

U.S. EPA Office of Research & Development
Publications—CERI
Cincinnati, OH 45268
513/684-7562

U.S. EPA Office of Solid Waste (WH-562)
Superfund Hotline
401 M St., S.W.
Washington, DC 20460
800/424-9346

U.S. Mine Safety and Health Administration
Department of Labor
4015 Wilson Blvd, Room 600
Arlington, VA 22203
703/235-1452

U.S. National Oceanic and Atmospheric Administration
Hazardous Materials Response Branch N/CMS 34
7600 Sand Point Way NE
Seattle, WA 98115

B-4 Computer Data Base Systems.

Hazardous Materials Information Exchange (HMIX)
Federal Emergency Management Agency
State and Local Programs Support Directory
Technological Hazards Division
500 C Street, S.W.
Washington, DC 20472

Supplements

These supplements are not part of the code text of this *Handbook*, but are included for information purposes only. They are printed in black for ease of reading.

Introduction

The subject of hazardous materials is too broad to be contained in any single source. The following Supplements to the *Hazardous Materials Response Handbook* have been included in an attempt to touch on some of the subject areas most important to emergency responders as they begin preparing to handle hazardous materials incidents.

These supplements do not by any means exhaust the information available even on these topics, but have been selected as helpful guidelines and excellent references for anyone beginning emergency response planning.

The Technical Committee on Hazardous Materials Response Personnel, which prepared NFPA 471 and 472, intends to develop other documents that will be of benefit to the response community. Future editions of this Handbook will likewise feature an expanded Supplement section covering more of the diverse topics relevant to hazardous materials response.

I Hazardous Materials Definitions

U.S. Department of Transportation, Research and Special Programs Administration, Office of Hazardous Materials Transportation, Washington, DC 20590.

The following supplement includes the definition, hazard class, and UN class of many hazardous substances.

Hazardous Materials Definitions

The following definitions have been abstracted from the Code of Federal Regulations, Title 49, Transportation, Parts 100–177. Refer to the referenced sections for complete details. Note: In column (1), Sec. 172.101, Hazardous Materials Table, the plus (+) *fixes* the proper *shipping name* and *hazard class*. The name and class do not change whether the material meets or does not meet the definition of that class. [Sec. 172.101(b)(1)]

Hazardous Material—A substance or material, including a hazardous substance, which has been determined by the Secretary of Transportation to be capable of posing an unreasonable risk to health, safety, and property when transported in commerce and which has been so designated. (See Sec. 171.8)

Multiple Hazards—A material meeting the definition of *more than one hazard class* is classed according to its position in the lists in Sec. 173.2(a) and (b).

DOT HAZARD CLASS	UN CLASS	DEFINITION
		An Explosive—Any chemical compound, mixture, or device which is designed to function by explosion, that is substantially instantaneous with the release of gas and heat. Exception—such compound, mixture, or device which is otherwise specifically classified in Parts 170–189. (See Sec. 173.50)
CLASS A	1	Detonating or otherwise of **maximum hazard**. The nine types of Class A explosives are defined in Sec. 173.53.
CLASS B	1	**Flammable Hazard**—In general, functions by rapid burning rather than detonation. Includes some explosive devices such as special fireworks, flash powders, etc. (Sec. 173.88)
CLASS C	1	**Minimum Hazard**—Certain types of fireworks and certain types of manufactured articles containing *restricted quantities* of Class A and/or Class B explosives as components. (Sec. 173.100)
BLASTING AGENT		A material designed for blasting which has been tested in accordance with Sec. 173.114a(b). It must be so insensitive that there is very little probability of: (1) accidental explosion *or* (2) going from burning to detonation. [Sec. 173.114a(b)]
		Compressed Gas—Any material or mixture having in-the-container a pressure EXCEEDING 40 psia at 70°F, OR a pressure exceeding 104 psia at 130°F; or any liquid flammable material having a vapor pressure exceeding 40 psia at 100°F. [Sec. 173.300(a)]
		Non-Liquefied Compressed gas is a gas (other than gas in solution) which, under the charged pressure, is entirely gaseous at a temperature of 70°F.

DOT HAZARD CLASS	UN CLASS	DEFINITION
		Liquefied Compressed Gas is a gas which, under the charged pressure, is partially liquid at a temperature of 70°F.
		Compressed Gas in Solution is a compressed gas which is dissolved in a solvent.
FLAMMABLE GAS	2	Any compressed gas meeting criteria as specified in Sec. 173.300(b). This includes: lower flammability limit, flammability limit range, flame projection, or flame propagation.
NONFLAMMABLE GAS	2	Any compressed gas *other than* a flammable compressed gas.
COMBUSTIBLE LIQUID	3	Any liquid having a flash point *at or above* 100°F and below 200°F. Authorized flash point methods are listed in Sec. 173.115(d). Exceptions are found in Sec. 173.115(b).
FLAMMABLE LIQUID	3	Any liquid having a flash point below 100°F. Authorized flash point methods are listed in Sec. 173.115(d). For exception, see Sec. 173.115(a).
		Pyroforic Liquid — Any liquid that ignites spontaneously in dry or moist air *at or below* 130°F. [Sec. 173.115(c)]
FLAMMABLE SOLID	4	Any solid material (other than an explosive) which is liable to cause fires through friction or retained heat from manufacturing or processing. It can be ignited readily and burns so vigorously and persistently, as to create a serious transportation hazard. Included in this class are spontaneously combustible and *water-reactive* materials. (Sec. 173.150)
		Spontaneously Combustible Material (Solid) — A solid substance (including sludges and pastes) which may undergo spontaneous heating or self-burning under normal transportation conditions. These materials may increase in temperature and ignite when exposed to air. (Sec. 171.8)
		Water-Reactive Material (Solid) — Any solid substance (including sludges and pastes) which react with water by igniting or giving off *dangerous quantities* of flammable or toxic gases. (Sec. 171.8)
ORGANIC PEROXIDE	5	An organic compound containing the bivalent -O-O structure. It may be considered a derivative of hydrogen peroxide where one or more of the hydrogen atoms have been replaced by organic radicals. It must be classed as an organic peroxide unless it meets certain criteria listed in Sec. 173.151(a).
OXIDIZER	5	A substance, such as chlorate, permanganate, inorganic peroxide, or a nitrate, that yields oxygen readily. It accelerates the combustion of organic matter. (See Sec. 173.151)
POISON A	2	**Extremely Dangerous Poisons** — Poisonous gases or liquids — a *very small amount* of the gas, or vapor of the liquid, mixed with air is *dangerous to life*. (Sec. 173.326)
POISON B	6	**Less Dangerous Poisons** — Substances, liquids or solids (including pastes and semi-solids), other than Class A or Irritating Materials, so toxic (or presumed to be toxic) to man that they are a hazard to health during transportation. (Sec. 173.381)
IRRITATING MATERIAL	6	A liquid or solid substance which, upon contact with fire or air, gives off dangerous or intensely irritating fumes. They do *not include any poisonous material, Class A*. (Sec. 173.381)
ETIOLOGIC AGENT	6	An **etiologic agent** means a living microorganism (or its toxin) which causes (or may cause) human disease. (Sec. 173.386)

DOT HAZARD CLASS	UN CLASS	DEFINITION
RADIOACTIVE MATERIAL	7	Any material, or combination of materials, that spontaneously gives off ionizing radiation. It has a specific activity greater than 0.002 microcuries per gram. (Sec. 173.403) [See Sec. 173.403(a) through (z) for details]
CORROSIVE MATERIAL	8	Any liquid or solid that causes visible destruction or irreversible damage to human skin tissue. Also, it may be a liquid that has a severe corrosion rate on steel. [See Sec. 173.240(a) and (b) for details]
ORM-OTHER REGULATED MATERIALS		(1) Any material that may pose *an unreasonable risk* to health and safety or property when transported in commerce; *and* (2) does not meet any of the definitions of the other hazard classes specified in this subpart; *or* (3) has been reclassed an ORM (specifically or permissively) according to this subchapter. [Sec. 173.500(a)]
ORM-A	9	A material which has an anesthetic, irritating, noxious, toxic, or other similar property. If the material leaks during transportation, passengers and crew would experience extreme annoyance and discomfort. [Sec. 173.500(b)(1)]
ORM-B	9	A material (including a solid when wet with water) the leakage of which could cause significant damage to the vehicle transporting it. Materials meeting one or both of the following criteria are ORM-B materials: (1) specifically designated by name in Sec. 172.101 and/or (2) a liquid substance that has a corrosion rate exceeding 0.250 inch per year (IPY) on non-clad aluminum. An acceptable test is described in NACE Standard TM-01-69. [Sec. 173.500(b)(2)]
ORM-C	9	A material which has other inherent characteristics not described as an ORM-A or ORM-B. It is unsuitable for shipment, unless properly identified and prepared for transportation. Each ORM-C material is specifically named in Sec. 172.101. [Sec. 173.500(b)(3)]
ORM-D	9	A material such as a consumer commodity which presents a limited hazard during transportation *due to its form, quantity, and packaging*. They must be materials for which *exceptions* are provided in Sec. 172.101. A shipping description applicable to ORM-D material is found in Sec. 172.101. [Sec. 173.500(b)(4)]
ORM-E	9	A material that is not included in any other hazard class, but is subject to the requirements of this subchapter. Materials in this class include (1) HAZARDOUS WASTE and (2) HAZARDOUS SUBSTANCE, as defined in Sec. 171.8 [Sec. 173.500(b)(5)]

THE FOLLOWING ARE OFFERED TO EXPLAIN SOME OF THE ADDITIONAL TERMS USED IN PREPARATION OF HAZARDOUS MATERIALS FOR SHIPMENT. (Sec. 171.8)

DOT TERM	EXPLANATION
CONSUMER COMMODITY (SEE ORM-D ABOVE)	A material that is packaged or distributed in a form intended and suitable for sale through retail sales-type agencies. The material is for use by individuals for personal care or household use. This term also includes drugs and medicines. (Sec. 171.8)
FLASH POINT	The minimum temperature at which the flammable vapors of a substance (in contact with a spark or flame) will ignite. For liquids, see Sec. 173.115.

DOT TERM	EXPLANATION
FORBIDDEN	Material is prohibited from being offered or accepted for transportation. This prohibition *does not* apply if these materials are diluted, stabilized, or incorporated in devices AND they are classed in accordance with Sec. 172.101(d)(1).
HAZARDOUS SUBSTANCE	A material, including its mixtures and solutions, that: (1) is listed in the Appendix to Sec. 172.101; (2) is in a quantity, in one package, which equals or exceeds the reportable quantity (RQ) listed in the Appendix to Sec. 172.101: (3) when in a mixture or solution, is in a concentration by weight which equals or exceeds the concentration corresponding to the RQ of the material as shown in the table of the "hazardous substance" definition in Sec. 171.8. This definition does not apply to petroleum products that are lubricants or fuels. (See 40 CFR 300.6)
HAZARDOUS WASTE	Any material that is (1) subject to the hazardous waste manifest requirements of the Environmental Protection Agency specified in the CFR, Title 40, Part 262; or (2) would be subject to these requirements (in the absence of an interim authorization to a State) see Title 40, CRF, Part 123, Subpart F; 49 CFR 171.8. Questions regarding EPA hazardous waste regulations, call (202) 554-1404, 554-1405, or 554-1406 in Washington, D.C.
LIMITED QUANTITY	The maximum amount of a hazardous material authorized for specific labeling and packaging exceptions. Consult the sections applicable to the particular hazard class. See Sec. 173.118, 173.118(a), 173.153, 173.244, 173.306, 173.345, 173.364, and 173.391.
REPORTABLE QUANTITY	The quantity specified in Column 3 of the Appendix to Sec. 172.101 for any material identified in Column 1 of the Appendix.

*This handout is designed as a training aid for all interested parties who may become involved with hazardous materials. It does not relieve persons from complying with the Department of Transportation's hazardous materials regulations. Specific criteria for hazard classes and related definitions are found in the Code of Federal Regulations (CFR), Title 49, Parts 100-177.**

II So You Want To Start a Haz Mat Team!

William J. Keffer, EPA Region VII

The following supplement identifies major areas of concern for a community forming a new hazardous materials response team. The author covers team make-up, resources and protective clothing and presents an example of personnel protection measures in the form of heat stress monitoring of responders wearing Level C protective clothing.

Special thanks are given to William J. Keffer, the author of this paper, for permitting its inclusion as a supplement to the *Hazardous Materials Response Handbook*.

So You Want To Start a Haz Mat Team!

Many local government and nonmanufacturing industry groups are becoming increasingly concerned about the risk and expense of dealing with hazardous materials releases. Plans are being made to develop the special capabilities needed to handle these incidents, but little practical material is available to assist groups in forming hazardous materials (haz mat) teams.

The following article is designed to provide some insight into what a haz mat response program would entail, as well as information on how to form a haz mat team.

INCIDENT EXPECTATIONS

According to the records collected from larger cities in Region VII over the past decade and discussions with members of various operational teams, it appears that haz mat releases account for 0.5 to 1.0% of the fire calls made by a local jurisdiction. Approximately half of the incidents will involve very small quantities (less than 55 gallons), and 75% occur at fixed facilities. (See Supplement III of this publication for further information on how to determine what hazardous materials are most likely to be released in a given community.)

HAZ MAT TEAM MAKE-UP

Composition of a haz mat team should be multi-disciplinary. Suggested membership for a small local team would have a minimum of 10 persons, possibly distributed as follows:

3 to 6 fire service personnel

A minimum of 2 health authorities; local department, poison control, or medical

A minimum of 1 law enforcement representative — especially for security and evacuation coordination

2 to 4 industry or academic types for special advice and assistance

A minimum of 1 sponsoring agency administration type for legal management of storage and disposal

These team members should have ample opportunity to train, cross-train, and practice haz mat skills in a variety of exercises including hands-on. Development of the training program should rely heavily on the Occupational Safety & Health Administration (OSHA) Haz Mat Worker Protection Rule and the recently published documents, NFPA 472 and 471, included in this publication.

RESOURCES

If a community or facility plans a group around the 10-person base, discounts the regular salaries of the individuals, and assumes they already have respiratory gear (SCBAs), a team can be developed and maintained for about $12,000 to $15,000. This team could present a highly effective Level B (non-encapsulated) response posture that will provide personnel safety for more than 90% of all incidents. This $12,000 to $15,000 will be spent in a variable manner after the first year, but for the first year these monies would break down as follows:

Comprehensive physical exams at about $500 each	$5000
Out-of-service training (2 weeks)	$2000
Communications for non-fire service (pagers)	$2000
References	$1000
Personal protective gear	$3000
Mitigation and screening supplies	$1500

Items such as transportation, in-service training, and premium pay have not been costed. Properly managed programs could generally rely on vehicles readily available within a municipality and in-service training needs still vary.

PROTECTIVE CLOTHING

Protective clothing, and the training needed for its proper use, are important considerations when preparing a newly formed haz mat team. Without proper protection, a haz mat team member is as helpless as the general public when dealing with the release of a hazardous material. Even becoming actively involved in effective size-up or simple mitigation procedures for a haz mat incident often requires the responder to wear protective equipment.

The purpose of personal protective gear is to minimize or eliminate the routes of exposure to the individual wearer, namely via inhalation and dermal (skin) exposure. In most hazardous materials incidents, respiratory protection is essential. Without adequate monitoring instruments, the only respiratory protection recommended is the self-contained breathing apparatus (SCBA). The SCBA is expensive (approximately $1400) and requires regular maintenance. It is impractical to assume that every first responder from law enforcement, fire, and health agencies will have a unit always available. A more feasible option would be a cooperative agreement between fire service and other responder agencies. Most fire sevice organizations that usually work on-scene at a haz mat incident are provided with SCBAs. An SCBA, as with any protective gear, should never be worn without advance formal training, prior experience, and regular fit testing and practice. Local fire departments may be able to provide the opportunity for the necessary training.

To determine the appropriate type of protection, the following must first be determined:

1. The positive identification of the material involved through:

    a. Bill of lading, consist, manifest, or other shipping papers or labels;

    b. Placards, UN or NA ID number;

    c. Talking with the driver or conductor.

2. The hazardous properties of the materials through:

    a. Data available from the various handbooks and guides. Examples: DOT *Emergency Response Guidebook*, NIOSH/OSHA *Pocket Guide to Chemical Hazards*, CHRIS Manual, NFPA *Fire Protection Guide on Hazardous Materials*.

    b. Effects of the materials on bystanders — respiratory problems, skin or mucous membrane irritation, etc.

    c. Environmental indicators — dead birds or other animals, discolored or burned vegetation.

There are more than a dozen types of material used for protective clothing. The quality of assembly can also vary, making acquisition a confusing process. Also, it is essential that sizes of suits, gloves, and boots are appropriate to each individual since a breach in the clothing eliminates the protection offered.

Table 1 is a simplistic chart to guide clothing selection for common chemicals. This chart is very conservative since personnel are rarely exposed to 100% concentrations of the chemicals encountered.

A basic splash protection kit is small enough to be kept in the trunk of a patrol car. The individual kits can be made up for under $150 and can be packaged and stored in a large, heavy PVC bag, which can also be used for the materials after they are contaminated. Each kit would contain one each of the items listed in Table 2.

Following is information on protective clothing for some of the more commonly released materials.

ANHYDROUS AMMONIA — The equipment should prevent any possibility of skin or eye contact with the spilled product. This may include rubber boots,

Table 1

CHEMICAL	MATERIALS FOR ≥ 2 HOURS	MATERIALS FOR 1/2 HOUR
Gasoline	Viton/Neoprene, Nitrile PVA	Butyl, PE, PVC
Chlorine	Vitron, Neoprene Saranex	PE, PVC
Ammonia		
Methyl chloride	Viton/Neoprene	Butyl, Rubber Neoprene, Nitrile PE, PVC
Methanol	Butyl, CPE, Viton, Teflon, Saranex	Rubber, Neoprene, Nitrile, PVA, PVC
Toluene	Viton, Teflon	Butyl, Rubber, Neoprene, Nitrile, PE, PVA, PVC
Hydrochloric acid	Neoprene, Nitrile Viton, CPE, Rubber	PVC
Sulfuric acid	CPE, Saranex, Viton	PVC, Nitrile
Nitric acid	Viton, Saranex	Nitrile, PVC
Sodium hydroxide	CPE, Nitrile	
PCBs	Neoprene, PVA, Viton	Rubber, Polyethylene

gloves, face shield, splash-proof goggles, and other impervious and resistant clothing. Fully encapsulating suits with self-contained breathing apparatus (SCBA) may be advisable in some cases to prevent contact with vapor or fume concentrations in the air. Compatible materials may include butyl rubber, natural rubber, neoprene, nitrile rubber, and polyvinyl chloride.

RESPIRATORY PROTECTION — For unknown concentrations, fire fighting, or high concentrations (above 500 ppm), an SCBA with full facepiece should be worn. For lesser concentrations, a gas mask with chin-style or front- or back-mounted ammonia canister (500 ppm or less) should be used within the use limitations of these devices.

CHLORINE — The equipment used should prevent any possibility of skin or eye contact with the spilled product. This may include rubber boots, gloves, face shield, splash-proof safety goggles, and other impervious and resistant clothing. Fully encapsulating suits with SCBA may be necessary to prevent contact with vapor or fume concentrations in the air. Compatible materials may include neoprene, chlorinated polyethylene, polyvinyl chloride, viton, and saranex.

RESPIRATORY PROTECTION — For unknown concentrations, fire fighting, or high concentrations (above 25 ppm), a full facepiece SCBA should be worn. For lesser concentrations, either a gas mask with chin-style or front- or back-

Table 2

ITEMS	COST	SOURCE
Coverall, Tyvek or Polypropylene	$2.80 each $70/case of 25	Saf-T-Glove, KCMO Kimberly-Clark Co. Grove, IL
Coverall, Saranex, or CPE	$10 to $25 each $225/case of 25	Saf-T-Glove, KCMO Chemron, Buffalo, NY
Acid Suit Coverall PVC/Nylon	$50 each	Sijal, Oreland, PA
Surgical Gloves Vinyl	$3.00/box of 100	Edmont, Coshocton, OH
Vinyl Outer Gloves	$15.72/cs of 12 pair	Edmont, Coshocton, OH
PVC-lined Gloves	$21.00/cs of 12 pair	Edmont, Coshocton, OH
Neoprene-lined Gloves	$31.44/cs of 12 pair	Edmont, Coshocton, OH
Buna-Nitrile coated Gloves	$24.96/cs of 12 pair	Edmont, Coshocton, OH
PVA/Organic Solvent Gloves	$113.16/cs of 12 pair	Edmont, Coshocton, OH
Butyl Gloves	$17.28/cs of 12 pair	Edmont, Coshocton, OH
Rubber Overboots	$19.22/pair	Ranger, Endicott, NY
Safety Goggles	$3.96/pair	Obtain locally
Duct Tape	$1.59/roll	Obtain Locally
Large Plastic Bags	$3.99/box	Obtain Locally
DOT Emergency Response Guidebook	$4.00 each	DOT, Kansas City, MO

mounted chlorine canister (25 ppm or less) or a chlorine cartridge respirator with a full facepiece may be used.

HYDROCHLORIC ACID — The equipment used should prevent any possibility of skin or eye contact with the spilled product. This may include rubber boots, gloves, face shield, splash-proof safety goggles, and other impervious and resistant clothing. Fully encapsulating suits with SCBAs may be necessary to prevent contact with vapor or fume concentrations in the air. Compatible materials may include butyl rubber, natural rubber, neoprene, nitrile rubber, nitrile rubber/polyvinyl chloride, chlorinated polyethylene, polyvinyl chloride, styrene-butadiene rubber, viton, nitrile-butadiene rubber, saranex and polycarbonate.

RESPIRATORY PROTECTION — For unknown concentrations, fire fighting, or high concentrations (above 100 ppm) an SCBA with full facepiece should be worn. For lesser concentrations, a gas mask with chin style or front- or back-mounted acid gas canister (100 ppm or less) or an acid cartridge respirator with a full facepiece may be used.

METHANOL — The equipment should prevent repeated or prolonged contact and any reasonable probability of eye contact with the spilled product. This may include rubber boots, gloves, face shield, splash-proof safety goggles, and other impervious and resistant clothing. Compatible materials may include butyl rubber, natural rubber, neoprene, neoprene/styrene-butadiene rubber, nitrile rubber, nitrile rubber/polyvinyl chloride, polyethylene, chlorinated polyethylene, polyurethane, styrene-butadiene, viton, and nitrile-butadiene rubber.

RESPIRATORY PROTECTION — For unknown concentrations, fire fighting, or high concentrations (above 200 ppm) an SCBA with full facepiece or the equivalent should be worn.

METHYL CHLORIDE — The equipment should prevent skin contact with cold methyl chloride or the cold containers and any reasonable probability of eye contact with the cold product. This may include rubber boots, gloves, face shields, splash-proof goggles, and other impervious and resistant clothing. Compatible materials may include neoprene and nitrile-butadiene rubber.

RESPIRATORY PROTECTION — For unknown concentrations, fire fighting, or high concentrations (above 100 ppm), an SCBA with full facepiece should be worn.

NITRIC ACID — The equipment should prevent any possibility of skin or eye contact with the spilled product. This may include rubber boots, gloves, face shields, splash-proof safety goggles, and other impervious and resistant clothing. Fully encapsulated suits with an SCBA may be advisable in some cases to prevent contact with high vapor or fume concentrations in the air. Compatible materials may include natural rubber, neoprene, nitrile rubber, polyethylene, chlorinated polyethylene, polyvinyl chloride, viton, nitrile-butadiene rubber, and saranex for concentrations more than 70% nitric acid.

RESPIRATORY PROTECTION — For unknown concentrations, fire fighting, or high concentrations (above 250 mg/m^3), an SCBA with a full facepiece should be worn. For lesser concentrations, a gas mask with chin-style or front- or back-mounted canister or a chemical cartridge respirator with a full facepiece should be used within the limitations of these devices. The canister or cartridge should provide protection against nitric acid and should not contain oxidizable materials such as activated charcoal.

SODIUM HYDROXIDE — The equipment should prevent any possibility of skin or eye contact with the spilled material. This may include rubber boots, gloves, face shield, safety goggles, and other impervious and resistant clothing for solids or liquids. Fully encapsulating suits with SCBAs may be advisable in some cases to prevent contact with high dust or mist concentrations in the air. Compatible materials may include butyl rubber, natural rubber, neoprene,

neoprene/styrene-butadiene rubber, nitrile rubber, polyethylene, chlorinated polyethane, polyurethane, polyvinyl alcohol, and polyvinyl chloride for the solid and its solutions, as well as nitrile rubber/PVC, styrene-butadiene, viton, nitrile-butadiene rubber, and saranex for 30-70% solutions.

RESPIRATORY PROTECTION—For unknown concentrations, fire fighting, or high concentrations (above 100 mg/m^3) an SCBA with full facepiece should be worn. For lesser concentrations, a high efficiency mist and particulate filter respirator with full facepiece should be worn.

SULFURIC ACID—The equipment should prevent any possibility of skin or eye contact with the spilled product. This may include rubber boots, gloves, face shield, splash-proof goggles, and other impervious and resistant clothing. Fully encapsulating suits with SCBA may be advisable in some cases to prevent contact with high vapor or fume concentrations in the air. Compatible materials may include butyl rubber, neoprene, nitrile rubber, chlorinated polyethylene, polyvinyl chloride, styrene-butadiene rubber, viton, and nitrile-butadiene rubber for concentrated (more than 70%) acid as well as less concentrated solutions.

RESPIRATORY PROTECTION—For unknown concentrations, fire fighting, or high concentrations (above 50 mg/m^3) an SCBA with full facepiece should be worn. For lesser concentrations, a gas mask with chin-style or front- or back-mounted acid gas canister and high efficiency particulate filter or a high efficiency particulate filter with a facepiece should be worn.

TOLUENE—The equipment should prevent repeated or prolonged contact and any reasonable probability of eye contact with the spilled product. This may include rubber boots, gloves, face shield, splash-proof goggles, and other impervious and resistant clothing. Compatible materials may include polyurethane, polyvinyl alcohol, viton, nitrile-butadiene rubber, saranex, and fluorine/chloroprene.

RESPIRATORY PROTECTION—For unknown concentrations, fire fighting, or high concentrations (above 200 ppm) an SCBA with full facepiece should be worn. For lesser concentrations, a gas mask with chin-style or front- or back-mounted organic vapor canister (2000 ppm or less) or an organic vapor cartridge respirator with a full facepiece (1000 ppm or less) should be used within the limitations of these devices.

POLYCHLORINATED BIPHENYLS—The equipment used should prevent any possibility of skin or eye contact with the spilled product. This may include rubber boots, gloves, face shield, splash-proof goggles, and other impervious and resistant clothing, rubber boots, boot covers, viton gloves, saranex coated tyvek suits.

RESPIRATORY PROTECTION—For unknown concentrations, SCBA or approved air purifying respirator, full-face, canister-equipped (MSHA/NIOSH-approved), and a knowledge of the limitations of such equipment.

PERSONAL PROTECTIVE CLOTHING

1. *Before Donning Protective Clothing*

   a. Determine identification of material and associated hazardous properties.

   b. Decide if it is essential for you to enter the contaminated area.

   c. Determine if the personal protective equipment available is adequate protection for the hazards involved.

2. *How to Determine Appropriate Level of Protection*

   a. If the hazardous material involved has an inhalation hazard (i.e., toxic vapors or toxic combustion products) *respiratory protection is required*.

      1) Indications:

         a) Data from literature on material
         b) Visible fire
         c) Visible cloud of vapors
         d) Irritation of eyes, nose, throat, nausea of bystanders
         e) Environmental Indicators – i.e., dead birds, animals, etc.

   b. If the hazardous material involved is a skin contact hazard, *dermal splash protection is required*.

      1) Indications:
         a) Data from literature on material
         b) Chance of contact with material while mitigating incident
         c) Irritation of skin from material contact by bystanders

3. *Decontamination Procedure*—Must be done in contamination reduction area or warm zone

   a. Open heavy-duty plastic bag.

   b. Remove booties and dispose in bag.

   c. Remove outer gloves and dispose in bag.

d. Remove respiratory equipment, if utilized.

e. Remove protective coveralls and dispose in bag.

f. Remove inner gloves and dispose in bag.

g. Seal bag with duct tape.

h. Call State Department of Natural Resources or EPA for disposal or specific decontamination instructions.

WARNING: Attempting to mitigate a hazardous material incident and utilizing personal protective equipment should never be attempted without *formal training* and *prior familiarization with equipment*.

HEAT STRESS MONITORING IN LEVEL C

This section has been prepared to update recent experiences and improve current heat stress monitoring (HSM) at Superfund cleanup activities where EPA protective ensembles known as Level C are in use. It is hoped that a uniform procedure can be adopted that will allow for maximum productivity, while educating workers and monitoring effectively for heat stress.

As the title states, this section applies to Level C protection. Once specific guidelines (i.e., vital signs) are in place, they will apply to all levels of protection. However, some of the methods and equipment may vary. Below are the issues to be addressed:

1. Temperature at which HSM will be implemented.

2. Vital signs to be measured and their ceiling values.

3. Work to rest ratios.

4. Equipment needs.

5. Fluid replacement.

6. Cooling devices.

7. Acclimation of workers.

8. Recordkeeping.

9. Education of workers in recognition.

Temperature

Following is a formula to be used in the field for the purpose of determining the necessity of heat stress monitoring. Ambient temperature, humidity, and cloud cover (solar load) are the three factors requiring consideration during the estimation of the actual thermal load inflicted on the body's cooling system. Acclimation is also important in considering the body's ability to effectively cool itself. The problem is that all people do not react to heat loads in the same manner or at the same rate.

Adjusted Temperature = *Tdb + 13 (% Cloud cover factor)

No Clouds = 1.00

25 % Clouds = 0.75

50 % Clouds = 0.50

75 % Clouds = 0.25

100 % Clouds = 0.00

The Wet Bulb Globe Temperature (WBGT) may also be used to determine the radiant heat load. Both the WBGT and the adjusted temperature should be measured and the higher value utilized. For Level C work, monitoring should begin at the adjusted temperature of 75 °F, with work to be halted at a temperature of 110 °F, unless effective auxiliary personal cooling devices are in use and only acclimated workers are used.

Example: Tdb = 70 °F

Cloud Cover = 50%

WBGT = 75 °F

Tdb + 13 (% cloud cover)

70 + 13 (.50) = 70 + 6.5 = 76.5 °F

Adjusted Temperature = 76.5 °F

WBGT = 77 °F

WBGT (77 °F) would be used as the actual temperature and heat stress monitoring would be initiated.

Vital Signs

Monitoring of the individual worker's vital signs is the most effective way to prevent heat illnesses. It is important to assign values that will identify the

---

*Tdb = dry bulb temperature

heat illness symptoms before they become serious. Previously, temperature, heart rate, and blood pressure were used as indicators of heat stress. It is now believed that skin temperature and body water loss are the most important vitals to monitor. Heart rate should also be monitored as it reflects the heart's reaction to the added thermal load. It is believed that blood pressure is not affected by heat stress and should no longer be routinely monitored. The following are suggested recommendations for vital signs monitoring:

Body Weight: When wearing impermeable clothing it is possible to have a sweat rate as high as 3.5 L/hr. These lost fluids must be replaced intermittently throughout the day. Thirst alone is not a good indicator of proper fluid replacement. The amount of weight loss should be replaced by the equivalent weight in replacement fluid. Body Water Loss (BWL) should not be allowed to exceed 1.5% of total body weight. A scale should be kept in the break area and should be accurate to plus or minus $\frac{1}{4}$ pound. The color of excreted urine is also an indicator of the need for fluid replacement. If fluids are not properly replaced, the urine becomes a deep yellow color. There are some interferences; for instance, taking vitamins will darken the color. These factors can be monitored by field personnel themselves as a general indication of their fluid intake as compared to their body water loss.

Temperature: The deep body core temperature (rectal) is the most representative of actual temperature in the body; however, in the field it has not been practical to use this measurement. As a substitute, the oral temperature has been used. The average oral temperature is 37 °C (98.6 °F), but is not necessarily a norm for every individual. Current standards indicate that a maximum rise in temperature to 99–99.6 °F is all that should be tolerated for civilian workers. A problem with this type of cutoff is that it does not allow for adjustment for workers who have abnormal base temperatures. A baseline should be established for each individual from data collected over a two-week period. As a guideline, the maximum rise in temperature should not exceed 1.5 °F, and should return within 0.5 °F before the worker is allowed to return to the hot zone. It is very important that initial temperature readings, as well as other measurements, be made promptly (within a very few minutes) on exit from the work zone for break periods.

Heart Rate: The heart rate is probably the best indicator of overall stress applied to the body. The pulse becomes more rapid as the body tries to cool itself, as well as reflecting aerobic exercise. A maximum heart rate should be established using a method called the age adjusted heart rate. This maximum should not be exceeded at any point, but can be maintained during Level C activities. Aerobic exercise is generally performed at 70 to 85% of the maximum attainable heart rate, which is generally considered to be 220 beats per minute. The age adjusted heart rate is figured as follows:

(0.7) (220–age)

A 20-year old would have an age adjusted heart rate of 140 beats per minute and is calculated as follows:

(0.7) (220–20)

(0.7) (200) = 140 beats per minute

Work to Rest Ratios

A significant change in the vital signs measurements can signify heat stress candidates. Careful monitoring by a qualified person (i.e., nurse, EMT, safety officer) will aid in the reduction of heat stress injuries. The individual assigned these duties should be trained to interpret these readings and advise the OSC and response manager to make the proper adjustments in work and rest periods.

The less severe cases of heat illnesses can be mitigated by rest and cooling of the body. In order to focus attention to the possible victims of heat stress, the following guidelines should be met:

1. If, at any time, the maximum age adjusted heart rate is exceeded, or the heart rate has not returned to within 10% of the baseline measurement at the end of the rest period, the next work period should be shortened by 33% and the affected individual not return to work until the heart rate is reduced.

2. If, at any time, the oral temperature rises 1.5 °F above the baseline, the subsequent work period should be reduced by 33%.

3. If at any time, the BWL exceeds 1.5% of the total body weight, the worker shall be instructed to increase his/her fluid intake.

Fluid Replacement

Fluid replacement and rest periods are the most effective deterrent of heat stress. Proper fluid replacement cannot be gauged by thirst because the sense of thirst is satisfied before the proper amounts of fluid are ingested. The following are suggestions that should be used to monitor and control the replacement of fluids:

1. Each worker should be assigned their own replacement fluids container with their name marked clearly on the outside. This procedure helps to gauge the amount of fluids consumed by each worker.

2. Water is excellent and highly recommended, but workers should be allowed to drink something they like. Alcoholic and caffeinated drinks are not good replacement fluids as they tend to increase the rate of body fluid loss. Fruit juices and electrolyte replacement drinks (Gatorade) should be diluted

3:1 if they are used. Water replacement is more important than the replacement of electrolytes since sweat contains only $\frac{1}{3}$ of the electrolyte balance of the blood. Whatever drink is used, it should be served cooled to between 50 and 60 °F.

3. Salt tablets are not suggested as they are under scrutiny at this time. It is suggested that food be salted generously at meals if workers have been exposed to excessive heat.

4. Fluid intake should begin before the work shift is started after the morning weight is recorded. This provides for body water that will be lost in the first work period.

5. For each $\frac{1}{2}$ pound lost, 8 oz. of replacement fluids should be consumed.

6. One and one half percent body water loss is acceptable, as long as it is replaced before the following morning.

Equipment Needs

Body Weight: As stated before, scales should be used that are accurate to ± $\frac{1}{4}$ pound. They should also be rugged and easily calibrated in the field.

Heart Rate Monitors: In most cases the workers can take their own heart rate after brief instruction. The heart rate should be measured by the radial pulse for 30 seconds. For more reliable readings there is a rugged, battery-powered commercial instrument that could be mounted in the break area which, when lightly held in the hand, gives the heart rate on a digital readout after approximately 30 seconds.

Temperature: Oral temperatures may be measured by digital thermometers currently used by the nurses at the Superfund cleanup sites.

Cooling Devices

Cooling devices generally should not be considered for temperatures under 95 °F. Available cooling vests (ice) are, in general, bulky and heavy and will contribute more to the stress load than the relief provided. The vests that are hooked to some type of cooling unit or compressed air are impractical for workers not performing tasks in a fixed location. Employees who have been asked to wear the cooling units have complained about the weight and uneven cooling and have frequently expressed a preference to just stay hot and get the work over with.

Acclimation

Acclimation to Level C is a very important step in the prevention of heat stress. The body needs time to adjust to the additional demands of wearing impermeable clothing in the heat, just as it does to adjust to cold weather. Impermeable clothing in hot weather prevents cooling by evaporation, which is the most effective mechanism that the body has to cool itself. With the added thermal stress and restriction of the natural cooling system the body uses, the first two weeks of Level C work can be the most dangerous, especially if the ambient temperatures are elevated and the work is physically demanding. Acclimation periods allow the body to adjust gradually to the added stress inflicted by donning impermeable protective gear.

Nonacclimated workers starting in Level C should participate in a program of increased personal monitoring and reduced work periods during the first two weeks of exposure. In the absence of additional monitoring by competent health personnel, the following suggested schedule should provide a safe working buildup for proper acclimation:

Day 1-3   Light work during the morning or late afternoon not to exceed two hours.

Day 4-6   Light work during the morning or late afternoon not to exceed three hours.

Day 6-8   Light work during the morning or late afternoon not to exceed four hours.

Day 8-10   Moderate work during the morning and afternoon, approximately four hours.

Day 10-12 Moderate work during the middle of the day, approximately five hours.

Day 12-14 Moderate work during the middle of the day, approximately six hours.

After day 14   Full days of moderate work.

Recordkeeping

Recordkeeping is vital in determining an individual's reaction to added thermal stress and will assist in job placement. A matrix considered to be a good system for heat stress tracking and recordkeeping is available from the Emergency Response Team of Edison, NJ. Using ERT's system of recordkeeping will produce an abundance of paperwork, but it can be incorporated into data using graphs or stored on computer disks. Any way that it is retained will provide valuable information for future references and make data available for future studies.

Worker Education

Education of workers required to wear personal protective gear is the most important preventative technique. Heat illness occurs more often through ignorance than any other contributing factor. It is imperative that personnel understand heat stress, its signs, symptoms, and contributing factors. The following two references should be mandatory reading for all personnel involved in wearing impermeable protective gear or in directing others who are. Daily safety briefings at job sites where protective gear is being worn are also important as a preventative measure.

- NIOSH Pub. No. 80-132, *Hot Environments*, U.S. Dept of Labor.

- NIOSH/OSHA/USCG/EPA, *Occupational Safety and Health Guidance Manual for Hazardous Waste Site Activities*, U.S. Dept. of Health and Human Services.

 **Beginning the Hazard
Analysis Process**

William J. Keffer, EPA Region VII

The first step in preparing to deal with hazardous materials incidents is to identify the types and possible sources of accidental releases. The following supplement outlines the steps involved in a community's analysis of its facilities using hazardous materials and the transportation corridors that bring hazardous materials into the community. Included by the author is an item-by-item review of the contents of a material safety data sheet and instructions on how to use the information presented therein.

Our thanks and appreciation are extended to author William J. Keffer for submitting his paper for inclusion as a supplement to the *Hazardous Materials Response Handbook*.

Beginning the Hazard Analysis Process

Several recent federal studies show that there are between 5 and 6 million chemicals known to man, with this number growing at the rate of about 6,000 chemicals per month. Furthermore, a recent computer review by the Chemical Abstract Service of the complete list of known chemicals indicates that a first responder might reasonably be expected to encounter any of 1.5 million of these chemicals in an emergency, with 33,000 to 63,000 of them considered hazardous. To complicate matters, these hazardous chemicals are known by 183,000 different names. Fortunately, not all of these chemicals are equally common.

The U.S. Department of Transportation (DOT) and the U.S. Environmental Protection Agency (EPA) have used several measures of toxicity and volume of production to develop a shortened list of chemicals that are considered hazardous when transported in commerce. This list is comprised of about 2,700 chemicals, all of which are listed in 40 CFR 172.101 and the 1987 *Emergency Response Guidebook for Hazardous Materials Incidents*.

The Occupational Safety and Health Administration (OSHA) regulates about 400 chemicals on the basis of occupational exposures. This list is found in the NIOSH (National Institute of Occupational Safety and Health) *Pocket Guide to Chemical Hazards*.

Even this abbreviated list can be intimidating to local response personnel hoping to develop a comprehensive hazard analysis for their community. Further complicating their job is the fact that, according to a recent study by the National Academy of Sciences, National Research Council (NRC), there is so little known about seven-eighths of the 63,000 hazardous chemicals that not even a partial assessment can be made of their health hazards. Some conclusions drawn from the NRC study are as follows:

1. Pesticides — Of the 3,350 pesticides classified as important chemicals, information sufficient to make a partial assessment of the health hazard is only available on about 1,100 to 1,200 (34%) of them.

2. Drugs — Of the 1,815 drugs or drug ingredients noted, about 36% have enough information for a partial assessment.

3. Food Additives — For the 8,627 food additives listed, there is partial information on 19%.

4. Other Chemicals — For the remaining 48,500 industrial chemicals, there is enough information on just 10% to develop a partial assessment.

These two points—the lack of a generally accepted name for chemicals considered hazardous and the lack of data for assessing the risk—create a stumbling block for emergency response personnel and community officials who are charged with developing a viable, effective local hazardous materials management system. Without a contingency plan, however, based on effective and accurate hazard analysis prior to an emergency, it would be difficult and time-consuming to develop the necessary information in the midst of an emergency. In order to bring what appears to be an insurmountable task into perspective, the local response community must get involved in the *hazard analysis* process.

Hazard Analysis is the process of identifying chemicals present in the community—either at fixed facilities or passing through transportation corridors—and evaluating the hazard, vulnerability, and risk they present. A hazard analysis can be performed by any individual or small group that understands the principles of hazard analysis.

The goal of this paper is to provide a summary of the sources of information and methods for use by local agencies and hazardous material teams to improve their understanding of the hazardous material problem in the community—information that can be used in both planning and emergency response operations. It will also introduce the Material Safety Data Sheet (MSDS) and its utility in the hazard analysis process.

Hazard Analysis Data Sources

The purpose of hazard analysis is to gather data on the locations, quantities, and health hazards of chemicals most likely to be released in a community. This process may seem monumental from two aspects: 1) the sheer numbers of chemicals out there, and 2) the lack of information in a usable form available on these chemicals. In attempting to pare the list down to a workable size and gather data on the location and identity of the chemicals found in and being transported through the community, a variety of methods may be used, including:

1. Historical records of chemicals having the most frequent instances of release on a national level. For example, all previous studies have shown that the most commonly released hazardous chemical is commercial vehicle fuel (gasoline). The EPA commissioned a national study in 1985 to look at hazardous chemicals and the sources of their release. This survey covered 6,928 incidents nationwide involving chemicals other than fuel. The source of releases study indicated:

74.8% were fixed facility (in-plant) incidents

25.2% were in-transit incidents

Acute Hazardous Events, Data Base Industrial Economics, Inc., Cambridge, MA 02140. December 1985.

The fixed facility incidents were distributed as follows:

20.7% storage

19.4% valves and pipes

14.1% process

17.9% unknown

27.8% other

The in-transit incidents were distributed by mode as follows:

54.5% truck

14.1% rail

3.8% water

3.1% pipeline

2.5% other

Perhaps the most useful data from this national study is the information on the chemicals most commonly involved in the 6,928 incidents — 49.5% of the incidents involved only 10 chemicals:

23.0% polychlorinated biphenyls (PCBs)

6.5% sulfuric acid

3.7% anhydrous ammonia

3.5% chlorine

3.1% hydrochloric acid

2.6% sodium hydroxide

1.7% methanol/methyl alcohol

1.7% nitric acid

1.4% toluene

1.4% methyl chloride

Of the 6,928 incidents, 468 involved human injury or death. The same 10 chemicals listed above accounted for 35.7% of the death and injury events, though not at the same rate as they occurred:

9.6% chlorine

6.8% anhydrous ammonia

5.6% hydrochloric acid

4.7% sulfuric acid

2.8% PCBs

2.4% toluene

1.9% sodium hydroxide

1.5% nitric acid

0.4% methyl alcohol

0.1% methyl chloride

When the data for occurrence and injury is viewed for chemicals like chlorine and PCBs, it is apparent that the release and injury data are different. That is, although PCBs were involved in more incidents, chlorine, when released, posed a greater threat to humans. Determining this potential for causing injury to humans is the area of hazard assessment that takes the most effort on the part of the local response community. Gathering this information in a systematic manner cannot be done in the midst of an incident.

2. Summaries of previous incidents from emergency management and environmental response organizations at the local, state, regional, and federal levels. For example, EPA Region VII and states in the region have comprehensive, computerized records of all reported incidents by county since 1977. These records are available to any jurisdiction on request.

3. Local fire and police department records may disclose many incidents involving hazardous materials.

4. Local yellow pages and the state industrial directory will show most local fixed facilities that manufacture, store, or use chemicals. As an aid to this search, the EPA has recently prepared a summary for 14 types of facilities that shows what types of hazardous chemicals may be encountered. Copies of the summaries may be obtained from EPA by calling the national RCRA (Resource Conservation and Recovery Act) Industry Hotline toll free at 800-424-9346.

Once the chemicals in a community have been identified in name and quantity (either for contingency planning or emergency response), there are several national data bases for evaluating the hazard, vulnerability, and risk presented by those chemicals, including:

A. Poison Control Centers. If the chemical is a consumer product, the quickest way to get comprehensive hazard information is through the regional Poison Control Center.

B. Manufacturer's technical medical staff. If the chemical is an industrial bulk chemical, effective assistance is generally available from CHEMTREC (Chemical Transportation Emergency Center) through the technical medical staff of the company that manufacturers the chemical. Dial 800-CMA-8200 for nonemergency situations; 800-424-9300 for emergencies.

C. Agency for Toxic Substances and Disease Registry (ATSDR). If the chemical is a mixture or a waste, if a second opinion is required, or if the chemical is unknown, a good source of information is the new agency at the Centers for Disease Control called the Agency for Toxic Substances and Disease Registry. It can be reached by dialing 404-452-4100 days, or 404-329-2888 nights and weekends.

When contacting any of these sources, remember there is a lack of full information on health assessment for many chemicals. Answers and information from any of the sources listed above may be qualified, and each local response group, as part of their contingency planning, should locate a competent medical authority to work with the response community and assist in obtaining and interpreting health effects data.

Material Safety Data Sheets (MSDS)

The recent state and federal legislation on hazard communication, right-to-know, and mandatory local notification for certain hazardous chemicals will assist in developing pre-emergency and on-scene hazard assessments of the chemicals in the community. This legislation is bound to make the MSDS one of the best sources of information on chemical hazards.

Personnel from the Chemical Manufacturers Association, the sponsor of CHEMTREC, recently stated that they have available some 190,000 different MSDS and are receiving more all the time. For most local communities, it will be necessary to work with hard copies of the MSDS and begin with the 10 most commonly released chemicals. Later one can add to that list based on the results of the local hazard analysis that determined additional unique or peculiar local hazards.

In the next few years, all local response groups will be provided with some MSDS sheets by local industry, as required by SARA, Title III, so it will be helpful to become familiar with them now. However, remember that MSDS are not a cure-all and require some systematic way to approach their use. In any case, becoming familiar with the type of information presented on the MSDS and how that information will be of assistance in making a hazard assessment is essential—whether for pre-emergency planning or when responding to an emergency.

Minimum content of an MSDS is mandated by OSHA. Each sheet must contain the following sections:

1. The chemical name, chemical formula, common synonyms, chemical family, and manufacturer's name and emergency telephone number.

2. Hazardous ingredients and regulatory exposure limits, if any.

3. Physical data.

4. Fire and explosion hazard data.

5. Health hazard data.

6. Reactivity data.

7. Spill or leak procedures.

8. Special protection information.

9. Special precautions.

Response and/or planning personnel are encouraged to review the MSDS for at least the 10 most commonly released chemicals (previously listed), plus gasoline. They should obtain current MSDS from companies in their community, even though there are several sources of generic MSDS including some where the information is computerized, to establish and maintain good working relationships with local companies.

In reviewing these 11 MSDS, take note of the different ways information is presented and the lack of uniform presentation. It is not required to use the standardized content in the format suggested by OSHA. From the varying presentations, you will gain some insights on the use of MSDS and factors to be considered in interpreting them. The depth of information furnished in MSDS varies depending on the extent of what is known about the chemical, as well as the management attitude of the company providing the information.

MSDS Section 1–Materials Identification–This section identifies the chemical by name, synonyms, and/or chemical family name. The manufacturer's name and emergency telephone number are listed in order to obtain additional data and assistance. Several preparers choose to emphasize the health hazards, precautionary measures, and emergency contacts at the top of the sheets.

MSDS Section 2–Ingredients and Hazards–Absolute clarity in describing all ingredients of a material and their hazardous components is essential; however, experience indicates that is not always the case. Most of the 11 reviewed MSDS address pure substances or aqueous dilutions, with the notable exception of that for gasoline. When discussing chlorine and its hazards, the MSDS preparer assumed chlorine to be the pure chemical in the gaseous form, whereas we are more likely to encounter chlorine as a solid (HTH–commonly used in swimming pool chemical control) or as commercial bleaches (a liquid that is fairly dilute). Most commercial bleaches for domestic use contain from 30,000 to 50,000 parts per million (ppm) sodium hypochlorite as a source of chlorine.

Upon reviewing the ingredients and hazards of anhydrous ammonia and ammonium hydroxide, we see that anhydrous ammonia is a colorless gas with an extremely pungent odor and that ammonium hydroxide is a clear, colorless liquid. Although their forms are different and their ability to impinge on exposures when released is different, the hazard is the same. Ammonia is intensely corrosive to human tissue, whether it is inhaled, contacts the skin, or is ingested. OSHA regulates workplace exposures of ammonia at 50 ppm (permissible exposure limit, PEL) while the American Conference of Governmental Industrial Hygienists (ACGIH) recommends a level of 25 ppm (threshold limit value, TLV). Additionally, OSHA regulations state that at concentrations of 500 ppm in air, the material becomes immediately dangerous to life and health (IDLH).

The OSHA system is designed to provide working conditions for reasonably healthy adult humans for 8-hour exposures for 40 hours per week for 40 years. This data is not directly applicable to general populations. Obviously, anyone with preexisting respiratory ailments would be expected to be more affected by irritants and by those chemicals that affect the central nervous system (CNS). The limits are not applicable to children, especially those in the first year of life, since their metabolism and nervous system responses are significantly different than adults or older children.

Let's look at the form and hazard information extracted from the 11 MSDS selected:

Name	Form	Exposure Limit	IDLH
Ammonium hydroxide	liquid	50 ppm	500 ppm
Anhydrous ammonia	gas	50 ppm	500 ppm
Chlorine	gas	1 ppm (ceiling)	25 ppm
Gasoline (unleaded)	liquid	300 ppm (ACGIH) 500 ppm (OSHA) petroleum distillate 10 ppm benzene	
Hydrochloric acid	liquid	5 ppm	100 ppm
Methyl alcohol	liquid	200 ppm	25,000 ppm
Nitric acid	liquid	2 ppm	100 ppm
Polychlorinated biphenyls	liquid	0.5 mg/m^3 (ACGIH)	
Sodium hydroxide	solid	2 mg/m^3	200 mg/m^3
Sulfuric acid	liquid	1 mg/m^3	80 mg/m^3
Toluene	liquid	200 ppm	2,000 ppm

All of the OSHA limits are for airborne concentrations and vary widely among the substances listed. It is important to note that the ratio of exposure limit to IDLH concentration also varies widely. The greater the range between these two numbers, the greater the safety factor for an exposed person to avoid permanent harm or death.

One important area of information not available from this section is how the liquids and the solid on the list above become airborne concentrations and how fast this occurs. For this information, we will have to look elsewhere on the sheets.

MSDS Section 3–Physical Data—In the process of hazard assessment, the ability to evaluate physical data combined with health hazard data is essential. The common physical properties provided on the reviewed MSDS include boiling point, freezing point, specific gravity, vapor pressure, vapor density, solubility, and appearance. Other parameters may be provided at the discretion of the company completing the sheets.

Let's look briefly at the range in characteristics of the chemicals from the 11 sheets reviewed and discuss the use or implications of each for the first responder:

1. *Boiling point*—the temperature at which a liquid turns to a vapor.

Chemical	Boiling Point
Ammonium hydroxide	36°C
Anhydrous ammonia	−33°C
Chlorine	−34°C
Gasoline (unleaded)	38° to 204°C
Hydrochloric acid (37%)	53°C
Hydrochloric acid (35%)	65.6° to 110°C
Methyl alcohol	64.5°C
Nitric acid (60–68%)	122°C (67%)
Polychlorinated biphenyls	360° to 390°C
Sodium hydroxide	1390°C
Sodium hydroxide solution (50%)	145°C
Sulfuric acid	310°C
Toluene	231° to 232°C

Since the ambient temperature of this planet ranges from around −20°C to 50°C (−10°F to 120°F), any chemical with a boiling point below the ambient temperature will rapidly become a gas when released from its container. This is certainly the case for chlorine and anhydrous ammonia. Other materials with boiling points only slightly above normal ambient temperature will, if confined in a container, rapidly expand and pressurize that container with the potential for a rapid release if heated even slightly. Other materials, such as PCBs and sodium hydroxide pellets, will be unaffected by the heat of normal structural fires, but could be affected by the application of water to that fire. Sodium hydroxide pellets, for example, will dissolve in water to form a corrosive liquid.

2. *Freezing point* — temperature at which the liquid form of a chemical will turn into the solid form.

3. *Melting point* — temperature at which the solid form of a chemical will turn into the liquid form.

The two physical parameters above may be of limited use to response personnel for most chemicals. There are several chemicals for which control measures such as freezing are effective and where dry ice, for example, may be used to mitigate a release. Similarly, there are some chemicals where the form change under structural fire temperatures may be significant and may seriously alter the hazard to response personnel. Exposure of low melting point solids and most liquids to fire temperatures may result in production of toxic materials in the smoke plume.

4. *Specific gravity* — density of a chemical compared to the density of water. If the specific gravity is less than one, the chemical will float on water. If the specific gravity is greater than one, the chemical will sink. In either case, it is important for response personnel to consider the property of solubility concurrently with specific gravity. These properties for the 11 chemicals are listed below:

Chemical	Exposure Limit	Specific Gravity	Solubility
Ammonium hydroxide	50 ppm	0.9	infinite
Anhydrous ammonia	50 ppm	0.68	soluble
Chlorine	1 ppm	2.4	0.7%
Gasoline (unleaded)	300 ppm	0.7 − 0.8	insoluble
Hydrochloric acid	5 ppm	1.18	infinite
Methyl alcohol	200 ppm	0.8	miscible
Nitric acid	2 ppm	1.41	complete
Polychlorinated biphenyls	0.5 mg/m^3	1.5	0.01 ppm
Sodium hydroxide	2 mg/m^3	2.13	111 gm/100 gm
Sulfuric acid	1 mg/m^3	1.84	infinite
Toluene	200 ppm	0.86	0.05 gm/100 gm

Toluene, gasoline, and methyl alcohol are all flammable or combustible liquids with similar TLV levels. A glance at their respective solubilities, however, shows that mitigation techniques would have to be substantially different due to their solubility; i.e., methyl alcohol is completely miscible in water whereas the others are relatively insoluble. Not only would fire fighting methods differ, but additional attention would have to be paid to the solubility when environmental damage is possible. Many chemicals that are listed as only slightly soluble can still cause significant environmental toxicity to plants or aquatic life. Toxicity of methyl alcohol is 250 ppm and toluene 1,180 ppm; therefore each of them presents a serious environmental hazard if significant runoff is allowed to occur.

Most MSDS do not provide environmental risk information; therefore, this data will have to be sought from other sources. One excellent source for environmental risk information for many common chemicals is the EPA's OHM-TADS (Oil & Hazardous Materials–Technical Assistance Data System) Access to this system can be gained through any EPA regional office.

5. *Vapor density*—density of a gas compared to the density of air. If the vapor density is less than one, the material will rise in still air and dissipate. If the vapor density is greater than one, the vapor will attempt to sink in still air and potentially collect in low spots and valleys.

6. *Vapor pressure*—pressure exerted by vapors against the sides of the container. Vapor pressure is very temperature dependent. The lower the boiling point of the liquid, the greater the vapor pressure it will exert at a given temperature. In more common terms, the higher the vapor pressure, the more rapidly the material will change from the liquid form to a vapor when released to the environment and the higher the equilibrium concentration with air will be.

Boiling point, vapor pressure, and vapor density for the compounds of interest are listed in Table 1.

This data was found on the actual MSDS reviewed and, if you were to glance at the MSDS for sulfuric acid, it would be apparent that errors do creep in. Also, the detail of the information furnished varies from rough estimates or general statements for some materials to multiple listings for others.

If you can picture a room in which a release occurs from its container, and then look at the range of vapor pressures for the most commonly released substances, it will be apparent that both chlorine and anhydrous ammonia will present an almost instantaneous vapor (inhalation) hazard. Since both of these chemicals are soluble to some extent, a fog line may be helpful in suppression of volatization or reduction of concentrations, even when the release is continuous. However, when the intent is to reduce vapor produc-

Table 1

Chemical	Boiling Point	Vapor Pressure	Vapor Density
Ammonium hydroxide	36°C	115 mm Hg @ 20°C	1.2
Anhydrous ammonia	−33°C	23 atm @ 20°C	0.6
Chlorine	−34°C	4,800 mm Hg @ 20°C	2.49
Gasoline (unleaded)	38° − 204°C	N/A	N/A
Hydrochloric acid (37%)	53°C	190 mm Hg @ 20°C	1.27
Methyl alcohol	64.5°C	97 mm Hg @ 20°C	1.1
Nitric acid (60–68%)	122°C (67%)	62 mm Hg @ 20°C	2 to 3
Polychlorinated biphenyls	360° − 390°C	<1 mm Hg @ 20°C	N/A
Sodium hydroxide	1390°C	negligible	
Sodium hyrodxide solution (50%)	145°C	6.3 mm Hg @ 104°F	N/A
Sulfuric acid (96%)	310°C	22 mm Hg @ 145°C	<0.3 @ 25°C*
Sulfuric acid (93.2%)		<0.3 mm Hg @ 25°C	3.4
Toluene	231° − 232°C	22 mm Hg @ 20°C	3.14

*Some MSDS contain incorrect data. The vapor density of sulfuric acid is 3.4 and not 0.3.

tion, the water from the hose lines should not enter pooled materials like ammonia or chlorine. For materials like sodium hydroxide and PCBs, a vapor hazard is not likely to exist under real-world conditions.

MSDS Section 4–Fire and Explosion Data — Most of the MSDS reviewed contain specific information for the fire fighter on the physical characteristics of the chemicals when involved in a fire. These characteristics, as summarized in Table 2, should be made familiar to most fire personnel.

In addition to these normal fire characteristics, the chemicals have other fire-related hazards, some of which are reported in the fire and explosion section. Examples of these are:

1. Chlorine and anhydrous ammonia are generally stored in pressure containers, and the violent rupture of these containers represents a significant hazard, in addition to their toxicity.

2. Many of the chemicals listed generate toxic vapors or mists when involved in a fire, thus representing an additional hazard.

3. Hydrochloric acid, nitric acid, sodium hydroxide, and sulfuric acid are such vigorous oxidizers or reducers that, although they are not flammable hazards themselves, they react with metals to produce hydrogen gas, which is extremely flammable.

Table 2

Chemical	Flash Point	Autoignition Temperature	Flammability Limits	Extinguishing Media
Ammonium hydroxide	—	—	—	—
Anhydrous ammonia	1208°C	651°C	16–27%	shut off gas
Chlorine	N/A	N/A	N/A	N/A
gasoline (unleaded)	−45°F	536°–853°F	1.5–7.6%	dry chemical water spray
Hydrochloric acid	N/A	N/A	N/A	N/A
Methyl alcohol	52°F	385°F	6–36.5%	water spray
Nitric acid	none	none	N/A	N/A
PCBs	N/A	N/A	N/A	N/A
Sodium hydroxide	none	none	N/A	N/A
Sulfuric acid	none	none	N/A	N/A
Toluene	40°F	480°F	1.3–7.1%	dry chemical

MSDS Section 5–Health Hazard Information — This section of the MSDS presents information on the routes of exposure (inhalation, ingestion, dermal) and, in some cases, the severity of these risks (low, moderate, high). This information is essential for the selection of appropriate personal protective equipment and safety procedures for response actions at incidents. Some of the MSDS reviewed highlighted the major hazards in section one on the sheets, while others give a more detailed formal listing of the hazards in this section. Some sheets list the NFPA 704 rating for the specific chemical (use of the 704 system by local industry should be encouraged as it provides local emergency response personnel a basis for quick judgments about the severity of personal exposure). A brief summary of the hazards for each chemical is listed in Table 3.

The most common hazard of the chemicals listed is their corrosive effect on nearly every part of the human body. The effects of chlorine and sulfuric acid are very similar. What makes chlorine a greater risk is the volatility of the liquid when released compared to the acid, which is already a liquid at ambient temperatures and volatilizes very slowly.

MSDS Section 6–Reactivity Data — Generally, four areas of information are presented in this section, and all are potentially useful to those responding to a hazardous chemical emergency.

Table 3

Chemical	Hazards
Ammonium hydroxide	corrosive – severe eye and skin irritant
Anhydrous ammonia	corrosive – severe eye and skin irritant
Chlorine	corrosive – life threatening toxic effects may occur at concentrations of 25 ppm on short exposures
Gasoline	flammable – irritant – CNS effects – some evidence of carcinogenicity – also numerous chronic effects
Hydrochloric acid	corrosive – may be fatal if ingested
Methyl alcohol	flammable – may be fatal if ingested
Nitric acid	corrosive – strong oxidizer at higher concentrations
PCBs	very long-lasting material – some evidence of liver damage – carcinogenic risk and adverse reproductive effects
Sodium hydroxide	corrosive – may be fatal if swallowed, causing severe burns
Sulfuric acid	corrosive – causes severe burns – may be fatal if swallowed – harmful if inhaled
Toluene	flammable – chronic skin irritant – various systemic effects on CNS, liver, and kidneys

1. Stability – Is the material stable at ambient temperature and pressure or at normal storage conditions? Most of the chemicals reviewed are stable and not liable to undergo spontaneous changes.

2. Polymerization – Will the chemical change through polymerization at normal conditions of storage and temperature? For chemicals that spontaneously polymerize, this frequently leads to generation of heat and potential container failure.

3. Decomposition – What new chemicals and what hazards will be created by the thermal decomposition of the chemical? Important information is included in this section for officials concerned about the exposures of response personnel and general population if the chemical is exposed to fire. For example, formaldehyde may be formed from fire involving methyl alcohol, oxides of nitrogen from anhydrous ammonia, and from most of the other chemicals, oxides of carbon that may increase the hazards from simple asphyxiation.

4. Incompatibles – What materials may cause violent reactions with the chemical? Note especially the MSDS for gasoline.

With six million chemicals in the world, many can have a large number of potentially violent reactions. It is important to have some idea of the likelihood of these chemicals coming in contact with each other. Many of the chemicals reviewed are potent acids or bases, and they will certainly be incompatible with chemicals of widely differing pH. For example, the sheet for sulfuric acid lists water as being incompatible. The mixing of sulfuric acid (96%) with water (at pH of 7) releases enough heat to cause a violent reaction.

MSDS Section 7–Spill or Leak Procedures — This section contains suggested steps for handling releases of the chemical in question. The information provided is usually similar to the 1987 DOT *Emergency Response Guide*. It is important to note the order in which the material is presented. If the material is extremely flammable, but not particularly toxic, initial advice will usually be in control ignition sources. If the material is extremely toxic, initial advice will generally be to evacuate.

MSDS Section 8–Special Protection Information — On many MSDS now in use, this section is not very specific. Hopefully, improvements will be made that recommend specific respiratory and clothing information. It is important to know that none of the impervious clothing is suitable for all chemicals. For example, polyethylene protective clothing is not recommended for concentrated sulfuric acid, but is suitable for more dilute solutions.

Special problems may be created for first responders by those materials that adversely affect normal fire fighter protective clothing. These materials (i.e., chlorobenzene, methyl iodide, etc.), for which breakthrough times are less than one tank of air, may not offer any useful protection to the responder. Once a hazard analysis is completed and response organizations are at the point of their contingency planning where they are selecting response equipment, it is suggested that they obtain a copy of "Guidelines for the Selection of Chemical Protective Clothing," Arthur D. Little Co., Cambridge, Massachusetts, February 1987, and check out the recommendations for protective clothing for the chemicals in their community.

MSDS Section 9–Special Precautions — Many MSDS do not contain any information in this area. In some cases, for extremely flammable materials, there is an additional warning about sparks and radiant heat; for chlorine, there is a warning about igniting other combustible materials on contact; and for many other chemicals there is a reiteration of standard storage and handling procedures.

One important area that may be covered on some MSDS is the hazard of the chemical to animal or aquatic life. This information is frequently based on scientific testing of the compound or chemical in controlled laboratory settings. The information from this testing is presented in terminology different than the regulatory TLV and PEL information and will take an additional effort on the responder's part to be able to evaluate the information.

Depending on the potential use of the chemical, a series of tests may be run over a period of time in a manner that resembles successive elimination. Tests are run for a variety of acute and chronic effects as well as exotic effects on more and more complex animals. Initial screening is done on bacteria, which allows the testing of large numbers of individuals and numerous generations in small spaces and in a short time. Subsequent tests may be conducted with any of a wide variety of rodents, pigs, dogs, and, sometimes, primates. The sequence of the tests is shown in the following charts.

CHRONOLOGY OF TESTING

Early   Acute toxicity—handling hazards

•   1st level screen—mutagen/carcinogen

•   Subchronic toxicity—target organ, toxic dose—rodent

•   Birth defects—teratology

•   2nd level screen—mutagen/carcinogen

•   Absorption/distribution/metabolism/excretion—lab animals (metabolism/pharmacokinetics)

•   Subchronic toxicology—non-rodent species

•   Reproduction study

•   Behavioral tests

•   Synergism/potentiation

•   Residue evaluation

•   Long-term studies—carcinogenesis—rodents

•   Definitive test for mutagenesis—rodents

•   Metabolism/pharmacokinetics—humans

Late   Epidemiology

Tests are done to evaluate the physical/chemical properties of the substance, note routes of entry into the organisms being tested, and document exposure variables. The tests are used to evaluate the biological fate of the chemicals and develop a dose/response curve for the specific effects being evaluated. A hypothetical dose/response curve is shown in Figure 1. The most common expression of the results of these tests is the dose or concentration at which 50% are affected, known as the TD_{50}. Toxicologists exhibit their skills by the accuracy with which they can extrapolate animal data to predict effects on man. In general, TD_{50} data is commonly given for pesticides and other chemicals developed for pest and weed control. An example of interpreting this data for responders is shown below:

Table Relative Index of Toxicology

Toxicity Rating	Probable Oral Dose	Lethal Dose for Average Adult Humans
Practically nontoxic	> 15g/kg	more than a quart
Slightly toxic	5–150g/kg	between pint and quart
Moderately toxic	0.5–5g/kg	between ounce and quart
Very toxic	50–500mg/kg	between teaspoon and ounce
Extremely toxic	5–50mg/kg	7 drops to teaspoonful
Supertoxic	< 5mg/kg	a taste (less than 7 drops)

Dose/response curves deal with acute exposures, but it is important to also consider the potential for repetitive exposures at lower doses, which may accumulate in the body. This situation is called chronic exposure and is shown diagrammed below. The two toluene sheets discussed earlier provide chronic exposure data indicating the potential for brain cell damage from long-term inhalation of toluene vapor.

Summary

Local government emergency response personnel can avoid much confusion if they concentrate on utilizing MSDS in pre-emergency planning. That is, identify in advance the buildings or processes that store or utilize extremely

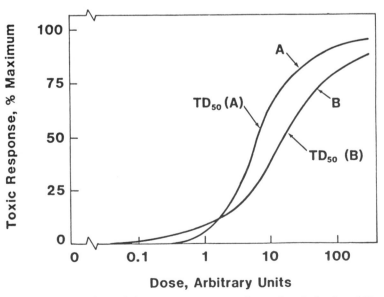

Figure 1 Hypothetical dose-response curve of two chemicals, A and B.

toxic materials and make basic decisions regarding "no attack" before an emergency occurs.

Once response community personnel have completed the hazard analysis and have evaluated the collected MSDS in the manner listed above, numerous additional questions need to be answered to complete the contingency planning portion of the work. Some of the questions that the hazard analysis and contingency planning process will help answer include:

1. What risks will be encountered by first responders?

2. What protective equipment (respiratory and clothing) will be needed by the responders?

3. What type of support resources are likely to be necessary for mitigation supplies or medical assistance?

One useful way to clarify these questions and to provide practical guidance for the response community is to use the highest risk situation from the hazard analysis as a scenario for a hands-on drill/exercise for the participating members of the response community.

IV

Hazardous Materials Emergency Response Plan

Sacramento Fire Department, 1231 I Street, Suite 401, Sacramento, CA 95814

The following supplement contains an excellent example of a working response plan for hazardous materials emergencies. Our particular thanks are given to the Sacramento Fire Department, and especially to the author Jan Dunbar, Battalion Chief, for their generous consent to our including this document as a supplement to the *Hazardous Materials Response Handbook*.

Sacramento Fire Department
Hazardous Materials Emergency Response Plan

I. PLANNING BASIS

A. Purpose

The purpose of this Hazardous Materials Emergency Response Plan is to establish an organization to mitigate hazardous materials incidents within the city limits of Sacramento, and on any mutual aid hazardous materials incident outside the city limits of Sacramento.

B. Objectives

1. To describe operational concepts, organization, and support systems required to implement the plan.

2. Identify authority, responsibilities, and actions of federal, state, local, and private industry agencies necessary to minimize damage to human health, natural systems, and property, and to aid in mitigating the hazard.

3. Establish an operational structure that has the ability to function not only within the city limits or Sacramento, but also on any mutual aid call where Sacramento Fire Department equipment responds to a hazardous materials incident outside the city limits of Sacramento.

4. To utilize fire department officers and members who have been trained to handle hazardous materials incidents.

5. To establish lines of authority and management for a hazardous materials incident.

II. ADMINISTRATION

A. Scope

1. Geographical Factors

This plan is concerned with hazardous materials incidents that occur within the territory of the city of Sacramento and on mutual aid calls.

2. The Hazard

The hazard shall include actual or potential fires, spills, leaks, rupture, or contamination, and any threat to life safety involving hazardous materials.

3. The Hazardous Material

The material itself may include explosives, flammables, combustibles, compressed gases, cryogenics, poisons and toxics, reactive and oxidizing agents, radioactive materials, corrosives, carcinogenics, or etiological agents, or any combination thereof.

4. The Incident

This plan is for any hazardous material incident associated with any mode of transportation, industrial processing and/or storage sites, waste disposal procedures, and illegal usage and disposal.

B. Authority

1. City Charter of the City of Sacramento.

2. Uniform Fire Code: Article 10, Section 10.101.

3. California Government Code: Chapter 7, Division 1, Title 2.

4. California Health and Safety Code: Sections 25115 and 25117, and Section 25600 through 25610.

5. California Vehicle Code: Article 4, Chapter 2, Division 2.

6. California State Office of Emergency Services Fire and Rescue Mutual Aid Plan.

C. References

1. California State Hazardous Materials Incident Contingency Plan, 1982.

2. California State Oil Spill Contingency Plan.

3. California State Radiological Emergency Assistance Plan.

4. Federal Response Plan.

5. Sacramento County Hazardous Materials Emergency Response Plan.

III. HAZARDOUS MATERIALS INCIDENT CLASSIFICATION

There are three (3) levels of hazardous materials incident classifications. The basis used for determining the level of a hazardous material incident are:

1. Level of technical expertise required to abate the incident.

2. Extent of municipal, county, and state government involvement.

3. Extent of evacuation of civilians.

4. Extent of injuries and/or deaths.

5. Extent and involvement of decontamination procedures.

A. Level I Incident (Known as a LEVEL I H.M.I.)

1. Spills, leaks, ruptures, and/or fires involving hazardous materials that can be contained, extinguished, and/or abated utilizing equipment, supplies, and resources immediately available to the fire department having jurisdiction; *and*

2. Hazardous material incidents that do not require evacuation of civilians.

B. Level II Incident (Known as Level II H.M.I.)

Any Sacramento Fire Department officer can upgrade a Level I HMI to a Level II HMI.

1. Hazardous materials incidents that can only be identified, tested, sampled, contained, extinguished, and/or abated utilizing the resources of the Sacramento Fire Department Hazardous Material Response Team; a hazardous materials incident that requires the use of chemical-protective gear and specialized equipment.

2. Hazardous materials incidents that require evacuation of civilians within the area of the fire department having jurisdiction; *and/or*

3. Fires involving hazardous materials that are permitted to burn for a controlled period of time, or are allowed to consume themselves.

C. Level III Incident (Known as Level III H.M.I.)

The officer of the HMRT, or the incident commander, can upgrade a LEVEL II HMI to a LEVEL III HMI.

1. Spills, leaks, and/or ruptures that can be contained and/or abated utilizing the highly specialized equipment and supplies available to environmental or industrial response personnel; *and/or*

2. Fires involving hazardous materials that are allowed to burn due to ineffectiveness or dangers of the use of extinguishing agents, or the unavailability of water; and/or there is a real threat of large container failure; and/or an explosion, detonation, BLEVE, or container failure has occurred; *and/or*

3. Hazardous material incident that requires evacuation of civilians extending across jurisdictional boundaries and/or there are serious civilian injuries and/or deaths as a result of the hazardous material incident; *and/or*

4. Hazardous material incident that requires at least two Sacramento Fire Department Hazardous Materials Response Teams; and/or decontamination

of civilians or personnel is required on scene; and/or the Sacramento Fire Department mobile decontamination unit is required on scene; *and/or*

5. Hazardous material incident that has become one of multi-agency involvement of large proportions.

IV. INCIDENT COMMAND AND SCENE MANAGEMENT

A. Incident Commander

The *INCIDENT COMMANDER* (I.C.) shall be the designated fire department officer responsible for mitigating the hazards at the scene of hazardous materials incident. Upon his arrival, he shall secure and maintain immediate control until the situation has been abated.

1. The Sacramento Fire Department shall accept and provide the position of INCIDENT COMMANDER for the scene of all hazardous materials incidents within the *city limits of Sacramento*. The fire department shall coordinate and direct within its control all fire department activities within its jurisdiction and responsibility to include, but not be limited to, rescue and first aid, product identification, scene stabilization and management, suppression activities, protection of exposures, containment, agency notification, scene isolation, personnel protection, and decontamination.

2. The captain of the Sacramento Fire Department's Hazardous Materials Response Team shall report to and function through the fire department INCIDENT COMMANDER.

3. The INCIDENT COMMANDER shall report to and function through the SCENE MANAGER.

B. Scene Manager

The *SCENE MANAGER* (S.M.) shall employ overall management and coordination of a hazardous materials incident. The SCENE MANAGER shall be responsible for the identification of the incident resources and needs, the procurement of these resources, and the coordination of the resources, so as to abate the incident and protect life, property, and the environment.

The SCENE MANAGER shall not be responsible for the detailed direction of technical or specialized procedures, but shall oversee that these procedures are carried out when needed. Scene management decisions are to be made with assistance of expert advisors and specialists.

1. *Freeways, State Roads*

For all hazardous materials incidents that occur on any freeway or state road, including those within the city of Sacramento, the SCENE MANAGER shall be the *California Highway Patrol*, in accordance with Section 2454 of the California Vehicle Code.

2. *Within Sacramento City Limits*

a. Streets and Roads

For all hazardous materials incidents that occur within the city limits of Sacramento, the *Sacramento Fire Department* shall function as the SCENE MANAGER in accordance with the "Memorandum of Understanding for Hazardous Materials Incidents" as signed by the police department and the fire department.

b. Public and Private Property (Off-Road)

For all other hazardous materials incidents that occur within the city limits of Sacramento, the *Sacramento Fire Department* shall function as the SCENE MANAGER. The SCENE MANAGER'S position shall be provided by the ranking chief officer of the fire department on the scene.

NOTE: Under all of the above conditions, the INCIDENT COMMANDER of the fire department shall provide direct control and authority of all fire department related activities at the scene of any hazardous materials incident.

3. *Outside the Sacramento City Limits*

a. Streets and Roads

For all hazardous materials incidents that occur on state roads, all freeways, and all surface streets in unincorporated portions of a county, the *California Highway Patrol* shall function as SCENE MANAGER in accordance with Section 2454 of the California Vehicle Code.

b. Public and Private Property (Off-Road)

For hazardous materials incidents that occur on public and private property in the County of Sacramento, the *Sacramento County Sheriff* shall function as SCENE MANAGER in accordance with the Sacramento County Hazardous Materials Response Plan.

V. HAZARDOUS MATERIALS RESPONSE TEAM (HMRT)

A. The Sacramento Fire Department shall maintain three (3) specially trained four-man teams for the specific purpose of responding to chemical emergencies. The HMRT can provide expertise and equipment especially developed to help control and abate a hazardous material incident.

The equipment, instruments, protective clothing, and kits assigned to a HMRT are *not* to be loaned to or used by any other fire fighter, individual, or agency, without the express consent of the HMRT officer on scene.

B. It shall be the responsibility of the HMRT officer to:

1. Determine what county, state, and federal agencies need to be notified of the incident, and what county, state, and federal agencies need to send a representative to the scene, and convey this information to the INCIDENT COMMANDER and SCENE MANAGER.

NOTE: Notification of all STATE and FEDERAL agencies can be accomplished by making one call to the *California State Office of Emergency Services* (OES).

NOTE: Review *APPENDIX TWO* for a detailed explanation of services and responsibilities of governmental agencies.

2. Insure that the *California State Office of Emergency Services* (OES) always be notified of the following regarding each incident.

a. Any incident involving spillage of a suspected chemical or hazardous material that has contacted the ground.

b. Any incident involving leakage or venting of a toxic or flammable vapor or gas to the atmosphere.

c. Any incident that involves civilians who become injured or contaminated due to the nature of the incident.

d. Any incident that involves a wildlife kill or a wildlife threat.

3. Identify and establish a RESTRICTED ZONE when necessary, and enforce it.

4. Upgrade a LEVEL II HMI to a LEVEL III HMI through proper dispatch procedures when:

a. The incident is beyond the capabilities of that HMRT (not to include clean up procedures).

b. The HMRT officer wants a second HMRT to respond.

c. The HMRT officer wants a DUTY CHIEF to respond.

d. The HMRT officer has determined a need for the DECONTAMINATION UNIT to respond.

5. Work with, and be subordinate to, the Incident Commander of the fire department having jurisdiction.

VI. OPERATIONAL BUILDUP CHARTS

A. Level I Hazardous Materials Incident

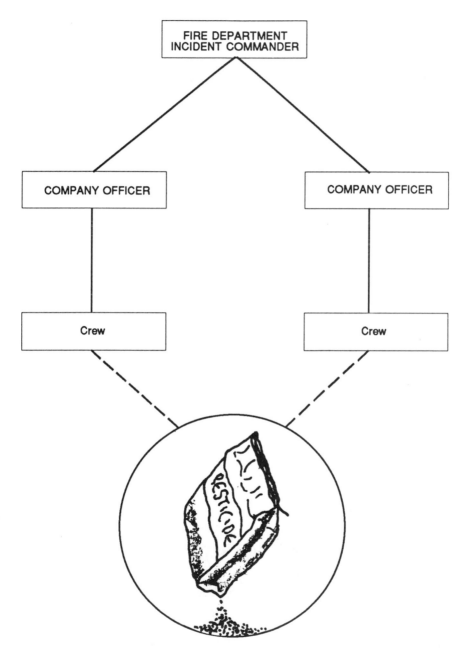

B. Level II Hazardous Materials Incident

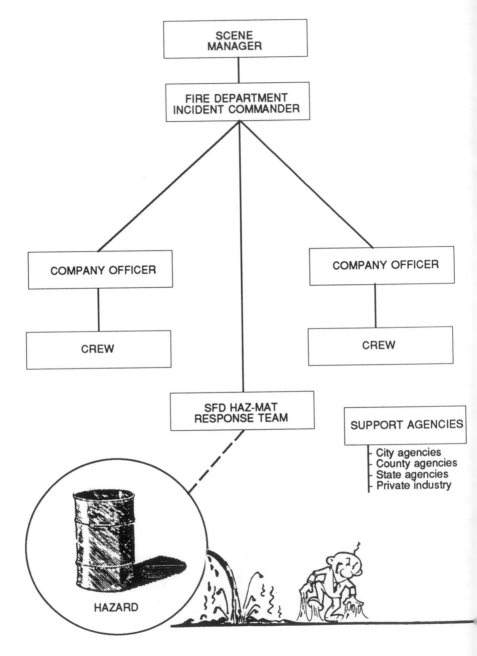

C. Level III Hazardous Materials Incident (Is in keeping with National Incident Management System)

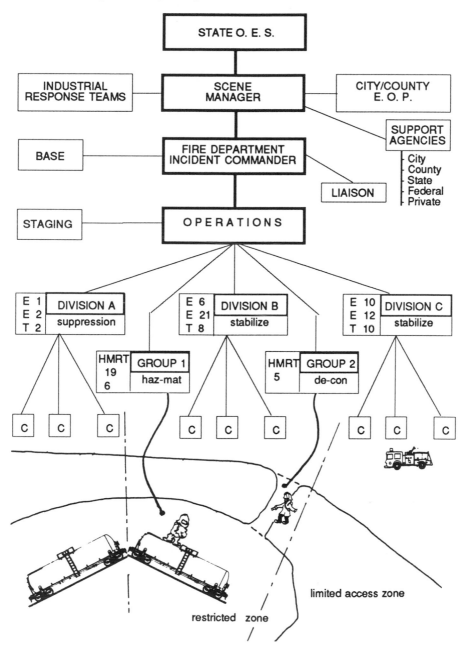

D. Level III Hazardous Materials Incident (Alternate I.C. build-up)

(Is in keeping with National Incident Management System.)

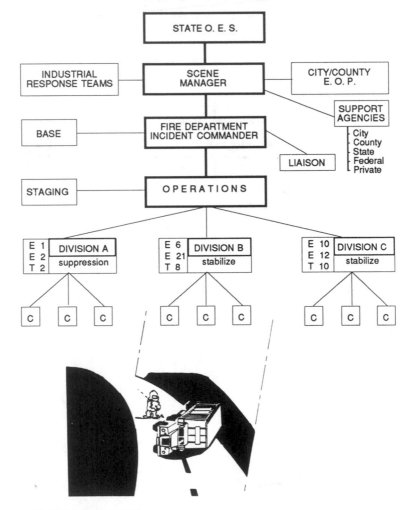

VII. CONTROL ZONES

A. Limited Access Zone

1. The LIMITED ACCESS ZONE (Yellow Zone, Warm Zone) shall be designated to define an area where some potential or real danger exists to the public or the environment.

2. Identification of a LIMITED ACCESS ZONE shall be done by the first arriving fire department company officer.

Access shall be limited to those members of agencies on scene who are appropriately protected and directly engaged in rescue, control, and preliminary stabilization measures.

B. Restricted Zone

1. The RESTRICTED ZONE (Red Zone, Hot Zone) shall be designated as necessary to identify and define an area of exceptional danger, including extreme threat to life safety.

2. Identification of a RESTRICTED ZONE shall be done by the HMRT.

Access shall be controlled by the HMRT. Only personnel of the HMRT and other designated personnel of necessity will be allowed access.

C. Decontamination Zone (Decontamination Corridor)

1. The DECONTAMINATION ZONE (or CORRIDOR) shall be designated as necessary to establish a procedure to decontaminate personnel, civilians, and equipment in an effort to reduce or stop the spread of suspected contaminants.

2. Identification and the set-up of a DECON ZONE or CORRIDOR shall be done by the HMRT.

   a. Access into the DECON ZONE of contaminated people shall be coordinated by HMRT. Only an officer of the HMRT may allow anyone to exit the DECON ZONE.

   b. Workers entering the DECON ZONE to assist in procedures shall do so only as directed by the HMRT, and only when appropriately protected.

3. DECON procedures shall be effected only by HMRT personnel.

Sacramento Fire Department
Hazardous Materials Emergency Response Plan

Annex A
The Operational Plan

I. ACTIVATION

This plan will become operational when the fire department receives notification of any hazardous materials incident.

II. NOTIFICATION

A. H.M.R.T. and Battalion Chief

1. The Comm Center shall notify through proper dispatch procedure the closest Hazardous Materials Response Team, and Battalion Chief, of a LEVEL II H.M.I.

2. The city of Sacramento has three (3) on-duty HMRTs daily, each with a four (4) man crew. They are:

Consists of			Station	
HMRT 5:		T- 5, LP- 5	5	(1st Battalion)
HMRT 6:		T- 6, LP- 6	6	(2nd Battalion)
HMRT 19:		T-19, LP-19	19	(3rd Battalion)

3. Response to any LEVEL II H.M.I. shall be Code 3 unless specified otherwise during dispatch.

B. Dispatch

1. Whenever a company officer upgrades an incident to a LEVEL II H.M.I., or when the Comm Center dispatches a LEVEL II H.M.I., the Comm Center shall *notify* the on-duty *DUTY CHIEF* that a LEVEL II is in progress.

2. The DUTY CHIEF shall be advised of the *nature* and *type* of the incident, the *location*, and the *units* responding.

3. The dispatch for a LEVEL II H.M.I. shall consist of: one engine, one H.M.R.T. (closest to the incident), and one battalion chief.

4. Upon arrival of the H.M.R.T. and the battalion chief, it may be determined that the incident should be upgraded to a LEVEL III H.M.I. A LEVEL III H.M.I. can be made only by the H.M.R.T. officer or the battalion chief on scene (Incident Commander). The dispatch for a LEVEL III shall consist of:

> one additional engine
>
> one additional H.M.R.T.
>
> one on-duty deputy chief

5. The DECONTAMINATION UNIT shall be dispatched only when it is called for from the incident scene.

C. Agency Call Lists

1. Check lists shall be maintained of the necessary and required agencies that can provide technical assistance or that may become involved in the hazardous materials incident.

2. The Comm Center shall make state agency, federal agency, industrial, and mutual aid notification *only* as requested and directed from the field.

III. IMMEDIATE ON-SCENE ACTIONS

A. Fire Department First on Scene

1. *Identification*

   a. The first fire department officer to arrive on the scene will attempt to identify:

   1) The type of material involved.

   2) The quantity of material involved.

   3) The possibility of contamination.

   4) The immediate exposure problem.

   b. This information is to be transmitted to Comm Center immediately.

(LEVEL I)

   c. If the fire department officer determines that the incident is Level I, he will proceed to mitigate the problem.

(LEVEL II)

   d. If the fire department officer determines that the incident is LEVEL II, he is to *upgrade the incident to a LEVEL II./H.M.I.* and take whatever action is necessary to control the incident within the capability of his men, resources, and equipment.

   e. The fire department officer first on the scene will inform incoming companies of his evaluation and the actions he is taking, and shall direct the responding H.M.R.T. and Battalion Chief on which access routes they should use when approaching the incident.

   f. The first arriving company officer shall initiate steps to identify and establish the LIMITED ACCESS ZONE.

2. *Command Post*

a. The H.M.R.T. officer or battalion chief, whichever arrives first on the scene, shall establish a command post in the most strategically desirable and safe location.

b. The battalion chief shall assume the position of Incident Commander upon his arrival.

3. *Base*

a. A base shall be established as necessary outside of the anticipated hazard area in a removed and safe location. This area, and its resources, will be coordinated by assigned personnel.

b. The base location is to be transmitted to the Comm Center as soon as possible, and all responding units dispatched to this incident shall be directed to report to base.

4. *Communication*

a. At the direction of the INCIDENT COMMANDER, all units on scene shall be directed to switch to a FIRE TAC radio channel.

b. The INCIDENT COMMANDER shall continue to monitor the main fire dispatch channel.

5. *Incident on State Roads or City Streets*

a. If the hazardous materials incident is on a freeway, state road, or city street, the appropriate law enforcement agency shall be summoned to the scene. They will establish communication with the INCIDENT COMMANDER.

b. It is critical to the successful mitigation of the hazardous materials incident that good communications be established and maintained between the INCIDENT COMMANDER and the SCENE MANAGER.

6. *Incident on Public or Private Property*

a. If the hazardous materials incident occurs on public or private property, and not on freeways or state roads, then the highest ranking Chief Officer of the fire department on scene shall assume the position of SCENE MANAGER.

b. The SCENE MANAGER shall assure the designation of the INCIDENT COMMANDER.

7. *Hazardous Material Response Team Operation*

a. It shall be the responsibility of the H.M.R.T. to identify the problem, to determine the possible course of action that could happen as a result of this incident, and to relay this information to the INCIDENT COMMANDER.

b. The INCIDENT COMMANDER shall evaluate this information, and will determine the best course of action to take to mitigate the hazard. The INCIDENT COMMANDER shall then convey this determination to the SCENE MANAGER, along with any needs he may have to accomplish this action.

c. The H.M.R.T. Officer may request that the incident be upgraded to a LEVEL III HMI.

8. *Evacuation*

a. If the evacuation is necessary, the INCIDENT COMMANDER and/or the SCENE MANAGER shall notify the appropriate law enforcement agencies.

b. The law enforcement agency having jurisdiction, assisted by other appropriate agencies, shall plan and conduct an orderly evacuation within a specified geographical area.

c. Utilization of the news media to assist in notifying the public of an evacuation shall be coordinated through the All Hazard Warning System (E.B.S. — Emergency Broadcasting System).

9. *Clean-up and Disposal*

a. The primary responsibility for the assumption of all costs for the clean-up and disposal of a chemical shall be:

1) The person or persons whose negligent or willful act caused such spill or release.

2) The person or persons who own or had custody of the chemical or hazardous material or waste at the time of the spill or release.

3) The person or persons who owned or had custody or control of the container or transport vehicle that held such chemical or hazardous material or waste.

b. The SCENE MANAGER, INCIDENT COMMANDER, and HMRT Officer shall work together to identify the responsible party. When, in the opinion of the HMRT Officer, the substance must be cleaned up according to Cal/OSHA and EPA regulations, the responsible party or a representative of his agency must call a reputable and licensed hazardous waste hauler.

c. The Sacramento Fire Department is authorized to clean up or abate any chemical, hazardous material, or waste released or deposited upon or into any property within the city of Sacramento. In the event the responsible party cannot be located or identified, or refuses to cooperate:

1) On freeways and state roads — the SCENE MANAGER (CHP) shall ask Cal/Trans to make the call.

2) On Sacramento city streets — the SCENE MANAGER (SFD Chief Officer) shall ask City Street Department to make the call.

3) On private property — the SCENE MANAGER (SFD Chief Officer) shall inform the on-duty DUTY CHIEF (Deputy Chief). Only the deputy chief can give the approval to make the call.

4) On state property — the SCENE MANAGER shall ask State Police to make the call.

d. The persons described in subsection (a) above shall be liable to the Sacramento Fire Department for all costs incurred as a result of clean-up and disposal efforts.

e. It shall not be the responsibility of the Sacramento Fire Department to remove any hazardous material. The HMRT shall remain on scene until arrival of the appropriate and approved agency. The HMRT shall oversee the clean-up operation in an advisory capacity to insure that removal of product and containers are done so correctly and safely.

10. *Cost Recovery*

a. In the event the Sacramento Fire Department ultimately must pay for clean-up and removal of a hazardous material, the Sacramento Fire Department may pursue all appropriate legal avenues to initiate action against the spiller to recover all costs incurred.

b. The agency acting in the capacity of SCENE MANAGER, when it is not the Sacramento Fire Department, shall assist the Sacramento Fire Department in initiating action to recover all costs incurred in the handling of the hazardous material incident.

B. Fire Department Not First on Scene

1. When another agency is on the scene of a hazardous material incident prior to the arrival of the fire department, it generally will be a law enforcement agency. Upon arrival of the first department officer, he will establish contact and communication with the agency first on scene.

2. The fire department officer first on the scene shall ascertain, if possible, what level of an incident is needed. If it is a LEVEL II INCIDENT, the fire officer shall call for the proper dispatch. The fire department officer shall then

proceed to gather as much information as possible about the incident to pass on to the HMRT.

3. Until a higher ranking chief officer of the fire department arrives, the fire department officer on the scene shall be the INCIDENT COMMANDER and will establish the necessary incident command structure to deal with the hazardous material incident.

4. A LEVEL II INCIDENT shall follow the same operational procedure as outlined for LEVEL II. It will be understood that a LEVEL III INCIDENT will by its nature involve a large number of outside agencies.

C. Mutual Aid

1. The HMRT units shall be available for response to a hazardous materials incident outside the city limits of Sacramento, when the request is made through the established mutual aid procedure.

2. The HMRT and the decontamination unit will be available for mutual aid response to any part of the Sacramento County, and for a distance of 50 miles radius from the boundaries of the city of Sacramento.

3. When the HMRT unit is dispatched, the responsible battalion chief will also be dispatched.

4. The battalion chief will become the liaison officer between the agency requesting the assistance of the HMRT and the officer of the HMRT unit.

5. The HMRT unit can function as an informational and/or operational unit on mutual aid dispatched calls.

APPENDIX ONE

Definitions

Assisting Agencies. Any outside agency (not a Sacramento Fire Department agency) that assists at the scene of a hazardous materials incident and provides supporting services not within the responsibility or capability of the Sacramento Fire Department. Such services would include, but not be limited to, road closures and detours, technical advice, sampling and monitoring capabilities, cleanup, offloading, disposal, and other supportive tasks as requested by the INCIDENT COMMANDER and SCENE MANAGER.

B.L.E.V.E. An acronym for Boiling Liquid Expanding Vapor Explosion.

Cleanup. Incident scene activity directed to removing the hazardous material and all contaminated debris, including dirt, water, road surfaces, containers, vehicles, contaminated articles, and extinguishment tools and materials, and returning the scene to as near normal as it existed prior to the incident. Cleanup is not the function of the Sacramento Fire Department, but overseeing and observing cleanup operations would be the responsibility of the INCIDENT COMMANDER and the SCENE MANAGER. Technical guidance for cleanup can be given by the HMRT Officer.

Command. To direct and delegate authoritatively through an organization that provides effective implementation of departmental control procedures.

Command Post—Location. When positioned in a safe strategic location, provides a base for the SCENE MANAGER when managing the overall incident. Representatives of all agencies involved at the incident should provide liaison officers to the command post.

Command Post—Vehicle. A vehicle, located in a safe and strategic location, that provides for the SCENE MANAGER a facility for tactical planning and includes such resources as multiple channels, resource and reference books, maps, reports, etc.

Containment. Includes all activities necessary to bring the scene of a hazardous materials incident to a point of stabilization, and to the greatest degree of safety as possible.

Coordination. The administering and management of several tasks so as to act together in a smooth concerted way. To bring together in a uniform manner the function of several agencies.

Cost Recovery.   A process that enables an agency to be reimbursed for costs incurred at a hazardous materials incident.

Explosion.   A sudden release of a large amount of energy in a destructive manner. It is as a result of powders, mists, or gases undergoing instantaneous ignition, liquids or solids undergoing sudden decomposition, or a pressurized vessel undergoing overpressure rupture, with such force as to generate tremendous heat, cause severe structural damage, occasionally generating a shock wave, and propelling shrapnel.

Hazardous Material.   A material or substance in a quantity or form that, when not properly controlled or contained, may pose an unreasonable risk to health, safety, property, and the environment, is of such a nature as to require implementation of special control procedures supplementing standard departmental procedures, and may require the use of specialized equipment and reference material. For the purpose of this plan, "hazardous material," "hazardous substance," "dangerous material," and "dangerous chemical" are synonymous.

Categories of Hazardous Material:

a. *Explosive*–Any chemical compound, mixture, or device the primary or common purpose of which is to function by explosion, with substantial instantaneous release of gas and heat.

b. *Flammable Liquid*–Any liquid having a flash point below 100 degrees Fahrenheit as determined by tests listed in 49CFR Sec. 173.115(d).

c. *Combustible Liquid*–Any liquid having a flash point at or above 100 degrees Fahrenheit and below 200 degrees Fahrenheit as determined by tests listed in 49CFR Sec. 173.115.

d. *Flammable Gas*–Any gas that, in a mixture of 13 percent or less by volume with air, is flammable at atmospheric pressure; or its flammable range with air at atmospheric pressure is wider than 12 percent (by volume) regardless of a lower flammability limit.

e. *Nonflammable Gas*–Any compressed gas other than a flammable gas.

f. *Flammable Solid*–Any solid material, other than an explosive, that is liable to cause fires through friction or retained heat from manufacturing or processing or that can be ignited readily and, when ignited, burns so vigorously and persistently as to create a serious transportation hazard.

g. *Oxidizer*–A substance that yields oxygen readily to stimulate the combustion of other material.

h. *Organic Peroxide*–An organic compound that may be considered a derivative of hydrogen peroxide where one or more of the hydrogen atoms have been replaced by organic radicals. It readily releases oxygen to stimulate the combustion of other materials.

i. *Poison A*–A poison gas. Extremely dangerous. Gases or liquids of such nature that a very small amount of the gas, or vapor from the liquid, is dangerous or lethal to life when mixed with air.

j. *Poison B*–Liquids or solids, including pastes, semisolids, and powders other than Class A or Irritating Materials, that are known to be so toxic to man as to afford a hazard to health.

k. *Irritating Material*–A liquid or solid substance that upon contact with fire or when exposed to air gives off dangerous or intensely irritating fumes, but not including any Class A poisonous material.

l. *Radioactive Material*–Also known as Radiological Material, it is any material or combination of materials that spontaneously emits ionizing radiation and has a specific gravity greater than 0.002 microcuries per gram.

m. *Corrosive Material*–Any liquid or solid, including powders, that cause visible destruction of human skin tissue; or a liquid that has a severe corrosion rate on steel or aluminum.

n. *Etiological Agent*–An Etiological Agent is a viable microorganism, or its toxin, that causes or may cause human disease.

o. *Consumer Commodity*–A material that is packaged or distributed in a form intended and suitable for sale through retail sales agencies for use or consumption by individuals for purposes of personal care or household use. This term also includes drugs and medicines.

HAZMAT.   An abbreviation for *Hazardous Material*.

H.M.R.T.   An abbreviation for Hazardous Materials Response Team.

Hazardous Material Incident.   Any spill, leak, rupture, fire, or accident that results, or has the potential to result, in the loss or escape of a hazardous material from its container.

Incident Command.   A system of command and control designed to assure the smooth implementation of immediate and continued operational procedures until the incident has been contained or abated.

Incident Commander.   A representative of the Sacramento Fire Department that is responsible for overall direction and control of immediate on-scene fire department functions. Incident Commander shall report to the Scene Manager and is abbreviated I.C.

Leak.   A leak will be considered to be the release or generation of a toxic, poisonous, or noxious liquid or gas in a manner that poses a threat to air and ground quality and to health and safety.

Rupture.   A rupture will be considered to be the physical failure of a container, releasing or threatening to release a hazardous material. Physical failure may be due to forces acting upon the container in such a manner as to cause punctures, creases, tears, corrosion, breakage, or collapse.

Scene Manager.   The Scene Manager shall be the representative of the agency that is responsible for overall management and coordination of all activities at the scene of a hazardous materials incident, until the scene has been abated of the hazard. Scene Manager is abbreviated S.M.

Spill.   A spill will be considered the release of a liquid, powder, or solid form of a hazardous material out of its original container.

Stabilization.   Incident scene activities directed to channel, restrict, and/or halt the spread of hazardous material; to control the flow of a hazardous material to an area of lesser hazard; to implement procedures to insure against ignition; or to control a fire in such a manner as to be safe, such as a controlled burn, flaring off, or extinguishment by consumption of the fuel.

Transportation.   Methods of transporting or moving commodities and materials, including highway, railroad, pipeline, waterborne vessels, aircraft, and other means.

APPENDIX TWO

I. RESPONSIBILITY OF AGENCIES

A. City Municipal Agencies

1. *The Sacramento Fire Department* shall assume the role of Incident Commander (I.C.) at the scene of a hazardous material incident. The Fire Department shall coordinate and effect rescue efforts, first aid, containment, and hazard reduction activities. The Incident Commander shall work with, and coordinate his department's activities through, the Scene Manager (S.M.).

2. *The Sacramento Police Department* shall have responsibility for crowd control, traffic control and relocation, and scene security and shall coordinate and control evacuation activities. For hazardous materials incidents that occur on state roads within the city of Sacramento, the California Highway Patrol shall assume the role of Scene Manager (S.M.).

3. *Traffic and Engineering* (Public Works) shall be responsible for necessary surface road closures, detours, and establishing traffic control zones. This agency shall also assist the fire department in appropriate scene stabilization and abatement of hazard for incidents on surface roads. They can also provide sand and barricades.

4. *Water and Sewer Departments* shall be responsible for maintaining clear and flowing water distribution systems and monitoring remedial actions when a hazardous material may effect water sources and distribution systems. They may also assist in product analysis utilizing their chemistry lab.

5. *The City Office of Emergency Planning* shall be used to coordinate procurement of needed resources and efforts of other agencies within their jurisdiction. This office also has the ability to obtain resources of the state O.E.S.

B. County Municipal Agencies

Notification of and requests for assistance from the following Sacramento County Agencies *must* be made through *COUNTY DISPATCH*.

1. *The County Health Department* shall provide assistance and information regarding environmental health dangers and can provide information regarding clean-up and disposal procedures. This agency must be notified through County Dispatch if there are any civilians injured or contaminated as a result of a hazardous materials incident. This agency must also be notified of any hazardous materials incident involving any type or intensity of radioactive

material. The County Health Department can send specialists to the scene to assist in radiological monitoring.

2. *The County Air Pollution Control Board* shall provide information on atmosphere levels of contamination and shall monitor meteorological information and wind drift. They shall be notified through County Dispatch of any hazardous materials incident that is releasing, or could imminently release, a toxic or poisonous substance to the atmosphere. This agency may elect to respond to very serious air pollution related incidents and provide air monitoring capabilities.

3. *The County Department of Public Works* shall be responsible for providing traffic control zones and detours. They can, upon request of the Scene Manager, send sand and barricades to county incidents.

4. *The Sacramento County Sheriff* shall provide for the position of SCENE MANAGER for hazardous materials incidents off road in the county of Sacramento. This agency shall have the responsibility for providing crowd control and scene security and shall coordinate and control evacuation activities.

5. *The Fire Department* having jurisdiction shall assume the role of INCIDENT COMMANDER (I.C.) of the scene of a hazardous materials incident. The Fire Department shall coordinate and effect rescue efforts, first aid, containment, and hazard reduction activities. The INCIDENT COMMANDER shall work with, and coordinate his department's activities through, the Scene Manager (S.M.).

6. *Sacramento County Office of Emergency Planning* shall be used to coordinate procurement of needed resources on hazardous materials incidents in the county of Sacramento, and will coordinate efforts to acquire services of other Sacramento county agencies. Only this agency can grant expenditure of funds for clean-up of spills in Sacramento County. The Sacramento County Office of Emergency Planning may elect and send a representative to the scene of serious hazardous materials incidents in Sacramento County.

7. *Sacramento County Department of Agriculture* has the responsibility to regulate and enforce agricultural use of pesticides. Any hazardous materials incident involving an agricultural pesticide must be reported to this agency through County Dispatch. This agency can do some soil testing. They may elect to send a representative to the scene of any hazardous materials incident in the county that poses a threat to the community or the environment. They can provide some information on pesticide products and control measures.

C. State Government

Notification of and requests for assistance from any California state agency must be made through State Office of Emergency Services.

1. *Office of Emergency Services* (OES) is responsible for the notification and coordination of all state agencies that may become involved in the response to hazardous materials incidents. They shall coordinate with CHP to notify and alert state and federal agencies of incidents that occur on state roads. They shall coordinate state radiological monitoring of areas, personnel, and equipment in support of local authority.

2. *California Highway Patrol* (CHP) has the responsibility for traffic supervision and control on all state roads, state owned bridges, and highways within unincorporated areas. They shall provide traffic control, traffic rerouting, road closure, prevention of unauthorized entry into restricted and limited access areas, and assist local authorities as requested. They shall function as the overall Scene Manager (S.M.) for any hazardous material incident occurring on highways and roads within their primary responsibility.

3. *California Department of Transportation* (Cal/Trans) has the responsibility for maintaining the state highway system. They can assist as requested in the identification, containment, and clean-up of hazardous materials on these state highways. They can provide as requested highway closure, traffic management, and restoration of orderly traffic flow. They shall assist as requested to monitor contamination and shall repair and restore contaminated highways. They shall coordinate their on-scene activities through the Scene Manager.

4. *Department of Fish and Game* (DFG) has the responsibility for protecting the state's natural living and wildlife resources and their habitat. They can provide recommendations and guidelines when a hazardous material has contaminated or may contaminate streams or waterways. They shall coordinate their on-scene activities through the Scene Manager. Whenever there is a kill of wildlife or the threat of such, they must be notified through OES. On request they will send a representative to the scene.

5. *Department of Water Resources* (DWR) is responsible for protecting the State Water Project, including aqueducts, reservoirs, pumps, and natural channels, from pollutants. They can provide as requested advice and assistance concerning corrective actions to mitigate any incident affecting the state water system. They shall coordinate their on-scene activities through the Scene Manager.

6. *State Water Resources Control Board* (SWRCB) is responsible for protecting surface and ground water throughout the state. They can provide advice concerning the potential impact of a hazardous material incident involving or threatening water resources. They can conduct water sampling and analysis, and advise critical water users of the situation. They shall coordinate their on-scene activities through the Scene Manager.

7. *Division of Oil and Gas* (DOG) of the Department of Conservation, is responsible for all oil and gas well operations within the territory of California. They can assist in determining the appropriate actions to be taken to

control and secure the scene in the event of a hazardous material incident emanating from a well or drilling operation. They shall coordinate their on-scene activities through the Scene Manager.

8. *Department of Food and Agriculture* (Dept. of Ag.) is responsible for providing proper and safe pesticide control and protecting the public and environment from potential adverse effects due to agricultural chemicals. This agency shall be notified of any incident involving agricultural chemicals. They can, upon request, provide technical assistance on pesticide related incidents and can advise other state and local authorities of the potential of contamination to farm lands, feed, animals, and the environment. They can advise and counsel the Scene Manager of suggested corrective actions to be taken or contemplated to mitigate the causes of the hazard or pollutions.

9. *Department of Forestry* (CDF) performs fire prevention and suppression duties to California state forested lands and to local jurisdictions on contract. Emergency support activities that they can provide the Scene Manager include: feeding operations of state and local workers, communication support as requested, environmental monitoring as requested, and support of local fire fighting activities in accordance to the State Mutual Aid Plan.

10. *Department of Parks and Recreation* (DPR) is responsible for the maintenance and upkeep of all state parks and recreation lands. They can provide assistance to local authorities in evacuation of state parks, assist the CDF in setting up emergency feeding stations, and provide emergency living facilities for evacuees and emergency workers. They shall coordinate their on-scene activities with the Scene Manager.

11. *Department of Industrial Relations* (DIR) has responsibility for investigating and compiling information regarding industrial accidents where workers are seriously injured or killed. They can assist in providing technical advice as to how toxic materials may be safely handled at the scene of an accident.

They can assist the Scene Manager in evaluating the health hazards of toxic materials at incidents involving industrial locations.

12. *Department of Health Services* (DOHS) is responsible for protecting public health from low level radioactivity and hazardous materials. Their further responsibilities include protecting food and water supplies from the effects of hazardous materials incidents and designating a location for the disposal of hazardous waste. All incidents involving radioactive materials should be reported to the DOHS via the OES.

   a. *Hazardous Materials Management Section* (HMMS) can provide technical advice regarding protective measures for use by response personnel and advice regarding suitable disposal of the hazardous waste.

b. *Disaster Medical Services* (DMS) can identify hospitals and medical facilities capable of handling contaminated patients.

c. *Radiological Health Section* (RHS) can assist local authority in monitoring, provide laboratory for analysis of food and feed, and establish and direct measures to mitigate the radiological impact on public health. They can recommend measures to limit the spread of the radioactivity.

The functions and services that DOHS can provide shall be coordinated with the Scene Manager.

13. *Air Resources Board* (ARB) is responsible for air quality management. Regional air quality boards should be notified in the event a toxic or hazardous material has the potential to affect air quality. They can provide technical advice and field monitoring capabilities of airborne contaminants. They can provide predictions to future ambient air movement and potential wind drift information to the Scene Manager and to the OES.

14. *Department of General Services* (DGS) is responsible for providing security to all state buildings. They can ensure security of evacuated state buildings and assist local law enforcement agencies as requested. They shall coordinate their on-scene activities through the Scene Manager.

D. Federal Government

1. *Environmental Protection Agency* (EPA) has the responsibility to assure protection of the environment from all types of contamination and must be notified of incidents of hazardous materials resulting in contamination. The National Contingency Plan specifies the federal on-scene Incident Commander for inland waters and ground to the EPA.

2. *Department of Energy* (DOE) has the responsibility and capability of assisting and providing technical information in the handling and disposal of radiological sources and nuclear materials.

3. *Department of Transportation* (DOT) is responsible for regulating the transportation of hazardous materials.

4. *U.S. Coast Guard* (USCG) has a responsibility encompassing the nation's coastline and major navigable waterways within the state. They can provide for the decontamination and clean-up of any material that enters and affects the waters. The National Contingency Plan specifies the federal on-scene Incident Commander for coastal waters as the USCG.

APPENDIX THREE

Notification Information Flow

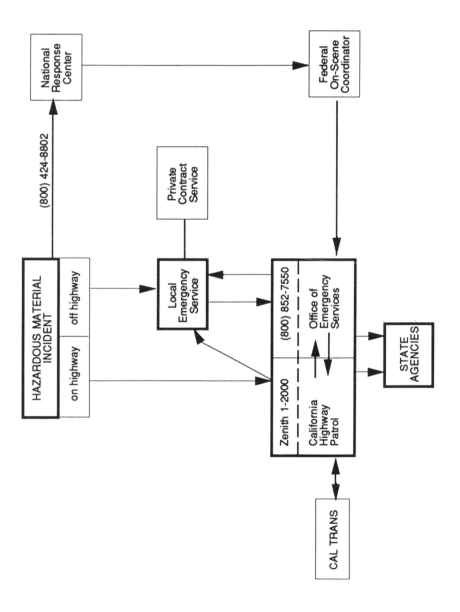

APPENDIX FOUR

Responsibility Matrix

STATE AGENCIES	Notification	Identification, Analysis, Technical Assistance	Coordination	Law Enforcement, Traffic Control	Rescue, Suppression, Containment	Cleanup & Disposal	Evacuation, Area Control	Emergency Medical	Public Health & Sanitation	Education & Public Information	Recovery	Training & Exercises	Natural Resources Damage Assessment
Office of Emergency Services	x	x	x						x	x	x	x	
California Highway Patrol	x	x	x	x				x		x		x	
State Water Resources Control Board		x	x			x			x		x	x	x
Department of Fish & Game		x	x	x		x						x	x
Department of Conservation (Oil & Gas)		x	x		x	x						x	
State Lands Commission		x	x		x	x						x	
Department of Transportation		x			x	x					x	x	
Department of Health Services		x			x	x		x	x	x		x	
Department of Food & Agriculture		x				x			x		x	x	
Department of Industrial Relations		x						x	x		x	x	
Department of Water Resources		x			x				x			x	x
Office of Fire Marshal		x								x		x	
Air Resources Board		x							x			x	
Department of Forestry					x	x						x	x
Department of Parks & Recreation						x	x					x	x
Military Department							x	x				x	
Public Utilities Commission		x										x	
Attorney General		x									x		
Department of General Services			x			x							
Emergency Medical Service Authority								x	x				
Department of Social Services			x								x		

	Notification	Identification, Analysis, Technical Assistance	Coordination	Law Enforcement, Traffic Control	Rescue, Suppression, Containment	Cleanup & Disposal	Evacuation, Area Control	Emergency Medical	Public Health & Sanitation	Education & Public Information	Recovery	Training & Exercises
LOCAL AGENCIES												
Emergency Services Coordinator	x	x	x							x	x	x
Fire Service		x	x		x			x				x
Law Enforcement			x	x			x					x
County Health Officer		x			x				x			x
Agriculture Commission		x			x				x			x
Air Pollution Control Officer		x							x			x
Public Works		x			x	x			x			x
Supervisors/Councilors										x	x	x
Schools							x	x				x
Hospitals		x						x				x
FEDERAL AGENCIES												
FEMA		x					x			x	x	x
U.S. Coast Guard	x	x	x	x	x	x						x
EPA	x	x	x			x						x
Others		x							x			x
NON-GOVERNMENT												
Private Facility Owners	x	x	x		x	x	x			x	x	x
Industry Co-Ops		x			x	x						x
Private Hazardous Waste Services		x			x	x						x
American National Red Cross							x	x			x	x
Salvation Army							x	x			x	x
Civil Air Patrol							x					x
Hospitals, Ambulances		x						x				x

APPENDIX FIVE

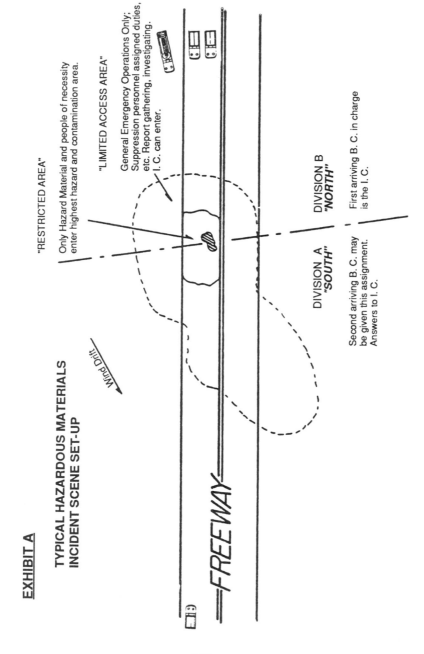

EXHIBIT A

TYPICAL HAZARDOUS MATERIALS
INCIDENT SCENE SET-UP

Wind Drift

"RESTRICTED AREA"

Only Hazard Material and people of necessity enter highest hazard and contamination area.

"LIMITED ACCESS AREA"

General Emergency Operations Only; Suppression personnel assigned duties, etc. Report gathering, investigating. I. C. can enter.

DIVISION B
"NORTH"

First arriving B. C. in charge is the I. C.

DIVISION A
"SOUTH"

Second arriving B. C. may be given this assignment. Answers to I. C.

FREEWAY

APPENDIX SIX

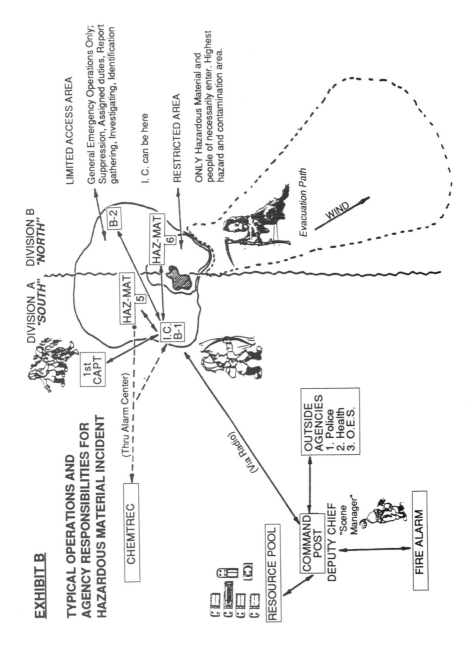

311

V Hazardous Materials Emergency Planning Guide

**National Response Team
of the National Oil and Hazardous
Substances Contingency Plan
2100 2nd Street SW
Washington, DC 20593**

The following document is reprinted from the US government publication and details the organization and planning of a community hazardous materials incident response plan.

The Background of This Guidance

This Haz Mat Emergency Planning Guide has been developed cooperatively by 14 federal agencies. It is being published by the National Response Team in compliance with Section 303(f) of the "Emergency Planning and Community Right-to-Know Act of 1986," Title III of the "Superfund Amendments and Reauthorization Act of 1986" (SARA).

This guide replaces the Federal Emergency Management Agency's (FEMA) Planning Guide and Checklist for Hazardous Materials Contingency Plans (popularly known as FEMA-10).

This guide also incorporates material from the U.S. Environmental Protection Agency's (EPA) interim guidance for its Chemical Emergency Preparedness Program (CEPP) published late in 1985. Included are Chapters 2 ("Organizing the Community"), 4 ("Contingency Plan Development and Content"), and 5 ("Contingency Plan Appraisal and Continuing Planning"). EPA is revising and updating CEPP technical guidance materials that will include site-specific guidance, criteria for identifying extremely hazardous substances, and chemical profiles and a list of such substances. Planners should use this general planning guide in conjunction with the CEPP materials.

In recent years, the U.S. Department of Transportation (DOT) has been active in emergency planning. The Research and Special Programs Administration (RSPA) has published transportation-related reports and guides and has contributed to this general planning guide. The U.S. Coast Guard (USCG) has actively implemented planning and response requirements of the National Contingency Plan (NCP), and has contributed to this general planning guide.

The U.S. Occupational Safety and Health Administration (OSHA) and the U.S. Agency for Toxic Substances and Disease Registry (ATSDR) have assisted in preparing this general planning guide.

In addition to its FEMA-10, FEMA has developed and published a variety of planning-related materials. Of special interest here is *Guide for Development of State and Local Emergency Operations Plans* (known as CPG 1-8) that encourages communities to develop multi-hazard emergency operations plans (EOPs) covering all hazards facing a community (e.g., floods, earthquakes, hurricanes, as well as hazardous materials incidents). This general planning guide complements CPG 1-8 and indicates in Chapter 4 how hazardous materials planners can develop or revise a multi-hazard EOP. Chapter 4 also describes a sample outline for an emergency plan covering only hazardous materials, if a community does not have the resources to develop a multi-hazard EOP.

315

The terms "contingency plan," "emergency plan," and "emergency operations plan" are often used interchangeably, depending upon whether one is reading the NCP, CPG 1-8, or other planning guides. This guide consistently refers to "emergency plans" and "emergency planning."

This guide will consistently use "hazardous materials" when generally referring to hazardous substances, petroleum, natural gas,* synthetic gas, acutely toxic chemicals, and other toxic chemicals. Title III of SARA uses the term "extremely hazardous substances" to indicate those chemicals that could cause serious irreversible health effects from accidental releases.

The major differences between this document and other versions proposed for review are the expansion of the hazards analysis discussion (Chapter 3) and the addition of Appendix A explaining the planning provisions of Title III of SARA.

---

*We recognize that natural gas is under a specific statute, but because this is a general planning guide (and because criteria for the list of extremely hazardous substances under Title III of SARA may be expanded to include flammability), local planners may want to consider natural gas.

1 Introduction

1.1 The Need for Hazardous Materials Emergency Planning

Major disasters like that in Bhopal, India, in December 1984, which resulted in 2,000 deaths and over 200,000 injuries, are rare. Reports of hazardous materials spills and releases, however, are increasingly commonplace. Thousands of new chemicals are developed each year. Citizens and officials are concerned about accidents (e.g., highway incidents, warehouse fires, train derailments, industrial incidents) happening in their communities. Recent evidence shows that hazardous materials incidents are considered by many to be the most significant threat facing local jurisdictions. Ninety-three percent of the more than 3,100 localities completing the Federal Emergency Management Agency's (FEMA) Hazard Identification, Capability Assessment, and Multi-Year Development Plan during fiscal year 1985 identified one or more hazardous materials risks (e.g., on highways and railroads, at fixed facilities) as a significant threat to the community. Communities need to prepare themselves to prevent such incidents and to respond to the accidents that do occur.

Because of the risk of hazardous materials incidents, and because local governments will be completely on their own in the first stages of almost any hazardous materials incident, communities need to maintain a continuing preparedness capacity. A specific tangible result of being prepared is an emergency plan. Some communities might have sophisticated and detailed written plans but, if the plans have not recently been tested and revised, these communities might be less prepared than they think for a possible hazardous materials incident.

1.2 Purpose of This Guide

The purpose of this guide is to assist communities in planning for hazardous materials incidents.

"Communities" refers primarily to local jurisdictions. There are other groups of people, however, that can profitably use this guide. Rural areas with limited resources may need to plan at the county or regional level. State officials seeking to develop a state emergency plan that is closely coordinated with local plans can adapt this guidance to their purposes. Likewise, officials of chemical plants, railroad yards, and shipping and trucking companies can use this guidance to coordinate their own hazardous materials emergency planning with that of the local community.

317

"Hazardous materials" refers generally to hazardous substances, petroleum, natural gas, synthetic gas, acutely toxic chemicals, and other toxic chemicals. "Extremely hazardous substances" is used in Title III of the Superfund Amendments and Reauthorization Act of 1986 to refer to those chemicals that could cause serious health effects following short-term exposure from accidental releases. The U.S. Environmental Protection Agency (EPA) published an initial list of 402 extremely hazardous substances for which emergency planning is required. Because this list may be revised, planners should contact EPA regional offices to obtain information. This guidance deals specifically with response to hazardous materials incidents—both at fixed facilities (manufacturing, processing, storage, and disposal) and during transportation (highways, waterways, rail, and air). Plans for responding to radiological incidents and natural emergencies such as hurricanes, floods, and earthquakes are not the focus of this guidance, although most aspects of plan development and appraisal are common to these emergencies. Communities should see NUREG 0654/FEMA-REP-1 and/or FEMA-REP-5 for assistance in radiological planning. (See Appendix C.) Communities should be prepared, however, for the possibility that natural emergencies, radiological incidents, and hazardous materials incidents will cause or reinforce each other.

The objectives of this guide are to:

• Focus community activity on emergency preparedness and response;

• Provide communities with information useful in organizing the planning task;

• Furnish criteria to determine risk and to help communities decide whether they need to plan for hazardous materials incidents;

• Help communities conduct planning that is consistent with their needs and capabilities; and

• Provide a method for continually updating a community's emergency plan.

This guide will *not*:

• Give a simple "fill-in-the-blanks" model plan (because each community needs an emergency plan suited to its own unique circumstances);

• Provide details on response techniques; or

• Train personnel to respond to incidents.

Community planners will need to consult other resources in addition to this guide. Related programs and materials are discussed in Section 1.5.

1.3   How to Use This Guide

This guide has been designed so it can be used easily by both those communities with little or no planning experience and those communities with extensive planning experience.

All planners should consult the decision tree in Exhibit 1 for assistance in using this guide.

Chapter 2 describes how communities can organize a planning team. Communities that are beginning the emergency planning process for the first time will need to follow Chapter 2 very closely in order to organize their efforts effectively. Communities with an active planning agency might briefly review

Exhibit 1 Overview of Planning Process.

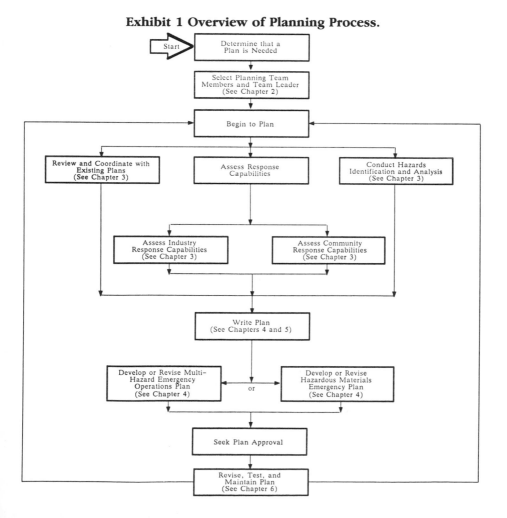

Chapter 2, especially to be sure that all of the proper people are included in the planning process, and move on to Chapter 3 for a detailed discussion of tasks for hazardous materials planning. Planners should review existing emergency plans, perform a hazards identification and analysis, assess prevention and response capabilities, and then write or revise an emergency plan.

Chapter 4 discusses two basic approaches to writing an emergency plan: (a) incorporating hazardous materials planning into a multi-hazard emergency operations plan (EOP) (see Section 1.5.1); and (b) developing or revising a plan dealing only with hazardous materials. Incorporating hazardous materials planning into a multi-hazard approach is preferable. Some communities, however, have neither the capability nor the resources to do this immediately. Communities that choose to develop or revise an EOP should consult FEMA's CPG 1-8 for specific structure requirements for the plan in addition to the discussion in Section 1.5.1. Communities that choose to develop or revise a single-hazard plan for hazardous materials can use the sample outline of an emergency plan in Chapter 4 to organize the various hazardous materials planning elements. (Note: Communities receiving FEMA funds must incorporate hazardous materials planning into a multi-hazard EOP.)

Chapter 5 describes the elements to be considered when planning for potential hazardous materials incidents. All communities (both those preparing an EOP under the multi-hazard approach and those preparing a single-hazard plan) should carefully follow Chapter 5 to ensure that they consider and include the planning elements related to hazardous materials.

Chapter 6 describes how to review and update a plan. Experience shows that many communities mistakenly presume that completing an emergency plan automatically ensures adequate preparedness for emergency response. All communities should follow the recommendations in Chapter 6 to ensure that emergency plans will be helpful during a real incident.

Appendix A is a summary for implementing the "Emergency Planning and Community Right-to-Know Act of 1986." Appendix B is a list of acronyms and abbreviations used in this guidance. Appendix C is a glossary of terms used throughout this guide. (Because this guide necessarily contains many acronyms and technical phrases, local planners should regularly consult Appendices B and C.) Appendix D contains criteria for assessing state and local preparedness. Planners should use this appendix as a checklist to evaluate their hazards analysis, the legal authority for responding, the response organizational structure, communication systems, resources, and the completed emergency plan. Appendix E is a list of references on various topics addressed in this guidance. Appendix F is a listing of addresses of federal agencies at the national and regional levels. Planners should contact the appropriate office for assistance in the planning process.

1.4 Requirements for Planning

Planners should understand federal, state, and local requirements that apply to emergency planning.

1.4.1 Federal Requirements

This section discusses the principal federal planning requirements found in the National Contingency Plan; Title III of SARA; the Resource Conservation and Recovery Act; and FEMA's requirements for Emergency Operations Plans.

A. *National Contingency Plan*

The National Contingency Plan (NCP), required by Section 105 of the Comprehensive Environmental Response, Compensation, and Liability Act (CERCLA), calls for extensive preparedness and planning. The National Response Team (NRT), comprised of representatives of various federal government agencies with major environmental, transportation, emergency management, worker safety, and public health responsibilities, is responsible for coordinating federal emergency preparedness and planning on a nationwide basis.

A key element of federal support to local responders during hazardous materials transportation and fixed facility incidents is a response by U.S. Coast Guard (USCG) or Environmental Protection Agency (EPA) On-Scene Coordinators (OSCs). The OSC is the federal official predesignated to coordinate and direct federal responses and removals under the NCP. These OSCs are assisted by federal Regional Response Teams (RRTs) that are available to provide advice and support to the OSC and, through the OSC, to local responders.

Federal responses may be triggered by a report to the National Response Center (NRC), operated by the Coast Guard. Provisions of the federal Water Pollution Control Act (Clean Water Act), CERCLA ("Superfund"), and various other federal laws require persons responsible for a discharge or release to notify the NRC immediately. The NRC Duty Officer promptly relays each report to the appropriate Coast Guard or EPA OSC, depending on the location of an incident. Based on this initial report and any other information that can be obtained, the OSC makes a preliminary assessment of the need for a federal response.

This activity may or may not require the OSC or his/her representative to go to the scene of an incident. If an on-scene response is required, the OSC will go to the scene and monitor the response of the responsible party or state or local government. If the responsible party is unknown or not taking appropriate action, or the response is beyond the capability of state and local governments, the OSC may initiate federal actions. The Coast Guard has OSCs at 48 locations (zones) in 10 districts, and the EPA has OSCs in its 10 regional offices and in certain EPA field offices. (See Appendix F for appropriate addresses.)

Regional Response Teams (RRTs) are composed of representatives from federal agencies and a representative from each state within a federal region. During a response to a major hazardous materials incident involving transportation or a fixed facility, the OSC may request that the RRT be convened to provide advice or recommendations on specific issues requiring resolution.

An enhanced RRT role in preparedness activities includes assistance for local community planning efforts. Local emergency plans should be coordinated with any federal regional contingency plans and OSC contingency plans prepared in compliance with the NCP. Appendix D of this guide contains an adaptation of extensive criteria developed by the NRT Preparedness Committee to assess state and/or local emergency response preparedness programs. These criteria should be used in conjunction with Chapters 3, 4, and 5 of this guide.

B. *Title III of SARA ("Superfund Amendments and Reauthorization Act of 1986")*

Significant new hazardous materials emergency planning requirements are contained in Title III of SARA (also known as the "Emergency Planning and Community Right-to-Know Act of 1986"). (See Appendix A for a detailed summary on implementing Title III.)

Title III of SARA requires the establishment of state emergency response commissions, emergency planning districts, and local emergency planning committees. The governor of each state appoints a state emergency response commission whose responsibilities include: designating emergency planning districts; appointing local emergency planning committees for each district; supervising and coordinating the activities of planning committees; reviewing emergency plans; receiving chemical release notifications; and establishing procedures for receiving and processing requests from the public for information about and/or copies of emergency response plans, material safety data sheets, the list of extremely hazardous substances prepared as part of EPA's original Chemical Emergency Preparedness Program initiative (see Section 1.5.2), inventory forms, and toxic chemical release forms.

Forming emergency planning districts is intended to facilitate the preparation and implementation of emergency plans. Planning districts may be existing political subdivisions or multijurisdictional planning organizations. The local emergency planning committee for each district must include representatives from each of the following groups or organizations:

• Elected state and local officials;

• Law enforcement, civil defense, fire fighting, health, local environmental, hospital, and transportation personnel;

• Broadcast and print media;

• Community groups; and

• Owners and operators of facilities subject to the requirements of Title III of SARA.

Each emergency planning committee is to establish procedures for receiving and processing requests from the public for information about and/or copies of emergency response plans, material safety data sheets, and chemical inventory forms. The committee must designate an official to serve as coordinator of information.

Facilities are subject to emergency planning and notification requirements if a substance on EPA's list of extremely hazardous substances is present at the facility in an amount in excess of the threshold planning quantity for that substance. (See *Federal Register*, Vol. 51, No. 221, 41570 et seq.) The owner or operator of each facility subject to these requirements must notify the appropriate state emergency response commission that the facility is subject to the requirements.

Each facility must also notify the appropriate emergency planning committee of a facility representative who will participate in the emergency planning process as a facility emergency coordinator. Upon request, facility owners and operators are to provide the appropriate emergency planning committee with information necessary for developing and implementing the emergency plan for the planning district.

Title III provisions help to ensure that adequate information is available for the planning committee to know which facilities to cover in the plan. (See Appendix A for a discussion of how the local planning committee can use information generated by Title III.) Section 303 (d)(3) requires facility owners and operators to provide to the local emergency planning committee whatever information is necessary for developing and implementing the plan.

When there is a release of a chemical identified by Title III of SARA, a facility owner or operator, or a transporter of the chemical, must notify the community emergency coordinator for the emergency planning committee for each area likely to be affected by the release, and the state emergency response commission of any state likely to be affected by the release. (This Title III requirement does not replace the legal requirement to notify the National Response Center for releases of CERCLA Section 103 hazardous substances.)

Each emergency planning committee is to prepare an emergency plan by October 1988 and review it annually. The committee also evaluates the need for resources to develop, implement, and exercise the emergency plan; and makes recommendations with respect to additional needed resources and how to provide them. Each emergency plan must include: facilities and transportation routes related to specific chemicals; response procedures of facilities, and local emergency and medical personnel; the names of community and facility emergency coordinators; procedures for notifying officials and the public in the event of a release; methods for detecting a release and identifying areas and populations at risk; a description of emergency equipment and facilities in the community and at specified fixed facilities; evacuation plans; training programs; and schedules for exercising the emergency plan. (These plan requirements are listed in greater detail in Chapter 5.) The completed plan is to be reviewed by the state emergency response commission and, at the request of the local emergency planning committee, may be reviewed by the federal Regional Response Team.

(Note: Many local jurisdictions already have emergency plans for various types of hazards. These plans may only require modification to meet emergency plan requirements in Title III of SARA.)

Finally, with regard to planning, Title III of SARA requires the NRT to publish guidance for the preparation and implementation of emergency plans. This Hazardous Materials Emergency Planning Guide is intended to fulfill this requirement. Other Title III provisions supporting emergency planning are discussed in Appendix A.

C. Resource Conservation and Recovery Act

The Resource Conservation and Recovery Act (RCRA) established a framework for the proper management and disposal of all wastes. The Hazardous and Solid Waste Amendments of 1984 (HSWA) expanded the scope of the law and placed increased emphasis on waste reduction, corrective action, and treatment of hazardous wastes.

Under Subtitle C of RCRA, EPA identifies hazardous wastes, both generically and by listing specific wastes and industrial process waste streams; develops standards and regulations for proper management of hazardous wastes by the generator and transporter, which include a manifest that accompanies waste shipments; and develops standards for the treatment, storage, and disposal of the wastes. These standards are generally implemented through permits that are issued by EPA or an authorized state. To receive a permit, persons wishing to treat, store, or dispose of hazardous wastes are required to submit permit applications, which must include a characterization of the hazardous wastes to be handled at the facility, demonstration of compliance with standards and regulations that apply to the facility, and a contingency plan. There are required opportunities for public comment on the draft permits, through which local governments and the public may comment on the facility's contingency plan. It is important that local emergency response authorities be familiar with contingency plans of these facilities. Coordination with local community emergency response agencies is required by regulation (40 CFR 264.37), and EPA strongly encourages active community coordination of local response capabilities with facility plans.

When a community is preparing an emergency plan that includes underground storage tanks (containing either wastes or products), it should coordinate with EPA's regional offices, the states, and local governments. Underground storage tanks are regulated under Subtitle C or I of RCRA.

D. *FEMA Emergency Operations Plan Requirements*

Planning requirements for jurisdictions receiving FEMA funds are set forth in 44 CFR Part 302, effective May 12, 1986. This regulation calls for states and local governments to prepare an emergency operations plan (EOP) that conforms with the requirements for plan content contained in FEMA's CPG 1-3, CPG 1-8, and CPG 1-8A. These state and local government EOPs must identify the available personnel, equipment, facilities, supplies, and other resources in the jurisdiction, and state the method or scheme for coordinated actions to be taken by individuals and government services in the event of natural, man-made (e.g., hazardous materials), and attack-related disasters.

E. *OSHA Regulations*

Occupational Safety and Health Administration regulations require employers involved in hazardous waste operations to develop and implement an emergency response plan for employees. The elements of this plan must include: (1) recognition of emergencies; (2) methods or procedures for alerting employees on site; (3) evacuation procedures and routes to places of refuge or safe distances away from the danger area; (4) means and methods for emergency medical treatment and first aid for employees; (5) the line of authority for employees; (6) on-site decontamination procedures; (7) site control means; and (8) methods for evaluating the plan. Employers whose employees will be responding to hazardous materials emergency incidents from their regular work location or duty station (e.g., a fire department, fire brigade, or emergency medical service) must also have an emergency response plan. (See 29 CFR Part 1910.120.)

1.4.2 State and Local Requirements

Many states have adopted individual laws and regulations that address local government involvement in hazardous materials. Local authorities should investigate state requirements and programs before they initiate preparedness and planning activities. Emergency plans should include consideration of any state or local community right-to-know laws. When these laws are more demanding than the federal law, the state and local laws sometimes take precedence over the federal law.

1.5 Related Programs and Materials

Because emergency planning is a complex process involving a variety of issues and concerns, community planners should consult related public and private sector programs and materials. The following are selected examples of planning programs and materials that may be used in conjunction with this guide.

1.5.1 FEMA's Integrated Emergency Management System (CPG 1-8)

FEMA's Guide for Development of State and Local Emergency Operations Plans (CPG 1-8) provides information for emergency management planners and for state and local government officials about FEMA's concept of emergency operations planning under the Integrated Emergency Management System (IEMS). IEMS emphasizes the integration of planning to provide for all hazards discovered in a community's hazards identification process. CPG 1-8 provides extensive guidance in the coordination, development, review, validation, and revision of EOPs (see Section 4.2). (See Appendix F for FEMA's address and telephone number.)

This guide for hazardous materials emergency planning is deliberately meant to complement CPG 1-8. Chapter 4 describes how a community can incorporate hazardous materials planning into an existing multi-hazard EOP, or how it can develop a multi-hazard EOP while addressing possible hazardous materials incidents. In either case, communities should obtain a copy of CPG 1-8 from FEMA and follow its guidance carefully. All communities, even those with sophisticated multi-hazard EOPs, should consult Chapter 5 of this guide to ensure adequate consideration of hazardous materials issues.

1.5.2 EPA's Chemical Emergency Preparedness Program (CEPP)

In June 1985, EPA announced a comprehensive strategy to deal with planning for the problem of toxics released to the air. One section of this strategy, the Chemical Emergency Preparedness Program (CEPP), was designed to address accidental releases of acutely toxic chemicals. This program has two goals: to increase community awareness of chemical hazards and to enhance state and local emergency planning for dealing with chemical accidents. Many of the CEPP goals and objectives are included in Title III of SARA (see Section 1.4.1). EPA's CEPP materials (including technical guidance, criteria for identifying extremely hazardous substances, chemical profiles and list) are designed to complement this guidance and to help communities perform hazards identification and analysis as described in Chapter 3 of this guide. CEPP materials can be obtained by writing EPA. (See Appendix F.)

1.5.3 DOT Materials

The U.S. Department of Transportation's (DOT) *Community Teamwork* is a guide to help local communities develop a cost-effective hazardous materials transportation safety program. It discusses hazards assessment and risk analysis, the development of an emergency plan, enforcement, training, and legal authority for planning. Communities preparing an emergency plan for transportation-related hazards might use *Community Teamwork* in conjunction with this guide.

Lessons Learned is a report on seven hazardous materials safety planning projects funded by DOT. The projects included local plans for Memphis, Indianapolis, New Orleans, and Niagara County (NY); regional plans for Puget Sound and the Oakland/San Francisco Bay Area; and a state plan for Massachusetts. The *Lessons Learned* report synthesizes the actual experiences of these projects during each phase of the planning process. A major conclusion of this study was that local political leadership and support from both the executive and legislative branches are important factors throughout the planning process. Chapter 2 of this guide incorporates portions of the experiences and conclusions from *Lessons Learned*.

DOT's *Emergency Response Guidebook* provides guidance for fire fighters, police, and other emergency services personnel to help them protect themselves and the public during the initial minutes immediately following a hazardous materials incident. This widely used guidebook is keyed to the identification placards required by DOT regulations to be displayed prominently on vehicles transporting hazardous materials. All first responders should have copies of the *Emergency Response Guidebook* and know how to use it.

DOT has also published a four-volume guide for small towns and rural areas writing a hazardous materials emergency plan. DOT's objectives were to alert officials of those communities to the threat to life, property, and the environment from the transportation of hazardous materials, and to provide simplified guidance for those with little or no technical expertise. Titles of the volumes are: Volume I, A Community Model for Handling Hazardous Materials Transportation Emergencies; Volume II, Risk Assessment Users Manual for Small Communities and Rural Areas; Volume III, Risk Assessment/Vulnerability Model Validation; and Volume IV, Manual for Small Towns and Rural Areas to Develop a Hazardous Materials Emergency Plan. (See Appendix F for DOT's address and telephone number.)

1.5.4 Chemical Manufacturers Association's Community Awareness and Emergency Response Program (CMA/CAER)

The Chemical Manufacturers Association's (CMA) Community Awareness and Emergency Response (CAER) program encourages chemical plant managers to take the initiative in cooperating with local communities to develop integrated emergency plans for responding to hazardous materials incidents. Because chemical industry representatives can be especially knowledgeable during the planning process, and because many chemical plant officials are willing and able to share equipment and personnel during response operations, community planners should seek out local CMA/CAER participants. Even if no such local initiative is in place, community planners can approach chemical plant managers or contact CMA and ask for assistance in the spirit of the CAER program.

Users of this general planning guide might also purchase and use the following three CMA/CAER publications: "Community Awareness and Emergency Response Program Handbook," "Site Emergency Response Planning," and "Community Emergency Response Exercise Program." (See Appendix E for CMA's address.)

2

Selecting and Organizing the Planning Team

2.1 Introduction

This chapter discusses the selection and organization of the team members who will coordinate hazardous materials planning. The guidance stresses that successful planning requires community involvement throughout the process. Enlisting the cooperation of all parties directly concerned with hazardous materials will improve planning, make the plan more likely to be used, and maximize the likelihood of an effective response at the time of an emergency. Experience shows that plans are not used if they are prepared by only one person or one agency. Emergency response requires trust, coordination, and cooperation among responders who need to know who is responsible for what activities, and who is capable of performing what activities. This knowledge is gained only through personal interaction. Working together in developing and updating plans is a major opportunity for cooperative interaction among responders.

(As indicated in Section 1.4.1, Title III of SARA requires governors to appoint a state emergency response commission that will designate emergency planning districts and appoint local emergency planning committees for each district. The state commission might follow the guidance in this chapter when appointing planning committees.)

2.2 The Planning Team

Hazardous materials planning should grow out of a process coordinated by a team. The team is the best vehicle for incorporating the expertise of a variety of sources into the planning process and for producing an accurate and complete document. The team approach also encourages a planning process that reflects the consensus of the entire community. Some individual communities and/or areas that include several communities have formed hazardous materials advisory councils (HMACs). HMACs, where they exist, are an excellent resource for the planning team.

2.2.1 Forming the Planning Team

In selecting the members of a team that will bear overall responsibility for hazardous materials planning, four considerations are most important:

• The members of the group must have the ability, commitment, authority, and resources to get the job done;

• The group must possess, or have ready access to, a wide range of expertise relating to the community, its industrial facilities and transportation systems, and the mechanics of emergency response and response planning;

• The members of the group must agree on their purpose and be able to work cooperatively with one another; and

• The group must be representative of all elements of the community with a substantial interest in reducing the risks posed by hazardous materials.

A comprehensive list of potential team members is presented in Exhibit 2.

In those communities receiving FEMA funds, paid staff may already be in place for emergency operations planning and other emergency management tasks. This staff should be an obvious resource for hazardous materials planning. FEMA has two training courses for the person assigned as the planning team leader and for team members—Introduction to Emergency Management, and Emergency Planning. Another course, Hazardous Materials Contingency Planning, is an interagency "train-the-trainer" course presented cooperatively by EPA, FEMA, and other NRT agencies. Course materials and the schedule of offerings are available through state emergency management agencies.

2.2.2 Respect for All Legitimate Interests

While many individuals have a common interest in reducing the risks posed by hazardous materials, their differing economic, political, and social perspectives may cause them to favor different means of promoting safety. For example, people who live near a facility with hazardous materials are likely to be greatly concerned about avoiding any threat to their lives, and are likely to be less intensely concerned about the costs of developing accident prevention and response measures than some of the other groups involved. Others in the community are likely to be more sensitive to the costs involved, and may be anxious to avoid expenditures for unnecessarily elaborate prevention and response measures. Also, facility managers may be reluctant for proprietary reasons to disclose materials and processes beyond what is required by law.

There may also be differing views among the agencies and organizations with emergency response functions about the roles they should play in case of an incident. The local fire department, police department, emergency management agency, and public health agency are all likely to have some responsibilities in responding to an incident. However, each of these organizations might envision a very different set of responsibilities for their respective agencies for planning or for management on scene.

In organizing the community to address the problems associated with hazardous materials, it is important to bear in mind that all affected parties have a legitimate interest in the choices among planning alternatives. There-fore, strong efforts should be made to ensure that all groups with an interest in the planning process are included.

Some interest groups in the community have well-defined political identi-ties and representation, but others may not. Government agencies, private industry, environmental groups, and trade unions at the facilities are all likely to have ready institutional access to an emergency planning process. Nearby residents, however, may lack an effective vehicle for institutional representa-tion. Organizations that may be available to represent the residents' interests include neighborhood associations, church organizations, and ad hoc orga-nizations formed especially to deal with the risks posed by the presence of specific hazardous materials in a neighborhood.

Exhibit 2
Potential Members of an Emergency Planning Team

Part A: Experience shows that the following individuals, groups, and agencies should participate in order for a successful plan to be developed:

*Mayor/city manager (or representative)
*County executive (or representative)/board of supervisors
*State elected officials (or representative)
*Fire department (paid and volunteer)
*Police department
*Emergency management or civil defense agency
*Environmental agency (e.g., air and/or water pollution control agency)
*Health department
*Hospitals, emergency medical service, veterinarians, medical community
*Transportation agency (e.g., DOT, port authority, transit au-thority, bus company, truck or rail companies)
*Industry (e.g., chemical and transportation)
Coast Guard/EPA representative (e.g., agency response pro-gram personnel)
Technical experts (e.g., chemist, engineer)
*Community group representative
*Public information representative (e.g., local radio, TV, press)

Part B: Other groups/agencies that can be included in the planning process, depending on the community's individual priorities:

Agriculture agency

---

*Required by Title III of SARA

> Indian tribes within or adjacent to the affected jurisdiction
> Public works (e.g., waste disposal, water, sanitation, and roads)
> Planning department
> Other agencies (e.g., welfare, parks, and utilities)
> Municipal/county legal counsel
> Workers in local facilities
> Labor union representatives (e.g., chemical and transportation, industrial health units)
> Local business community
> Representatives from volunteer organizations (e.g., Red Cross)
> Public interest and citizens groups, environmental organizations, and representatives of affected neighborhoods
> Schools or school districts
> Key representatives from bordering cities and counties
> State representatives (Governor, legislator's office, State agencies)
> Federal agency representatives (e.g., FEMA, DOT/RSPA, ATSDR, OSHA)

2.2.3 Special Importance of Local Governments

For several reasons, local governments have a critical role to play in the development of emergency preparedness. First, local governments bear major responsibilities for protecting public health and safety; local police and fire departments, for example, often have the lead responsibility for the initial response to incidents involving hazardous materials. Second, one of the functions of local government is to mediate and resolve the sometimes competing ideas of different interest groups. Third, local governments have the resources to gather necessary planning data. Finally, local governments generally have the legislative authority to raise funds for equipment and personnel required for emergency response. Support from the executive and legislative branches is essential to successful planning. Appropriate government leaders must give adequate authority to those responsible for emergency planning.

2.2.4 Local Industry Involvement

Because fixed facility owners and operators are concerned about public health and safety in the event of an accidental release of a hazardous material, and because many facility employees have technical expertise that will be helpful to the planning team, the team should include one or more facility representatives. Title III of SARA requires facility owners or operators to notify the emergency planning committee of a facility representative who will participate in the emergency planning process as a facility emergency coordinator. In planning districts that include several fixed facilities, one or more representative facility emergency coordinators could be active members of the planning team. The planning team could consult with the other facility emergency coordinators and/or assign them to task forces or committees (see Section 2.3.2). Title III of SARA also requires facilities to submit to the local emergency planning committee any information needed to develop the plan.

2.2.5 Size of Planning Team

For the planning team to function effectively, its size should be limited to a workable number. In communities with many interested parties, it will be necessary to select from among them carefully so as to ensure fair and comprehensive representation. Some individuals may feel left out of the planning process. This can be offset by providing these individuals access to the process through the various approaches noted in the following sections, such as membership on a task force or advisory council. In addition, all interested parties should have an opportunity for input during the review process.

2.3 Organizing the Planning Process

After the planning team members have been identified, a team leader must be chosen and procedures for managing the planning process must be established.

2.3.1 Selecting a Team Leader

A community initiating a hazardous materials emergency planning process may choose to appoint an individual to facilitate and lead the effort, or may appoint a planning team and have the group decide who will lead the effort. Either approach can be used. It is essential to establish clear responsibility and authority for the project. The chief executive (or whoever initiates the process) should determine which course is better suited to local circumstances. (The emergency planning committee required by Title III of SARA is to select its own chairperson.) Regardless of how the team leader is selected, it is his or her primary responsibility to oversee the team's efforts through the entire planning process. Because the role of leader is so significant, a co-chair or back-up could also be named.

Five factors are of major importance in selecting a team leader:

• The degree of respect held for the person by groups with an interest in hazardous materials;

• Availability of time and resources;

• The person's history of working relationships with concerned community agencies and organizations;

• The person's management and communication skills; and

• The person's existing responsibilities related to emergency planning, prevention, and response.

Logical sources for a team leader include:

• *The chief executive or other elected official*. Leadership by a mayor, city or county council member, or other senior official is likely to contribute substantially to public confidence, encourage commitment of time and resources by other key parties, and expedite the implementation of program initiatives. Discontinuity in the planning process can result, however, if an elected official leaves office.

• *A public safety department.* In most communities, the fire department or police department bears principal responsibility for responding to incidents involving chemical releases and, typically, for inspecting facilities as well. A public safety department, therefore, may have personnel with past experience in emergency planning and present knowledge of existing responsibilities within the community.

• *The emergency management or civil defense agency.* In many communities, officials of such an agency will be knowledgeable and experienced in planning for major disasters from a variety of causes. One of the primary responsibilities of a community's emergency management coordinator is to guide, direct, and participate in the development of a multi-hazard emergency operations plan. In some states, existing laws require that this agency be the lead agency to prepare and distribute emergency plans.

• *The local environmental agency or public health agency.* Persons with expertise and legal responsibility in these areas will have special knowledge about the risks posed by hazardous materials.

• *A planning agency.* Officials in a planning agency will be familiar with the general planning process and with the activities and resources of the community.

• *Others.* Communities should be creative and consider other possible sources for a team leader, such as civic groups, industry, academic institutions, volunteer organizations, and agencies not mentioned above. Experience in leading groups and committees, regardless of their purpose, will prove useful in emergency planning.

Personal considerations as well as institutional ones should be weighed in selecting a team leader. For example, a particular organization may appear to have all the right resources for addressing hazardous materials incidents. But if the person in charge of that organization does not interact well with other local officials, it might be best to look for a different leader.

A response coordinator generally is knowledgeable about emergency plans and is probably a person who gets things done. Be aware, however, that a good response coordinator is not necessarily a good planner. He or she might make a good chief advisor to someone better suited for the team leader job.

2.3.2 Organizing for Planning Team Responsibilities

The planning team must decide who shall conduct the planning tasks and establish the procedures for monitoring and approving the planning tasks.

A. *Staffing*

There are three basic staffing approaches that may be employed to accomplish the tasks involved in emergency planning:

• *Assign staff*. Previous experience in related planning efforts demonstrates the usefulness of assigning one or more dedicated staff members to coordinate the planning process and perform specific planning tasks. The staff may be assigned within a "lead agency" having related responsibilities and/or expertise, or may be created separately through outside hiring and/or staff loans from government agencies or industry.

• *Assign task forces or committees*. Planning tasks can be performed by task forces or committees composed entirely or in part of members of the planning team. Adding knowledgeable representatives of government agencies, industry, environmental, labor, and other community organizations to the individual task forces or committees not only supplements the planning team expertise and resources, but also provides an opportunity for additional interested parties to participate directly in the process.

• *Hire contractors or consultants*. If the personnel resources available for the formation of a dedicated staff and task forces or committees are limited, and funds can be provided, the planning team may elect to hire contractors or consultants. Work assigned to a contractor can range from a specialized job, such as designing a survey, to performing an entire planning task (e.g., hazards identification and analysis). A disadvantage of hiring contractors or consultants is that it does not help build a community-centered capability or planning infrastructure.

The three approaches presented above are not mutually exclusive. A community may adopt any combination of the approaches that best matches its own circumstances and resources.

B. *Managing the Planning Tasks*

The monitoring and approval of planning assignments are the central responsibilities of the planning team. In order to have ongoing cooperation in implementing the plan, it is recommended that the planning team operate on a consensus basis, reaching general agreement by all members of the team. Achieving consensus takes more time than majority voting, but it is the best way to ensure that all represented parties have an opportunity to express their views and that the decisions represent and balance competing interests. If it is determined that a consensus method is inappropriate or impossible (e.g., because of the multi-jurisdictional nature of a group), the planning team should formally decide how issues will be resolved.

The team leader should work with the team members to establish clear goals and deadlines for various phases of the planning success. Progress toward these goals and deadlines should be monitored frequently.

Planning meetings, a necessary element of the planning process, often do not make the best use of available time. Meetings can be unnecessarily long and unproductive if planning members get bogged down on inappropriate side issues. Sometimes, when several agencies or groups sit down at one table, the meeting can become a forum for expressing political differences and other grievances fueled by long-standing interagency rivalries. For a team to be effective, a strong team leader will have to make sure that meeting discussions focus solely on emergency planning.

Another point to consider is that the team approach requires the melding of inputs from different individuals, each with a different style and sense of priorities. A team leader must ensure that the final plan is consistent in substance and tone. An editor may be used to make sure that the plan's grammar, style, and content all ultimately fit well together.

On critical decisions, it may be desirable to extend the scope of participation beyond the membership of the planning team. Approaches that might be used to encourage community consensus building through broadened participation in the process include invited reviews by key interest groups, or formation of an advisory council composed of interested parties that can independently review and comment on the planning team's efforts. Chapter 6 contains further guidance on consensus-building approaches.

The procedures to be used for monitoring and approving planning assignments should be carefully thought out at the beginning of the planning process; planning efforts work best when people understand the ground rules and know when and how they will be able to participate. The monitoring and approval process can be adjusted at any time to accommodate variations in local interest.

Planning committees formed according to Title III of SARA are to develop their own rules. These rules include provisions for public notification of committee activities; public meeting to discuss the emergency plan; public comments; response to public comments by the committee; and distribution of the emergency plan.

C. *The Use of Computers*

Computers are handy tools for both the planning process and for maintaining response preparedness. Because new technology is continually being developed, this guide does not identify specific hardware or software packages that planning teams and/or response personnel might use. Local planners should consult Regional FEMA or EPA offices (see Appendix F) for more detailed descriptions of how some communities are using computers.

The following list summarizes some ways in which computers are useful both in the planning process and for maintaining response preparedness.

- *Word processing*. Preparation and revision of plans is expedited by word processing. Of special interest to planners is the use of word processing to keep an emergency plan up to date on an annual or semiannual basis.

- *Modeling*. Planners might consider applying air dispersion models for chemicals in their community so that, during an emergency, responders can predict the direction, velocity, and concentration of plume movement. Similarly, models can be developed to predict the pathways of plumes in surface water and ground water.

- *Information access*. Responders can use a personal computer on site to learn the identity of the chemical(s) involved in the incident (e.g., when placards are partially covered), the effects of the chemical(s) on human health and the environment, and appropriate countermeasures to contain and clean up the chemical(s). Communities that intend to use computers on scene should also provide a printer on scene.

- *Data storage*. Communities can store information about what chemicals are present in various local facilities, and the availability of equipment and personnel that are needed during responses to incidents involving specific chemical(s). Compliance with Title III will generate large amounts of data (e.g., MSDS forms, data on specific chemicals in specific facilities, data on accidental releases). (See Appendix A.) Such data could be electronically stored and retrieved. These data should be reviewed and updated regularly. Area maps with information about transportation and evacuation routes, hospital and school locations, and other emergency-related information, can also be stored in computer disks.

State and local planners with personal computer communications capability can access the federally operated National Hazardous Materials Information Exchange (NHMIE) by dialing (312) 972-3275. Users can obtain up-to-date information on hazmat training courses, planning techniques, events and conferences, and emergency response experiences and lessons learned. NHMIE can also be reached through a toll-free telephone call (1-800-752-6367; in Illinois, 1-800-367-9592).

2.4 Beginning to Plan

When the planning team members and their leader have been identified and a process for managing the planning tasks is in place, the team should address several interrelated tasks. These planning tasks are described in the next chapter.

3

Tasks of the Planning Team

3.1 Introduction

The major tasks of the planning team in completing hazardous materials planning are:

• Review of existing plans, which prevents plan overlap and inconsistency, provides useful information and ideas, and facilitates the coordination of the plan with other plans;

• Hazards analysis, which includes hazards identification, vulnerability analysis, and risk analysis;

• Assessment of preparedness, prevention, and response capabilities, which identifies existing prevention measures and response capabilities (including mutual aid agreements) and assesses their adequacy;

• Completion of hazardous materials planning that describes the personnel, equipment, and procedures to be used in case of accidental release of a hazardous material; and

• Development of an ongoing program for plan implementation/maintenance, training, and exercising.

This chapter discusses the planning tasks that are conducted prior to the preparation of the emergency plan. Chapters 4 and 5 provide guidance on plan format and content. Chapter 6 discusses the team's responsibilities for conducting internal and external reviews, exercises, incident reviews, and training. This chapter begins with a discussion of the organizational responsibilities of the planning team.

3.2 Review of Existing Plans

Before undertaking any other work, steps should be taken to search out and review all existing emergency plans. The main reasons for reviewing these plans are (1) to minimize work efforts by building upon or modifying existing emergency planning and response information and (2) to ensure proper coordination with other related plans. To the extent possible, currently used plans should be amended to account for the special problems posed by hazardous materials, thereby avoiding redundant emergency plans. Even plans that are no longer used may provide a useful starting point. More general plans can also be a source of information and ideas. In seeking to identify existing plans, it will be helpful to consult organizations such as:

- State and local emergency management agencies;

- Fire departments;

- Police departments;

- State and local environmental agencies;

- State and local transportation agencies;

- State and local public health agencies;

- Public service agencies;

- Volunteer groups, such as the Red Cross;

- Local industry and industrial associations; and

- Regional offices of federal agencies such as EPA and FEMA.

When reviewing the existing plans of local industry and industrial associations, the planning team should obtain a copy of the CAER program handbook produced by CMA. (See Section 1.5.4.) The handbook provides useful information and encourages industry-community cooperation in emergency planning.

In addition to the above organizations, planning teams should coordinate with the RRTs and OSCs described in Section 1.4.1. Communities can contact or obtain information on the RRT and OSC covering their area through the EPA regional office or USCG district office. (See Appendix F for a list of these contacts.)

3.3 Hazards Analysis: Hazards Identification, Vulnerability Analysis, Risk Analysis

A hazards analysis is a critical component of planning for hazardous materials releases. The information developed in a hazards analysis provides both the factual basis to set priorities for planning and also the necessary documentation for supporting hazardous materials planning and response efforts.

There are several concepts involved in analyzing the dangers posed by hazardous materials. Three terms — hazard, vulnerability, risk — have different technical meanings but are sometimes used interchangeably. This guidance adopts the following definitions:

- *Hazard*. Any situation that has the potential for causing injury to life or damage to property and the environment.

- *Vulnerability*. The susceptibility of life, property, and the environment to injury or damage if a hazard manifests its potential.

- *Risk*. The probability that injury to life or damage to property and the environment will occur.

A hazards analysis may include vulnerability analysis and risk analysis, or it may simply identify the nature and location of hazards in the community. Developing a complete hazards analysis that examines all hazards, vulnerabilities, and risks may be neither possible nor desirable. This may be particularly true for smaller communities that have less expertise and fewer resources to contribute to the task. The planning team must determine the level of thoroughness that is appropriate. In any case, planners should ask local facilities whether they have already completed a facility hazards analysis. Title III requires facility owners or operators to provide to local emergency planning communities information needed for the planning process.

As important as knowing how to perform a hazards analysis is deciding how detailed an analysis to conduct. While a complete analysis of all hazards would be informative, it may not be feasible or practical given resource and time constraints. The value of a limited hazards analysis should not be underestimated. Often the examination of only major hazards is necessary, and these may be studied without undertaking an elaborate risk analysis. Thus, deciding what is really needed and what can be afforded is an important early step in the hazards analysis process. In fact, the screening of hazards and setting analysis priorities is an essential task of the planning team.

The costs of hazards analysis can and often should be reduced by focusing on the hazards posed by only the most common and/or most hazardous substances. A small number of types of hazardous materials account for the vast majority of incidents and risk. The experience from DOT's *Lessons Learned* is that the most prevalent dangers from hazardous materials are posed by common substances, such as gasoline, other flammable materials, and a few additional chemicals. The CEPP technical guidance presents a method that may be used to assist in ranking hazards posed by less prevalent but extremely hazardous substances, such as liquid chlorine, anhydrous ammonia, and hydrochloric and sulfuric acids.

A hazards analysis can be greatly simplified by using qualitative methods (i.e., analysis that is based on judgment rather than measurement of quantities involved). Smaller communities may find that their fire and police chiefs can provide highly accurate assessments of the community's hazardous materials problems. Other, larger communities may have the expertise and resources to utilize quantitative techniques but may decide to substitute qualitative methods in their place should it be cost effective to do so.

Simple or sophisticated, the hazards analysis serves to characterize the nature of the problem posed by hazardous materials. The information that is developed in the hazards analysis should then be used by the planning team to orient planning appropriate to the community's situation. Do not commit valuable resources to plan development until a hazards analysis is performed.

3.3.1 Developing the Hazards Analysis

The procedures that are presented in this section are intended to provide a simplified approach to hazards analysis for both facility and transportation hazards. Communities undertaking a hazards analysis should refer to CEPP technical guidance for fixed facilities and to *Lessons Learned* and *Community Teamwork* for transportation.

The components of a hazards analysis include the concepts of hazard, vulnerability, and risk. The discussion that follows summarizes the basic procedures for conducting each component.

A. *Hazards Identification*

The hazards identification provides information on the facility and transportation situations that have the potential for causing injury to life or damage to property and the environment due to a hazardous materials spill or release. The hazards identification should indicate:

• The types and quantities of hazardous materials located in or transported through a community;

• The location of hazardous materials facilities and routes; and

• The nature of the hazard (e.g., fire, explosions) most likely to accompany hazardous materials spills or releases.

To develop this information, consider hazardous materials at fixed sites and those that are transported by highway, rail, water, air, and pipeline. Examine hazardous materials at:

• Chemical plants;

• Refineries;

• Industrial facilities;

• Petroleum and natural gas tank farms;

• Storage facilities/warehouses;

• Trucking terminals;

• Railroad yards;

• Hospital, educational, and governmental facilities;

• Waste disposal and treatment facilities;

- Waterfront facilities, particularly commercial marine terminals;

- Vessels in port;

- Airports;

- Nuclear facilities; and

- Major transportation corridors and transfer points.

For individual facilities, consider hazardous materials:

- Production;

- Storage;

- Processing;

- Transportation; and

- Disposal.

Some situations will be obvious. To identify the less obvious ones, interview fire and police chiefs, industry leaders, and reporters; review news releases and fire and police department records of past incidents. Also, consult lists of hazardous chemicals that have been identified as a result of compliance with right-to-know laws. (Title III of SARA requires facility owners and operators to submit to the local emergency planning committee a material safety data sheet for specified chemicals, and emergency and hazardous chemical inventory forms. Section 303 (d)(3) of Title III states that "upon request from the emergency planning committee, the owner or operator of the facility shall promptly provide information ... necessary for developing and implementing the emergency plan.") Use the CEPP technical guidance for help in evaluating the hazards associated with airborne releases of extremely hazardous substances.

The hazards identification should result in compilation of those situations that pose the most serious threat of damage to the community. Location maps and charts are an excellent means of depicting this information.

B. *Vulnerability Analysis*

The vulnerability analysis identifies what in the community is susceptible to damage should a hazardous materials release occur. The vulnerability analysis should provide information on:

- The extent of the vulnerable zone (i.e., the significantly affected area) for a spill or release and the conditions that influence the zone of impact (e.g., size of release, wind direction);

• The population, in terms of size and types (e.g., residents, employees, sensitive populations—hospitals, schools, nursing homes, day care centers), that could be expected to be within the vulnerable zone;

• The private and public property (e.g., homes, businesses, offices) that may be damaged, including essential support systems (e.g., water, food, power, medical) and transportation corridors; and

• The environment that may be affected, and the impact on sensitive natural areas and endangered species.

Refer to the CEPP technical guidance or DOT's *Emergency Response Guidebook* to obtain information on the vulnerable zone for a hazardous materials release. For information on the population, property, and environmental resources within the vulnerable zone, consider conducting:

• A windshield survey of the area (i.e., first hand observation by driving through an area);

• Interviews of fire, police, and planning department personnel; and

• A review of planning department documents, and statistics on land use, population, highway usage, and the area's infrastructure.

The vulnerability analysis should summarize information on all hazards determined to be major in the hazards identification.

C. *Risk Analysis*

The risk analysis assesses the probability of damage (or injury) taking place in the community due to a hazardous materials release and the actual damage (or injury) that might occur, in light of the vulnerability analysis. Some planners may choose to analyze worst-case scenarios. The risk analysis may provide information on:

• The probability that a release will occur and any unusual environmental conditions, such as areas in flood plains, or the possibility of simultaneous emergency incidents (e.g., flooding or fire hazards resulting in release of hazardous materials);

• The type of harm to people (acute, delayed, chronic) and the associated high-risk groups;

• The type of damage to property (temporary, repairable, permanent); and

• The type of damage to the environment (recoverable, permanent).

Use the Chemical Profiles in the CEPP technical guidance or a similar guide to obtain information on the type of risk associated with the accidental airborne release of extremely hazardous substances.

Developing occurrence probability data may not be feasible for all communities. Such analysis can require specialized expertise not available to a community. This is especially true of facility releases that call for detailed analysis by competent safety engineers and others (e.g., industrial hygienists) of the operations and associated risk factors of the plant and engineering system in question (refer to the American Institute of Chemical Engineers' *Guidelines for Hazard Evaluation Procedures*). Transportation release analysis is more straightforward, given the substantial research and established techniques that have been developed in this area (refer to *Community Teamwork* and *Lessons Learned*).

Communities should not be overly concerned with developing elaborate quantitative release probabilities. Instead, occurrence probabilities can be described in relative terms (e.g., low, moderate, high). The emphasis should be on developing reasonable estimates based on the best available expertise.

3.3.2   Obtaining Facility Information

The information that is needed about a facility for hazards analysis may already be assembled as a result of previous efforts. As indicated in Section 1.4.1, industry is required by Title III of SARA to provide inventory and release information to the appropriate emergency planning committee. Local emergency planning committees are specifically entitled to any information from facility owners and operators deemed necessary for developing and implementing the emergency plan. The EPA administrator can order facilities to comply with a local committee's requests for necessary information; local planning committees can bring a civil suit against a facility that refuses to provide requested information. Some state and local governments have adopted community right-to-know legislation. These community right-to-know provisions vary, but they generally require industry and other handlers of hazardous materials to provide information to state or local authorities and/or the public about hazardous materials in the community. Wisconsin, for example, requires all hazardous materials spills to be reported to a state agency. Such requirements provide a data base that the planning team can use to determine the types of releases that have occurred in and around the community.

Requesting information from a facility for a hazards analysis can be an opening for continuing dialogue within the community. The information should be sought in such a way that facilities are encouraged to cooperate and participate actively in the planning process along with governmental agencies and other community groups. Respecting a commercial facility's needs to protect confidential business information (such as sensitive process information) will encourage a facility to be forthcoming with the information necessary for the community's emergency planning. The planning team can learn what the facility is doing and what measures have been put in place to reduce risks, and also identify what additional resources such as personnel, training, and equipment are needed in the community. Because facilities use different kinds of hazard assessments (e.g., HAZOP, Fault-tree analysis), local planners need to indicate specifically what categories of information they are interested in receiving. These categories may include:

- Identification of chemicals of concern;

- Identification of serious events that can lead to releases (e.g., venting or system leaks, runaway chemical reaction);

- Amounts of toxic material or energy (e.g., blast, fire radiation) that could be released;

- Predicted consequences of the release (e.g., population exposure illustrated with plume maps and damage rings) and associated damages (e.g., deaths, injuries);

- Whether the possible consequences are considered acceptable by the facility; and

- Prevention measures in place on site.

The facilities themselves are a useful resource; the community should work with the facility personnel and utilize their expertise. The assistance that a facility can provide includes:

- Technical experts;

- Facility emergency plans;

- Cleanup and recycling capabilities;

- Spill prevention control and countermeasures (SPCC);

- Training and safe handling instructions; and

- Participation in developing the emergency plan, particularly in defining how to handle spills on company property.

Cooperative programs such as CMA's CAER program are also a source for hazard information. One of the major objectives of the CAER program is to improve local emergency plans by combining chemical plant emergency plans with other local planning to achieve an integrated community emergency plan. The planning team should ask the facility if it is participating in the CAER program; this may stimulate non-CMA members to use the CAER approach. If a facility is participating in the CAER program, the emergency plans developed by the facility will serve as a good starting point in information gathering and emergency planning. The CAER program handbook also encourages companies to perform hazards analyses of their operations. Local planners should ask facilities if they have adhered to this recommendation and whether they are willing to share results with the planning team.

3.3.3 Example Hazards Analysis

Exhibit 3 presents an example of a very simple hazards analysis for a hypothetical community. Hazards A, B, and C are identified as three among other major hazards in the community. Information for the exhibit could have been obtained from windshield surveys of the area; the CEPP technical guidance; information gained from facilities under Title III provisions; and/or interviews with fire, police, county planners, and facility representatives. These interviews also could have provided input into the exhibit's qualitative assessments of hazard occurrence.

Once completed, the hazards analysis is an essential tool in the planning process. It assists the planning team to decide:

- The level of detail that is necessary;

- The types of response to emphasize; and

- Priority hazards or areas for planning.

The examples presented in Exhibit 3 illustrate the basic fact that there are no hard and fast rules for weighing the relative importance of different types of hazards in the context of the planning process. Compare example hazards B and C in the exhibit. Hazard C involves a substance, methyl isocyanate (MIC), whose lethal and severe chronic effects were evident at Bhopal. As described in the example, an MIC release could affect 200 plant workers and 1000 children in a nearby school. By contrast, the ammonia in example hazard B is less lethal than MIC and threatens fewer people. With just this information in mind, a planner might be expected to assign the MIC a higher planning priority than he would the ammonia. Consider now the "probability of occurrence." In hazard C, plant safety and prevention measures are excellent, and an MIC incident is correspondingly unlikely to occur. On the other hand, poor highway construction and weather conditions that affect visibility make an ammonia incident (example hazard B) far more probable. Planners must balance all factors when deciding whether to give planning priority to B or C. Both situations are dangerous and require emergency planning. Some would argue that the lethality of MIC outweighs the presence of good safety and prevention procedures; others would argue that the frequency of highway interchange accidents is reason enough to place greater emphasis on planning to deal with an ammonia incident. Each planning team must make such judgments on priorities in light of local circumstances.

Before initiating plan development, the planning team should complete an assessment of available response resources, including capabilities provided through mutual aid agreements. Guidance for conducting such an assessment is presented in the following section.

Exhibit 3
Example Hazards Analysis for a Hypothetical Community

	Hazard A	Hazard B	Hazard C
1. HAZARDS IDENTIFICATION (MAJOR HAZARDS)			
a. Chemical	Chlorine	Ammonia	Liquid methyl isocyanate (MIC)
b. Location	Water treatment plant	Tank truck on local interstate highway	Pesticide manufacturing plant in nearby semi-rural area
c. Quantity	2000 lbs	5000 lbs	5000 lbs
d. Properties	Poisonous; may be fatal if inhaled. Respiratory conditions aggravated by exposure. Contact may cause burns to skin and eyes. Corrosive. Effects may be delayed.	Poisonous; may be fatal if inhaled. Vapors cause irritation of eyes and respiratory tract. Liquid will burn skin and eyes. Contact with liquid may cause frostbite. Effects may be delayed. Will burn within certain vapor concentration limits and increase fire hazard in the presence of oil or other combustible materials.	Causes death by respiratory distress after inhalation. Other health effects would include permanent eye damage, respiratory distress, and disorientation. Explosive. Extremely flammable.
2. VULNERABILITY ANALYSIS			
a. Vulnerable zone	A spill of 2000 lbs of chlorine from a storage tank could result in an area of radius 1650 feet (0.3 miles) where chlorine gas may exceed the level of concern.	A spill of 5000 lbs of ammonia resulting from a collision of a tank truck could result in an area of radius 1320 feet (0.25 miles) where ammonia exceeds its level of concern.	A spill of 5000 lbs of methyl isocyanate could affect an area of radius 3300 feet (0.6 miles) with MIC vapors exceeding the level of concern (assuming that the liquid is hot when spilled, the tank is not diked, and the MIC is at 100% concentration).
b. Population within vulnerable zone	Approximately 500 residents of a nursing home; workers at small factory.	Up to 700 persons in residences, commercial establishments, or vehicles near highway interchange. Seasonal influx of visitors to forest preserve in the fall.	Up to 200 workers at the plant and 1000 children in a school.

Exhibit 3 (Continued)
Example Hazards Analysis for a Hypothetical Community

	Hazard A	Hazard B	Hazard C
c. Private and public property that may be damaged	Facility equipment, vehicles, and structures susceptible to damage from corrosive fumes. Community's water supply may be temporarily affected given that the facility is its primary supplier. Mixture with fuels may cause an explosion.	25 residences, 2 fast food restaurants, one 30 room motel, a truck stop, a gas station and a mini-market. Highway and nearby vehicles may be susceptible to damage from a fire or explosion resulting from the collision.	Runoff to a sewer may cause an explosion hazard as MIC reacts violently with water.
d. Environment that may be affected	Terrestrial life.	Adjacent forest preserve is highly susceptible to forest fires especially during drought conditons.	Nearby farm animals.
3. RISK ANALYSIS			
a. Probability of hazard occurrence	Low — because chlorine is stored in an area with leak detection equipment in 24 hour service with alarms. Protective equipment is kept outside storage room.	High — Highway interchange has a history of accidents due to poor visibility of exits and entrances.	Low — facility has up to date containment facilities with leak detection equipment, and an emergency plan for its employees. There are good security arrangements that would deter tampering or accidents resulting from civil uprisings.
b. Consequences if people are exposed	High levels of chlorine gas in the nursing home and factory could cause death and respiratory distress. Bedridden nursing home patients are especially susceptible.	Release of vapors and subsequent fire may cause traffic accidents. Injured and trapped motorists are subject to lethal vapors and possible incineration. Windblown vapors can cause respiratory distress for nearby residents and business patrons.	If accident occurs while school is in session, children could be killed, blinded, and/or suffer chronic debilitating respiratory problems. Plant workers would be subject to similar effects at any time.
c. Consequences for property	Possible superficial damage to facility equipment structures from corrosive fumes (repairable).	Repairable damage to highway. Potential destruction of nearby vehicles due to fire or explosions.	Vapors may explode in a confined space causing property damage (repairable). Damage could result from fires (repairable).

Exhibit 3 (Continued)
Example Hazards Analysis for a Hypothetical Community

	Hazard A	Hazard B	Hazard C
d. Consequences of environmental exposure	Possible destruction of surrounding fauna and flora.	Potential for fire damage to adjacent forest preserve due to combustible material (recoverable in the long term).	Farm animals and other fauna could be killed or suffer health effects necessitating their destruction or indirectly causing death.
e. Probability of simultaneous emergencies	Low	High	Low
f. Unusual environmental conditions	None	Hilly terrain prone to mists, thus creating adverse driving conditions.	Located in a 500 year river flood plain.

3.4 Capability Assessment

This section contains sample questions to help the planning team evaluate preparedness, prevention, and response resources and capabilities. The section is divided into three parts. The first part covers questions that the planning team can ask a technical representative from a facility that may need an emergency plan. The second part includes questions related to transportation.

The third part addresses questions to a variety of response and government agencies, and is designed to help identify all resources within a community. This information will provide direct input into the development of the hazardous materials emergency plan and will assist the planning team in evaluating what additional emergency response resources may be needed by the community.

3.4.1 Facility Resources

What is the status of the safety plan (also referred to as an emergency or contingency plan) for the facility? Is the safety plan consistent with any community emergency plan?

- Is there a list of potentially toxic chemicals available? What are their physical and chemical characteristics, potential for causing adverse health effects, controls, interactions with other chemicals? Has the facility complied with the community right-to-know provisions of Title III of SARA?

- Has a hazards analysis been prepared for the facility? If so, has it been updated? Has a copy been provided to the local emergency planning committee?

- What steps have been taken to reduce identified risks?

- How does the company reward good safety records?

- Have operation or storage procedures been modified to reduce the probability of a release and minimize potential effects?

- What release prevention or mitigation systems, equipment, or procedures are in place?

- What possibilities are there for safer substitutes for any acutely toxic chemicals used or stored at the facility?

- What possibilities exist for reducing the volume of the hazardous materials in use or stored at the facility?

- What additional safeguards are available to prevent accidental releases?

- What studies have been conducted by the facility to determine the feasibility of each of the following approaches for each relevant production process or operation: (a) input change, (b) product reformulation, (c) production process change, and (d) operational improvements?

• Are on-site emergency response equipment (e.g., fire fighting equipment, personal protective equipment, communications equipment) and trained personnel available to provide on-site initial response efforts?

• What equipment (e.g., self-contained breathing apparatus, chemical suits, unmanned fire monitors, foam deployment systems, radios, beepers) is available? Is equipment available for loan or use by the community on a reimbursable basis? (Note: Respirators should not be lent to any person not properly trained in their use.)

• Is there emergency medical care on site?

• Are the local hospitals prepared to accept and provide care to patients who have been exposed to chemicals?

• Who is the emergency contact for the site (person's name, position, and 24-hour telephone number) and what is the chain of command during an emergency?

• Are employee evacuation plans in effect and are the employees trained to use them in the event of an emergency?

• What kinds of notification systems connect the facility and the local community emergency services (e.g., direct alarm, direct telephone hook-up, computer hook-up) to address emergencies on site?

• What is the mechanism to alert employees and the surrounding community in the event of a release at the facility?

• Is there a standard operating procedure for the personal protection of community members at the time of an emergency?

• Does the community know about the meaning of various alarms or warning systems? Are tests conducted?

• How do facility personnel coordinate with the community government and local emergency and medical services during emergencies? Is overlap avoided?

• What mutual aid agreements are in place for obtaining emergency response assistance from other industry members? With whom?

• Are there any contacts or other prearrangements in place with specialists for cleanup and removal of releases, or is this handled in-house? How much time is required for the cleanup specialists to respond?

- What will determine concentrations of released chemicals existing at the site? (Are there toxic gas detectors, explosimeters, or other detection devices positioned around the facility? Where are they located?)

- Are wind direction indicators positioned within the facility perimeter to determine in what direction a released chemical will travel? Where are they located?

- Is there capability for modeling vapor cloud dispersion?

- Are auxiliary power systems available to perform emergency system functions in case of power outages at the facility?

- How often is the safety plan tested and updated? When was it last tested and updated?

- Does the company participate in CHEMNET or the CAER program?

- Does the company have the capability and plans for responding to off-site emergencies? Is this limited to the company's products?

What is the safety training plan for management and employees?

- Are employees trained in the use of emergency response equipment, personal protective equipment, and emergency procedures detailed in the plant safety plan? How often is training updated?

- Are simulated emergencies conducted for training purposes? How often? How are these simulations evaluated and by whom? When was this last done? Are the local community emergency response and medical service organizations invited to participate?

- Are employees given training in methods for coordinating with local community emergency response and medical services during emergencies? How often?

- Is management given appropriate training? How frequently?

- Is there an emergency response equipment and systems inspection plan?

- Is there a method for identifying emergency response equipment problems? Describe it.

- Is there testing of on-site alarms, warning signals, and emergency response equipment? How often is this equipment tested and replaced?

3.4.2 Transporter Resources

What cargo information and response organization do ship, train, and truck operators provide at a release?

• Do transport shipping papers identify hazardous materials, their physical and chemical characteristics, control techniques, and interactions with other chemicals?

• Do transports have proper placards?

• Are there standard operating procedures (SOPs) established for release situations? Have these procedures been updated to reflect current cargo characteristics?

• Who is the emergency contact for transport operators? Is there a 24-hour emergency contact system in place? What is the transport operation's chain of command in responding to a release?

What equipment and cleanup capabilities can transport operations make available?

• What emergency response equipment is carried by each transporter (e.g., protective clothing, breathing apparatus, chemical extinguishers)?

• Do transports have first-aid equipment (e.g., dressings for chemical burns and water to rinse off toxic chemicals)?

• By what means do operators communicate with emergency response authorities?

• Do transport operations have their own emergency response units?

• What arrangements have been established with cleanup specialists for removal of a release?

What is the safety training plan for operators?

• Are operators trained in release SOPs and to use emergency response equipment? How often is training updated?

• How often are release drills conducted? Who evaluates these drills and do the evaluations become a part of an employee's file?

• Are safe driving practices addressed in operator training? What monetary or promotional incentives encourage safety in transport operation?

Is there a transport and emergency response equipment inspection plan?

• What inspections are conducted? What leak detection and equipment readiness tests are done? What is the schedule for inspections and tests?

• Are problems identified in inspections corrected? How are maintenance schedules established?

3.4.3 Community Resources

What local agencies make up the community's existing response preparedness network? Some examples include:

• Fire department;

• Police/sheriff/highway patrol;

• Emergency medical/paramedic service associated with local hospitals or fire and police departments;

• Emergency management or civil defense agency;

• Public health agency;

• Environmental agency;

• Public works and/or transportation departments;

• Red Cross; and

• Other local community resources such as public housing, schools, public utilities, communications.

What is the capacity and level of expertise of the community's emergency medical facilities, equipment, and personnel?

Does the community have arrangements or mutual aid agreements for assistance with other jurisdictions or organizations (e.g., other communities, counties, or states; industry; military installations; federal facilities; response organizations)? In the absence of mutual aid agreements, has the community taken liability into consideration?

What is the current status of community planning and coordination for hazardous materials emergency preparedness? Have potential overlaps in planning been avoided?

• Is there a community planning and coordination body (e.g., task force, advisory board, interagency committee)? If so, what is the defined structure and authority of the body?

• Has the community performed any assessments of existing prevention and response capabilities within its own emergency response network?

• Does the community maintain an up-to-date technical reference library of response procedures for hazardous materials?

• Have there been any training sessions, simulations, or mock incidents performed by the community in conjunction with local industry or other organizations? If so, how frequently are they conducted? When was this last done? Do they typically have simulated casualties?

Who are the specific community points of contact and what are their responsibilities in an emergency?

• List the agencies involved, the area of responsibility (e.g., emergency response, evacuation, emergency shelter, medical/health care, food distribution, control access to accident site, public/media liaison, liaison with federal and state responders, locating and manning the command center and/or emergency operating center), the name of the contact, position, 24-hour telephone number, and the chain of command.

• Is there any specific chemical or toxicological expertise available in the community, either in industry, colleges and universities, poison control centers, or on a consultant basis?

What kinds of equipment and materials are available at the local level to respond to emergencies? How can the equipment, materials, and personnel be made available to trained users at the scene of an incident?

Does the community have specialized emergency response teams to respond to hazardous materials releases?

• Have the local emergency services (fire, police, medical) had any hazardous materials training, and if so, do they have and use any specialized equipment?

• Are local hospitals able to decontaminate and treat numerous exposure victims quickly and effectively?

• Are there specialized industry response teams (e.g., CHLOREP, AAR/BOE), state/federal response teams, or contractor response teams available within or close to the community? What is the average time for them to arrive on the scene?

• Has the community sought any resources from industry to help respond to emergencies?

Is the community emergency transportation network defined?

• Does the community have specific evacuation routes designated? What are these evacuation routes? Is the general public aware of these routes?

• Are there specific access routes designated for emergency response and services personnel to reach facilities or incident sites? (In a real incident, wind direction might make certain routes unsafe.)

Does the community have other procedures for protecting citizens during emergencies (e.g., asking them to remain indoors, close windows, turn off air-conditioners, tune into local emergency radio broadcasts)?

Is there a mechanism that enables responders to exchange information or ideas during an emergency with other entities, either internal or external to the existing organizational structure?

Does the community have a communications link with an Emergency Broadcast System (EBS) station? Is there a designated emergency communications network in the community to alert the public, update the public, and provide communications between the command center and/or emergency operating center, the incident site, and off-scene support? Is there a back-up system?

• What does the communications network involve (e.g., special radio frequency, network channel, siren, dedicated phone lines, computer hook-up)?

• Is there an up-to-date list, with telephone numbers, of radio and television stations (including cable companies) that broadcast in the area?

• Is there an up-to-date source list with a contact, position, and telephone number for technical information assistance? This can be federal (e.g., NRC, USCG CHRIS/HACS, ATSDR, OHMTADS), state, industry associations (e.g., CHEMTREC, CHLOREP, AAR/BOE, PSTN), and local industry groups (e.g., local AIChE, ASME, ASSE chapters).

Is there a source list with a contact, position, and telephone number for community resources available?

• Does the list of resources include: wreck clearing, transport, cleanup, disposal, health, analytical sampling laboratories, and detoxifying agents?

Have there been any fixed facility or transportation incidents involving hazardous materials in the community? What response efforts were taken? What were the results? Have these results been evaluated?

3.5 Writing an Emergency Plan

When the team has reviewed existing plans, completed a hazards identification and analysis, and assessed its preparedness, prevention, and response capabilities, it can take steps to make serious incidents less likely. Improved warning systems, increased hazardous materials training of industry and local response personnel, and other efforts at the local level, can all make a community better prepared to live safely with hazardous materials. The team should also begin to write an emergency plan if one does not already exist, or revise existing plans to include hazardous materials. Chapter 4 describes two approaches to developing or revising an emergency plan. Chapter 5 describes elements related to hazardous materials incidents that should be included in whichever type of plan the community chooses to write.

4 Developing the Plan

4.1 Introduction

Most communities have some type of written plan for emergencies. These plans range from a comprehensive multi-hazard approach as described in FEMA's CPG 1-8 (Guide for Development of State and Local Emergency Operations Plans) to a single telephone roster for call-up purposes, or an action checklist. Obviously the more complete and thorough a plan is, the better prepared the community should be to deal with any emergency that occurs.

As noted in Chapter 1, the "Emergency Planning and Community Right-to-Know Act of 1986" requires local emergency planning committees to develop local plans for emergency responses in the event of a release of an extremely hazardous substance. Those communities receiving FEMA funds are required to incorporate hazardous materials planning into their multi-hazard emergency operations plan (EOP). Other communities are encouraged to prepare a multi-hazard EOP in accord with CPG 1-8 since it is the most comprehensive approach to emergency planning. Not every community, however, may be ready for or capable of such a comprehensive approach. Because each community must plan in light of its own situation and resources, a less exhaustive approach may be the only practical, realistic way of having some type of near-term plan. Each community must choose the level of planning that is appropriate for it, based upon the types of hazard found in the community.

This chapter discusses two basic approaches to writing a plan: (1) development or revision of a hazardous materials appendix (or appendices to functional annexes) to a multi-hazard EOP following the approach described in FEMA's CPG 1-8, and (2) development or revision of a plan covering only hazardous materials. Each approach is discussed in more detail below.

4.2 Hazardous Materials Appendix to Multi-Hazard EOP

The first responders (e.g., police, fire, emergency medical team) at the scene of an incident are generally the same whatever the hazard. Moreover, many emergency functions (e.g., direction and control, communications, and evacuation) vary only slightly from hazard to hazard. Procedures to be followed for warning the public of a hazardous materials incident, for example, are not that different from procedures followed in warning the public about other incidents such as a flash flood. It is possible, therefore, to avoid a great deal of unnecessary redundancy and confusion by planning for all hazards at the same time. A multi-hazard EOP avoids developing separate structures, resources, and plans to deal with each type of hazard. Addressing the general aspects of all hazards first and then looking at each potential hazard individually to see if any unique aspects are involved result in efficiencies and economies in the long run. Multi-hazard EOPs also help ensure that plans and systems are reasonably compatible if a large-scale hazardous materials incident requires a simultaneous, coordinated response by more than one community or more than one level of government.

A community that does not have a multi-hazard plan is urged to consider seriously the advantages of this integrated approach to planning. In doing so, the community may want to seek state government advice and support.

CPG 1-8 describes a sample format, content, and process for state and local EOPs. It recommends that a multi-hazard EOP include three components—a basic plan, functional annexes, and hazard-specific appendices. It encourages development of a basic plan that includes generic functional annexes applicable to any emergency situation, with unique aspects of a particular hazard being addressed in hazard-specific appendices. It stresses improving the capabilities for simultaneous, coordinated response by a number of emergency organizations at various levels of government. Local communities that receive FEMA funds must incorporate hazardous materials planning into their multi-hazard EOP. In most of these communities, there are paid staff to do emergency operations planning as well as related emergency management tasks.

CPG 1-8 provides flexible guidance, recognizing that substantial variation in planning may exist from community to community. A community may develop a separate hazardous material appendix to each functional annex where there is a need to reflect considerations unique to hazardous materials not adequately covered in the functional annex. On the other hand, a community may develop a single hazardous materials appendix to the EOP, incorporating all functional annex considerations related to hazardous materials in one document. The sample plan format used in CPG 1-8 is a good one, but it is not the only satisfactory one. It is likely that no one format is the best for all communities of all sizes in all parts of the country. Planners should, therefore, use good judgment and common sense in applying CPG 1-8 principles to meet their needs. The community has latitude in formatting the plan but should closely follow the basic content described in CPG 1-8.

CPG 1-8 should be used in preparing the basic plan and functional annexes. This guide should be used as a supplement to CPG 1-8 to incorporate hazardous materials considerations into a multi-hazard EOP. Communities that want to develop standard operating procedures (SOP) manuals could begin with information included in the functional annexes of a multi-hazard EOP.

A community that is incorporating hazardous materials into a multi-hazard EOP should turn to Chapter 5 of this guide for a discussion of those elements that need to be taken into account in hazardous materials planning.

4.3 Single-Hazard Emergency Plan

If a community does not have the resources, time, or capability readily available to undertake multi-hazard planning, it may wish to produce a single-hazard plan addressing hazardous materials.

Exhibit 4 identifies sections of an emergency plan for hazardous materials incidents. The sample outline is not a model. It is not meant to constrain any community. Indeed, each community should seek to develop a plan that is best suited to its own circumstances, taking advantage of the sample outline where appropriate.

The type of plan envisioned in the sample outline would affect all governmental and private organizations involved in emergency response operations in a particular community. Its basic purpose would be to provide the necessary data and documentation to anticipate and coordinate the many persons and organizations that would be involved in emergency response actions. As such, the plan envisioned in this sample outline is intended neither to be a "hip-pocket" emergency response manual, nor to serve as a detailed standard operating procedures (SOP) manual for each of the many agencies and organizations involved in emergency response actions, although it could certainly be used as a starting point for such manuals. Agencies that want to develop an SOP manual could begin with the information contained under the appropriate function in Plan Section C of this sample outline. If it is highly probable that an organization will be involved in a hazardous materials incident response, then a more highly detailed SOP should be developed.

Exhibit 4
Sample Outline of a Hazardous Materials Emergency Plan.

(NOTE: Depending upon local circumstances, communities will develop some sections of the plan more extensively than other sections. See 5.1 for how the sample outline relates to SARA Title III requirements.)

A. Introduction

1. Incident Information Summary

2. Promulgation Document

3. Legal Authority and Responsibility for Responding

4. Table of Contents

5. Abbreviations and Definitions

6. Assumptions/Planning Factors

7. Concept of Operations

    a. Governing Principles

    b. Organizational Roles and Responsibilities

    c. Relationship to Other Plans

8. Instructions on Plan Use

    a. Purpose

    b. Plan Distribution

9. Record of Amendments

B. Emergency Assistance Telephone Roster

C. Response Functions*

1. Initial Notification of Response Agencies

2. Direction and Control

3. Communications (among Responders)

4. Warning Systems and Emergency Public Notification

---

*These "Response Functions" are equivalent to the "functional annexes" of a multi-hazard emergency operations plan described in CPG 1-8.

5. Public Information/Community Relations

6. Resource Management

7. Health and Medical Services

8. Response Personnel Safety

9. Personal Protection of Citizens

   a. Indoor Protection

   b. Evacuation Procedures

   c. Other Public Protection Strategies

10. Fire and Rescue

11. Law Enforcement

12. Ongoing Incident Assessment

13. Human Services

14. Public Works

15. Others

D. Containment and Cleanup

1. Techniques for Spill Containment and Cleanup

2. Resources for Cleanup and Disposal

E. Documentation and Investigative Follow-up

F. Procedures for Testing and Updating Plan

1. Testing the Plan

2. Updating the Plan

G. Hazards Analysis (Summary)

H. References

1. Laboratory, Consultant, and Other Technical Support Resources

2. Technical Library

5 Hazardous Materials Planning Elements

5.1 Introduction

This chapter presents and discusses a comprehensive list of planning elements related to hazardous materials incidents. Communities that are developing a hazardous materials appendix/plan need to review these elements thoroughly. Communities that are revising an existing appendix/plan need to evaluate their present appendix/plan and identify what elements need to be added, deleted, or amended in order to deal with the special problems associated with the accidental spill or release of hazardous materials.

Title III of SARA requires each emergency plan to include at least each of the following. The appropriate section of the plan as indicated in Exhibit 4 is shown in parentheses after each required Title III plan element.

(1)   Identification of facilities subject to the Title III requirements that are within the emergency planning district; identification of routes likely to be used for the transportation of substances on the list of extremely hazardous substances; and identification of additional facilities contributing or subjected to additional risk due to their proximity to facilities, such as hospitals or natural gas facilities. (Exhibit 4, Sections A.6 and G)

(2)   Methods and procedures to be followed by facility owners and operators and local emergency and medical personnel to respond to any releases of such substances. (Exhibit 4, Section C)

(3)   Designation of a community emergency coordinator and facility emergency coordinators, who shall make determinations necessary to implement the plan. (Exhibit 4, Section A.7b)

(4)   Procedures providing reliable, effective, and timely notification by the facility emergency coordinators and the community emergency coordinator to persons designated in the emergency plan, and to the public, that a release has occurred. (Exhibit 4, Sections C.1 and C.4)

(5)   Methods for determining the occurrence of a release, and the area or population likely to be affected by such release. (Exhibit 4, Sections A.6 and G)

(6)   A description of emergency equipment and facilities in the community and at each facility in the community subject to Title III requirements, and an identification of the persons responsible for such equipment and facilities. (Exhibit 4, Section C.6)

(7)   Evacuation plans, including provisions for a precautionary evacuation and alternative traffic routes. (Exhibit 4, Section C.9b)

(8)   Training programs, including schedules for training of local emergency response and medical personnel. (Exhibit 4, Sections C.6 and F.1)

(9)   Methods and schedules for exercising the emergency plan. (Exhibit 4, Section F.1)

The various planning elements are discussed here in the same order as they appear in the sample outline for a hazardous materials emergency plan in Chapter 4. Community planners might choose, however, to order these planning elements differently in a multi-hazard plan following the model of CPG 1-8.

5.2   Discussion of Planning Elements

The remainder of this chapter describes in detail what sorts of information could be included in each element of the emergency plan. These issues need to be addressed in the planning process. In some cases, they will be adequately covered in SOPs and will not need to be included in the emergency plan.

Planning Element A: Introduction

Planning Element A.1: Incident Information Summary

- Develop a format for recording essential information about the incident:

  - Date and time

  - Name of person receiving call

  - Name and telephone number of on-scene contact

  - Location

  - Nearby populations

  - Nature (e.g., leak, explosion, spill, fire, derailment)

  - Time of release

  - Possible health effects/medical emergency information

  - Number of dead or injured; where dead/injured are taken

- Name of material(s) released; if known

  - Manifest/shipping invoice/billing label

  - Shipper/manufacturer identification

  - Container type (e.g., truck, rail car, pipeline, drum)

  - Railcar/truck 4-digit identification numbers

  - Placard/label information

- Characteristics of material (e.g., color, smell, physical effects), only if readily detectable

- Present physical state of the material (i.e., gas, liquid, solid)

- Total amount of material that may be released

- Other hazardous materials in area

- Amount of material released so far/duration of release

- Whether significant amounts of the material appear to be entering the atmosphere, nearby water, storm drains, or soil

- Direction, height, color, odor of any vapor clouds or plumes

- Weather conditions (wind direction and speed)

- Local terrain conditions

- Personnel at the scene

Comment:

Initial information is critical. Answers to some of these questions may be unknown by the caller, but it is important to gather as much information as possible very quickly in order to facilitate decisions on public notification and evacuation. Some questions will apply to fixed facility incidents and others will apply only to transportation incidents. Some questions will apply specifically to air releases, while other questions will gather information about spills onto the ground or into water. Identification numbers, shipping manifests, and placard information are essential to identify any hazardous materials involved in transportation incidents, and to take initial precautionary and containment steps. First responders should use DOT's Emergency Response Guidebook to help identify hazardous materials. Additional information about the identity and characteristics of chemicals is available by calling CHEMTREC (800-424-9300). CHEMTREC and the Hazard Information Transmission (HIT) program are described in Appendix C.

This emergency response notification section should be:

BRIEF — never more than one page in length.

EASILY ACCESSIBLE — located on the cover or first page of the plan. It should also be repeated at least once inside the plan, in case the cover is torn off.

SIMPLE — reporting information and emergency telephone numbers should be kept to a minimum.

Copies of the emergency response notification form could be provided to potential dischargers to familiarize them with information needed at the time of an incident.

Planning Element A.2:   Promulgation Document

• Statement of plan authority

Comment:

A letter, signed by the community's chief executive, should indicate legal authority and responsibility for putting the plan into action. To the extent that the execution of this plan involves various private and public-sector organizations, it may be appropriate to include here letters of agreement signed by officials of these organizations.

Planning Element A.3:   Legal Authority and Responsibility for Responding

• Authorizing legislation and regulations

• Federal (e.g., CERCLA, SARA, Clean Water Act, National Contingency Plan, and Disaster Relief Act)

• State

• Regional

• Local

• Mandated agency responsibilities

• Letters of agreement

Comment:

If there are applicable laws regarding planning for response to hazardous materials releases, list them here. Analyze the basic authority of participating agencies and summarize the results here. The community may choose to enact legislation in support of its plan. Be sure to identify any agencies required to respond to particular emergencies.

Planning Element A.4:   Table of Contents

Comment:

All sections of the plan should be listed here and clearly labeled with a tab for easy access.

Planning Element A.5:   Abbreviations and Definitions

Comment:

Frequently used abbreviations, acronyms, and definitions should be gathered here for easy reference.

Planning Element A.6:   Assumptions/Planning Factors

- Geography
  - Sensitive environmental areas
  - Land use (actual and potential, in accordance with local development codes)
  - Water supplies
  - Public transportation network (roads, trains, buses)
  - Population density
  - Particularly sensitive institutions (e.g., schools, hospitals, homes for the aged)
- Climate/weather statistics
- Time variables (e.g., rush hour, vacation season)
- Particular characteristics of each facility and the transportation routes for which the plan is intended
  - On-site details
  - Neighboring population
  - Surrounding terrain

- Known impediments (tunnels, bridges)

- Other areas at risk

- Assumptions

Comment:

This section is a summary of precisely what local conditions make an emergency plan necessary. Information for this section will be derived from the hazards identification and analysis. Appropriate maps should be included in this section. Maps should show: water intake, environmentally sensitive areas, major chemical manufacturing or storage facilities, population centers, and the location of response resources.

Assumptions are the advance judgments concerning what would happen in the case of an accidental spill or release. For example, planners might assume that a certain percentage of local residents on their own will evacuate the area along routes other than specified evacuation routes.

Planning Element A.7:   Concept of Operations

Planning Element A.7a: Governing Principles

Comment:

The plan should include brief statements of precisely what is expected to be accomplished if an incident should occur.

Planning Element A.7b: Organizational Roles and Responsibilities

- Municipal government

  - Chief elected official

  - Emergency management director

  - Community emergency coordinator (Title III of SARA)

  - Communications personnel

  - Fire service

  - Law enforcement

  - Public health agency

  - Environmental agency

  - Public works

- County government

- Officials of fixed facilities and/or transportation companies

  - Facility emergency coordinators (Title III of SARA)

- Nearby municipal and county governments

- Indian tribes within or nearby the affected jurisdiction

- State government

  - Environmental protection agency

  - Emergency management agency

  - Public health agency

  - Transportation organization

  - Public safety organization

- Federal government

  - EPA

  - FEMA

  - DOT

  - HHS/ATSDR

  - USCG

  - DOL/OSHA

  - DOD

  - DOE

  - RRT

- Predetermined arrangements

- How to use outside resources

  - Response capabilities

  - Procedures for using outside resources

Comment:

This section lists all those organizations and officials who are responsible for planning and/or executing the pre-response (planning and prevention), response (implementing the plan during an incident), and post-response (cleanup and restoration) activities to a hazardous materials incident. One organization should be given command and control responsibility for each of these three phases of the emergency response. The role of each organization/official should be clearly described. The plan should clearly designate who is in charge and should anticipate the potential involvement of state and federal agencies and other response organizations. (Note: The above list of organizations and officials is not meant to be complete. Each community will need to identify all the organizations/officials who are involved in the local planning and response process.)

This section of the plan should contain descriptions and information on the RRTs and the predesignated federal OSC for the area covered by the plan. (See Section 1.4.1 of this guidance.) Because of their distant location, it is often difficult for such organizations to reach a scene quickly; planners should determine in advance approximately how much time would elapse before the federal OSC could arrive at the scene.

This section should also indicate where other disaster assistance can be obtained from federal, state, or regional sources. Prearrangements can be made with higher-level government agencies, bordering political regions, and chemical plants.

Major hazardous materials releases may overwhelm even the best prepared community, and an incident may even cross jurisdictional boundaries. Cooperative arrangements are an efficient means of obtaining the additional personnel, equipment, and materials that are needed in an emergency by reducing expenditures for maintaining extra or duplicative resources. Any coordination with outside agencies should be formalized through mutual aid and Good Samaritan agreements or memoranda of understanding specifying delegations of authority, responsibility, and duties. These formal agreements can be included in the plan if desired.

Planning Element A.7c: Relationship to Other Plans

Comment:

A major task of the planning group is to integrate planning for hazardous materials incidents into already existing plans. In larger communities, it is probable that several emergency plans have been prepared. It is essential to coordinate these plans. When more than one plan is put into action simultaneously, there is a real potential for confusion among response personnel unless the plans are carefully coordinated. All emergency plans (including facility plans and hospital plans) that might be employed in the event of an accidental spill or release should be listed in this section. The community plan should include the methods and procedures to be followed by facility owners and operators and local emergency response personnel to respond to any releases of such substances. The NCP, the federal regional contingency plan, any OSC plan for the area, and any state plan should be referenced. Of special importance are all local emergency plans.

Even where formal plans do not exist, various jurisdictions often have preparedness capabilities. Planners should seek information about informal agreements involving cities, counties, states, and countries.

Planning Element A.8:   Instructions on Plan Use

Planning Element A.8a: Purpose

Comment:

This should be a clear and succinct statement of when and how the plan is meant to be used. It is appropriate to list those facilities and transportation routes explicitly considered in the plan.

Plan Section A.8b: Plan Distribution

- List of organizations/persons receiving plan

Comment:

The entire plan should be available to the public; it can be stored at a library, the local emergency management agency, or some other public place. The plan should be distributed to all persons responsible for response operations. The plan distribution list should account for all organizations receiving such copies of the plan. This information is essential when determining who should be sent revisions and updates to the plan.

Planning Element A.9:   Record of Amendments

- Change record sheet
  - Date of change
  - Recording signature
  - Page numbers of changes made

Comment:

Maintaining an up-to-date version of a plan is of prime importance. When corrections, additions, or changes are made, they should be recorded in a simple bookkeeping style so that all plan users will be aware that they are using a current plan.

All that is necessary for this page is a set of columns indicating date of change, the signature of the person making the change, and the page number for identifying each change made.

Planning Element B: Emergency Assistance Telephone Roster

- List of telephone numbers for:

  - Participating agencies

  - Technical and response personnel

  - CHEMTREC

  - Public and private sector support groups

  - National Response Center

Comment:

An accurate and up-to-date emergency telephone roster is an essential item. The name of a contact person (and alternate) and the telephone number should be listed. Briefly indicate the types of expertise, services, or equipment that each agency or group can provide. Indicate the times of day when the number will be answered; note all 24-hour telephone numbers. All phone numbers and names of personnel should be verified at least every six months. When alternate numbers are available, these should be listed. This section of the plan should stand alone so that copies can be carried by emergency response people and others. Examples of organizations for possible inclusion in a telephone roster are as follows:

Community Assistance
Police
Fire
Emergency Management Agency
Public Health Department
Environmental Protection Agency
Department of Transportation
Public Works
Water Supply
Sanitation
Port Authority
Transit Authority
Rescue Squad
Ambulance
Hospitals
Utilities
  Gas

Phone
Electricity
Community Officials
  Mayor
  City Manager
County Executive
Councils of Government

Volunteer Groups
Red Cross
Salvation Army
Church Groups
Ham Radio Operators
Off-Road Vehicle Clubs

State Assistance
State Emergency Response Commission (Title III of SARA)
State Environmental Protection Agency
Emergency Management Agency
Department of Transportation
Police
Public Health Department
Department of Agriculture

Response Personnel
Incident Commander
Agency Coordinators
Response Team Members

Bordering Political Regions
Municipalities
Counties
States
Countries
River Basin Authorities
Irrigation Districts
Interstate Compacts
Regional Authorities
Bordering International Authorities
Sanitation Authorities/Commissions

Industry
Transporters
Chemical Producers/Consumers
Spill Cooperatives
Spill Response Teams

Media
Television
Newspaper
Radio

Federal Assistance (Consult regional offices listed in Appendix F for appropriate telephone numbers.)

Federal On-Scene Coordinator		
U.S. Department of Transportation		
U.S. Coast Guard		
U.S. Environmental Protection Agency		
Federal Emergency Management Agency	24 hours	202-646-2400
U.S. Department of Agriculture		
Occupational Safety and Health Administration		
Agency for Toxic Substances and Disease Registry	24 hours	404-452-4100
National Response Center	24 hours	800-424-8802
in Washington, DC area		202-426-2675
or		202-267-2675
U.S. Army, Navy, Air Force		
Bomb Disposal and/or Explosive Ordnance Team, U.S. Army		
Nuclear Regulatory Commission	24 hours	301-951-0550
U.S. Department of Energy Radiological Assistance	24 hours	202-586-8100
U.S. Department of the Treasury Bureau of Alcohol, Tobacco, and Firearms		

Other Emergency Assistance

CHEMTREC	24 hours	800-424-9300
CHEMNET	24 hours	800-424-9300
CHLOREP	24 hours	800-424-9300
NACA Pesticide Safety Team	24 hours	800-424-9300
Association of American Railroads/ Bureau of Explosives	24 hours	202-639-2222
Poison Control Center		
Cleanup Contractor		

Planning Element C: Response Functions

Comment:

Each function should be clearly marked with a tab so that it can be located quickly. When revising and updating a plan, communities might decide to add, delete, or combine individual functions.

Each response "function" usually includes several response activities. Some communities prepare a matrix that lists all response agencies down the left side of the page and all response activities across the top of the page. Planners can then easily determine which response activities need interagency coordination and which, if any, activities are not adequately provided for in the plan.

Function 1:   Initial Notification of Response Agencies

- 24-hour emergency response hotline telephone numbers

  - Local number to notify area public officials and response personnel

  - Number to notify state authorities

  - National Response Center (800-424-8802; 202-426-2675 or 202-267-2675 in Washington, DC area)

- Other agencies (with telephone numbers) to notify immediately (e.g., hospitals, health department, Red Cross)

Comment:

The local 24-hour emergency response hotline should be called first and therefore should have a prominent place in the plan. Provision should be made for notifying nearby municipalities and counties that could be affected by a vapor cloud or liquid plumes in a water supply.

Normally, the organization that operates the emergency response hotline will inform other emergency service organizations (e.g., health department, hospitals, Red Cross) once the initial notification is made. The plan should provide a method for notifying all appropriate local, state, and federal officials and agencies, depending upon the severity of the incident. To ensure that the appropriate federal on-scene coordinator (OSC) is notified of a spill or release, the NRC operated by the U.S. Coast Guard should be included in the notification listing. CERCLA requires that the NRC be notified by the responsible party of releases of many hazardous materials in compliance with the reportable quantity (RQ) provisions. The NRC telephone number is 800-424-8802 (202-426-2675 or 202-267-2675 in the Washington, DC area). If there is an emergency notification number at the state or regional level, it should be called before the NRC, and then a follow-up call made to the NRC as soon as practicable.

The plan should indicate how volunteer and off-duty personnel will be summoned. Similarly, there should be a method to notify special facilities (e.g., school districts, private schools, nursing homes, day care centers, industries, detention centers), according to the severity of the incident.

Function 2:   Direction and Control

- Name of on-scene authority

- Chain of command (illustrated in a block diagram)

- Criteria for activating emergency operating center

- Method for establishing on-scene command post and communications network for response team(s)

- Method for activating emergency response teams

- List of priorities for response actions

- Levels of response based on incident severity

Comment:

Response to a hazardous materials spill or release will involve many participants: police, fire fighters, facility personnel, health personnel, and others. It is also possible to have more than one organization perform the same service; for example, local police, the county sheriff and deputies, as well as the highway patrol may respond to perform police functions. Because speed of response is so important, coordination is needed among the various agencies providing the same service. It is essential to identify (by title or position) the one individual responsible for each participating organization, and the one individual responsible for each major function and service. The plan might require that the responsible person establish an incident command system (ICS).

Work out, in advance, the following:

(1)   Who will be in charge (lead organization)

(2)   What will be the chain of command

(3)   Who will activate the emergency operating center, if required

(4)   Who will maintain the on-scene command post and keep it secure

(5)   Who will have advisory roles (and what their precise roles are)

(6)   Who will make the technical recommendations on response actions to the lead agency

(7)   Who (if anyone) will have veto power

(8)   Who is responsible for requesting assistance from outside the community

This chain of command should be clearly illustrated in a block diagram.

Response action checklists are a way of condensing much useful information. They are helpful for a quick assessment of the response operation. If checklists are used, they should be prepared in sufficient detail to ensure that all crucial activities are included.

Planners should consider whether to have categories of response actions based on severity. The severity of an incident influences decisions on the level (or degree) of response to be made. This will determine how much equipment and how many personnel will be called, the extent of evacuation, and other factors.

The following chart summarizes who and what are involved in three typical emergency conditions. Information about the three response levels should be provided to special facilities (e.g., school districts, private schools, day care centers, hospitals, nursing homes, industries, detention centers).

Response Level	Description	Contact:
I. Potential Emergency Condition	An incident or threat of a release which can be controlled by the first response agencies and does not require evacuation of other than the involved structure or the immediate outdoor area. The incident is confined to a small area and does not pose an immediate threat to life or property.	Fire Department Emergency Medical Services Police Department Partial EOC Staff Public Information Office CHEMTREC National Response Center
II. Limited Emergency Condition	An incident involving a greater hazard or larger area which poses a potential threat to life or property and which may require a limited evacuation of the surrounding area.	All Agencies in Level 1 HAZMAT Teams EOC Staff Public Works Department Health Department Red Cross County Emergency Management Agency State Police Public Utilities
III. Full Emergency Condition	An incident involving a severe hazard or a large area which poses an extreme threat to life and property and will probably require a large scale evacuation; or an incident requiring the expertise or resources of county, State, Federal, or private agencies/organizations.	All Level I and II Agencies plus the following as needed: Mutual Aid Fire, Police, Emergency Medical State Emergency Management Agency State Department of Environmental Resources State Department of Health EPA USCG ATSDR FEMA OSC/RRT

Function 3:   Communications (among Responders)

• Any form(s) of exchanging information or ideas for emergency response with other entities, either internal or external to the existing organizational structure.

Comment:

This aspect of coordination merits special consideration. Different response organizations typically use different radio frequencies. Therefore, specific provision must be made for accurate and efficient communication among all the various organizations during the response itself. Several states have applied for one "on-scene" command radio frequency that all communities can use. At a minimum, it may be beneficial to establish radio networks that will allow for communication among those performing similar functions. The plan might specify who should be given a radio unit and who is allowed to speak on the radio. In order to avoid possible explosion/ fire hazards, all communications equipment (including walkie-talkies) should be intrinsically safe.

Function 4:   Warning Systems and Emergency Public Notification

• Method for alerting the public

  • Title and telephone number of person responsible for alerting the public as soon as word of the incident is received

  • List of essential data to be passed on (e.g., health hazards, precautions for personal protection, evacuation routes and shelters, hospitals to be used)

Comment:

This section should contain precise information on how sirens or other signals will be used to alert the public in case of an emergency. This should include information on what the different signals mean, how to coordinate the use of sirens, and the geographic area covered by each siren. (If possible, a backup procedure should be identified.) While a siren alerts those who hear it, an emergency broadcast is necessary to provide detailed information about the emergency and what people should do.

Sample Emergency Broadcast System messages should be prepared with blank spaces that can be filled in with precise information about the accident. One sample message should provide fundamental information about the incident and urge citizens to remain calm and await further information and instructions. Another sample message should be for an evacuation. Another sample message should describe any necessary school evacuations so that parents will know where their children are. Another sample message should be prepared to tell citizens to take shelter and inform them of other precautions they may take to protect themselves. The message should clearly identify those areas in which protective actions are recommended, using familiar boundaries. Messages might be developed in languages other than English, if customarily spoken in the area.

This section could be of urgent significance. When life-threatening materials are released, speed of response is crucial. It is not enough to have planned for alerting the community; one organization must be assigned the responsibility of alerting the public as soon as word of the accidental release is received. Delay in alerting the public can lead to the loss of life. In addition to sirens and the Emergency Broadcast System, it may be necessary to use mobile public address systems and/or house-by-house contacts. In this case, adequate protection must be provided for persons entering the area to provide such help.

Function 5:   Public Information/Community Relations

- Method to educate the public for possible emergencies

- Method for keeping the public informed

  - Provision for one person to serve as liaison to the public

  - List of radio and T.V. contacts

Comment:

Many communities develop a public information program to educate citizens about safety procedures during an incident. This program could include pamphlets; newspaper stories; periodic radio and television announcements; and programs for schools, hospitals, and homes for the aged.

It is important to provide accurate information to the public in order to prevent panic. Some citizens simply want to know what is happening. Other citizens may need to be prepared for possible evacuation or they may need to know what they can do immediately to protect themselves. Because information will be needed quickly, radio and television are much more important than newspapers in most hazardous materials releases. In less urgent cases, newspaper articles can provide detailed information to enhance public understanding of accidental spills and procedures for containment and cleanup. One person should be identified to serve as spokesperson. It is strongly recommended that the individual identified have training and experience in public information, community relations, and/or media relations. The spokesperson can identify for the media individuals who have specialized knowledge about the event. The chain of command should include this spokesperson. Other members of the response team should be trained to direct all communications and public relations issues to this one person.

Function 6:   Resource Management

- List of personnel needed for emergency response

- Training programs, including schedules for training of local emergency response and medical personnel

- List of vehicles needed for emergency response

- List of equipment (both heavy equipment and personal protective equipment) needed for emergency response

Comment:

This section should list the resources that will be needed, and where the equipment and vehicles are located or can be obtained. A major task in the planning process is to identify what resources are already available and what must still be provided. For information on the selection of protective equipment, consult the *Occupational Safety and Health Guidance Manual for Hazardous Waste Site Activities* prepared by NIOSH, OSHA, USCG, and EPA; and the EPA/Los Alamos "Guidelines for the Selection of Chemical Protective Clothing" distributed by the American Conference of Governmental Industrial Hygienists (Building B-7, 6500 Glynway Ave., Cincinnati, OH 45211).

This section should also address funding for response equipment and personnel. Many localities are initially overwhelmed by the prospect of providing ample funding for hazardous materials response activities. In large localities, each response agency is usually responsible for providing and maintaining certain equipment and personnel; in such cases, these individual agencies must devise funding methods, sources, and accounting procedures. In smaller localities with limited resources, officials frequently develop cooperative agreements with other jurisdictions and/or private industries. Some communities stipulate in law that the party responsible for an incident should ultimately pay the cost of handling it.

For a more detailed discussion of response training, consult Chapter 6 of this guide.

Function 7:   Health and Medical

- Provisions for ambulance service

- Provisions for medical treatment

Comment:

This section should indicate how medical personnel and emergency medical services can be summoned. It may be appropriate to establish mutual aid agreements with nearby communities to provide backup emergency medical personnel and equipment. The community should determine a policy (e.g., triage) for establishing priorities for the use of medical resources during an emergency. Medical personnel must be made aware of significant chemical hazards in the community in order to train properly and prepare for possible incidents. Emergency medical teams and hospital personnel must be trained in proper methods for decontaminating and treating persons exposed to hazardous chemicals. Planners should include mental health specialists as part of the team assisting victims of serious incidents. Protective action recommendations for sanitation, water supplies, recovery, and reentry should be addressed in this section.

Function 8:   Response Personnel Safety

- Standard operating procedure for entering and leaving sites

- Accountability for personnel entering and leaving the sites

- Decontamination procedures

- Recommended safety and health equipment

- Personal safety precautions

Comment:

Care must be taken to choose equipment that protects the worker from the hazard present at the site without unnecessarily restricting the capacities of the worker. Although the emphasis in equipment choices is commonly focused on protecting the worker from the risks presented by the hazardous material, impaired vision, restricted movements, or excessive heat can put the worker at equal risk. After taking these factors into account, the planner should list the equipment appropriate to various degrees of hazard using the EPA Levels of Protection (A, B, C, and D). The list should include: the type of respirator (e.g., self-contained breathing apparatus, supplied air respirator, or air purifying respirator) if needed; the type of clothing that must be worn; and the equipment needed to protect the head, eyes, face, ears, hands, arms, and feet. This list can then be used as a base reference for emergency response. The specific equipment used at a given site will vary according to the hazard. In addition, the equipment list should be reevaluated and updated as more information about the site is gathered to ensure that the appropriate equipment is being used. Responders should receive ongoing training in the use of safety equipment.

This section can also address liability related to immediate and long term health hazards to emergency responders. State and local governments may want to consider insurance coverage and/or the development of waivers for employees and contractors who may be on site during a hazmat incident.

Function 9:   Personal Protection of Citizens

Function 9a: Indoor Protection

- Hazard-specific personal protection

Comment:

The plan should clearly indicate what protective action should be taken in especially hazardous situations. Evacuation is sometimes, but not always, necessary. (See Function 9b.) For some hazardous materials it is safer to keep citizens inside with doors and windows closed rather than to evacuate them. It is perhaps appropriate to go upstairs (or downstairs). Household items (e.g., wet towels) can provide personal protection for some chemical hazards. Frequently a plume will move quickly past homes. Modern housing has adequate air supply to allow residents to remain safely inside for an extended period of time. Because air circulation systems can easily transport airborne toxic substances, a warning should be given to shut off all air circulation systems (including heating, air conditioning, clothes dryers, vent fans, and fire places) both in private and institutional settings.

In order for an indoor protective strategy to be effective, planning and preparedness activities should provide:

- An emergency management system and decision-making criteria for determining when an indoor protection strategy should be used;

- A system for warning and advising the public;

- A system for determining when a cloud has cleared a particular area;

- A system for advising people to leave a building at an appropriate time; and

- Public education on the value of indoor protection and on expedient means to reduce ventilation.

Function 9b: Evacuation Procedures

- Title of person and alternate(s) who can order/recommend an evacuation

- Vulnerable zones where evacuation could be necessary and a method for notifying these places

- Provisions for a precautionary evacuation

- Methods for controlling traffic flow and providing alternate traffic routes

- Shelter locations and other provisions for evacuations (e.g., special assistance for hospitals)

- Agreements with nearby jurisdictions to receive evacuees

- Agreements with hospitals outside the local jurisdictions

- Protective shelter for relocated populations

- Reception and care of evacuees

- Re-entry procedures

Comment:

Evacuation is the most sweeping response to an accidental release. The plan should clearly identify under what circumstances evacuation would be appropriate and necessary. DOT's *Emergency Response Guidebook* provides suggested distances for evacuating unprotected people from the scene of an incident during the initial phase. It is important to distinguish between general evacuation of the entire area and selective evacuation of a part of the risk zone. In either case, the plan should identify how people will be moved (i.e., by city buses, police cars, private vehicles). Provision must be made for quickly moving traffic out of the risk zone and also for preventing outside traffic from entering the risk zone. If schools are located in the risk zone, the plan must identify the location to which students will be moved in an evacuation and how parents will be notified of this location. Special attention must also be paid to evacuating hospitals, nursing homes, and homes for the physically and mentally disabled.

Maps (drawn to the same scale) with evacuation routes and alternatives clearly identified should be prepared for each risk zone in the area. Maps should indicate precise routes to another location where special populations (e.g., from schools, hospitals, nursing homes, homes for the physically or mentally disabled) can be taken during an emergency evacuation, and the methods of transportation during the evacuation.

Consideration of when and how evacuees will return to their homes should be part of this section.

This section on evacuation should include a description of how other agencies will coordinate with the medical community.

Copies of evacuation procedures should be provided to all appropriate agencies and organizations (e.g., Salvation Army, churches, schools, hospitals) and could periodically be published in the local newspaper(s).

Function 9c: Other Public Protection Strategies

- Relocation

- Water supply protection

- Sewage system protection

Comment:

Some hazardous materials incidents may contaminate the soil or water of an area and pose a chronic threat to people living there. It may be necessary for people to move out of the area for a substantial period of time until the area is decontaminated or until natural weathering or decay reduce the hazard. Planning must provide for the quick identification of a threat to the drinking water supply, notification of the public and private system operators, and warning of the users. Planners should also provide sewage system protection. A hazardous chemical entering the sewage system can cause serious and long-term damage. It may be necessary to divert sewage, creating another public health threat and environmental problems.

Function 10:   Fire and Rescue

- Chain of command among fire fighters

- List of available support systems

- List of all tasks for fire fighters

Comment:

This section lists all fire fighting tasks, as well as the chain of command for fire fighters. This chain of command is especially important if fire fighters from more than one jurisdiction will be involved. Planners should check to see if fire fighting tasks and the chain of command are mandated by their state law. Fire fighters should be trained in proper safety procedures when approaching a hazardous materials incident. They should have copies of DOT's Emergency Response Guidebook and know how to find shipping manifests in trucks, trains, and vessels. Specific information about protective equipment for fire fighters should be included here. (See Function 6, "Resource Management," and the *Occupational Safety and Health Guidance Manual for Hazardous Waste Site Activities*.)

This section should also identify any mutual aid or Good Samaritan agreements with neighboring fire departments, haz mat teams, and other support systems.

Function 11:   Law Enforcement

- Chain of command for law enforcement officials

- List of all tasks for law enforcement personnel

Comment:

This section lists all the tasks for law enforcement personnel during an emergency response. Planners should check to see if specific law enforcement tasks are mandated by their state law. Because major emergencies will usually involve state, county, and local law enforcement personnel, and possibly the military, a clear chain of command must be determined in advance. Because they are frequently first on scene, law enforcement officials should be trained in proper procedures for approaching a hazardous materials incident. They should have copies of DOT's *Emergency Response Guidebook* and know how to find shipping manifests in trucks, trains, and vessels. Specific information about protective equipment for law enforcement officials should be included here. (See Function 6, "Resource Management," and the *Occupational Safety and Health Guidance Manual for Hazardous Waste Site Activities*.)

This section should include maps that indicate control points where police officers should be stationed in order to expedite the movement of responders toward the scene and of evacuees away from the scene, to restrict unnecessary traffic from entering the scene, and to control the possible spread of contamination.

Function 12:   Ongoing Incident Assessment

- Field monitoring teams

- Provision for environmental assessment, biological monitoring, and contamination surveys

- Food/water controls

Comment:

After the notification that a release has occurred, it is crucial to monitor the release and assess its impact, both on and off site. A detailed log of all sampling results should be maintained. Health officials should be kept informed of the situation. Often the facility at which the release has occurred will have the best equipment for this purpose.

This section should describe who is responsible to monitor the size, concentration, and movement of leaks, spills, and releases, and how they will do their work. Decisions about response personnel safety, citizen protection (whether indoor or through evacuation), and the use of food and water in the area will depend upon an accurate assessment of spill or plume movement and concentration. Similarly, decisions about containment and cleanup depend upon monitoring data.

Function 13:   Human Services

- List of agencies providing human services

- List of human services tasks

Comment:

This section should coordinate the activities of organizations such as the Red Cross, Salvation Army, local church groups, and others that will help people during a hazardous materials emergency. These services are frequently performed by volunteers. Advance coordination is essential to ensure the most efficient use of limited resources.

Function 14:   Public Works

- List of all tasks for public works personnel

Comment:

This section lists all public works tasks during an emergency response. Public works officials should also be familiar with Plan Section D ("Containment and Cleanup").

Function 15:   Others

Comment:

If the preceding list of functions does not adequately cover the various tasks to be performed during emergency responses, additional response functions can be developed.

Planning Element D: Containment and Cleanup

Planning Element D.1:   Techniques for Spill Containment and Cleanup

- Containment and mitigation actions

- Cleanup methods

- Restoration of the surrounding environment

Comment:

Local responders will typically emphasize the containment and stabilization of an incident; state regulatory agencies can focus on cleanup details. Federal RRT agencies can provide assistance during the cleanup process. It is the releaser's legal and financial responsibility to clean up and minimize the risk to the health of the general public and workers that are involved. The federal OSC or other government officials should monitor the responsible party cleanup activities.

A clear and succinct list of appropriate containment and cleanup countermeasures should be prepared for each hazardous material present in the community in significant quantities. This section should be coordinated with the section on "Response Personnel Safety" so that response teams are subjected to minimal danger.

Planners should concentrate on the techniques that are applicable to the hazardous materials and terrain of their area. It may be helpful to include sketches and details on how cleanup should occur for certain areas where spills are more likely.

It is important to determine whether a fire should be extinguished or allowed to burn. Water used in fire fighting could become contaminated and then would need to be contained or possibly treated. In addition, some materials may be water-reactive and pose a greater hazard when in contact with water. Some vapors may condense into pools of liquid that must be contained and removed. Accumulated pools may be recovered with appropriate pumps, hoses, and storage containers. Various foams may be used to reduce vapor generation rates. Water sprays or fog may be applied at downwind points away from "cold" pools to absorb vapors and/or accelerate their dispersal in the atmosphere. (Sprays and fog might not reduce an explosive atmosphere.) Volatile liquids might be diluted or neutralized.

If a toxic vapor comes to the ground on crops, on playgrounds, in drinking water, or other places where humans are likely to be affected by it, the area should be tested for contamination. Appropriate steps must be taken if animals (including fish and birds) that may become part of the human food chain are in contact with a hazardous material. It is important to identify in advance what instruments and methods can be used to detect the material in question.

Restoration of the area is a long-range project, but general restoration steps should appear in the plan. Specific consideration should be given to the mitigation of damages to the environment.

Planning Element D.2:   Resources for Cleanup and Disposal

- Cleanup/disposal contractors and services provided

- Cleanup material and equipment

- Communications equipment

- Provision for long-term site control during extended cleanups

- Emergency transportation (e.g., aircraft, four-wheel-drive vehicles, boats)

- Cleanup personnel

- Personal protective equipment

- Approved disposal sites

Comment:

This section is similar to the yellow pages of the telephone book. It provides plan users with the following important information:

- What types of resources are available (public and private);

- How much is stockpiled;

- Where it is located (address and telephone number); and

- What steps are necessary to obtain the resources.

Organizations that may have resources for use during a hazardous materials incident include:

- Public agencies (e.g., fire, police, public works, public health, agriculture, fish and game);

- Industry (e.g., chemical producers, transporters, storers, associations; spill cleanup contractors; construction companies);

- Spill/equipment cooperatives; and

- Volunteer groups (ham radio operators, four-wheel-drive vehicle clubs).

Resource availability will change with time, so keep this section of the plan up-to-date.

Hazardous materials disposal may exceed the capabilities of smaller cities and towns; in such cases, the plan should indicate the appropriate state and/or federal agency that is responsible for making decisions regarding disposal.

Disposal of hazardous materials or wastes is controlled by a number of federal and state laws and regulations. Both CERCLA and RCRA regulate waste disposal and it is important that this section reflect the requirements of these regulations for on-site disposal, transportation, and off-site disposal. The plan should include an updated list of RCRA disposal facilities for possible use during an incident.

Many states have their own regulations regarding transport and ultimate disposal of hazardous waste. Usually such regulations are similar and substantially equal to federal regulations. Contact appropriate state agency offices for information on state requirements for hazardous waste disposal.

Planning Element E: Documentation and Investigative Follow-Up

• List of required reports

• Reasons for requiring the reports

• Format for reports

• Methods for determining whether the response mechanism worked properly

• Provision for cost recovery

Comment:

This section indicates what information should be gathered about the release and the response operation. Key response personnel could be instructed to maintain an accurate log of their activities. Actual response costs should be documented in order to facilitate cost recovery.

It is also important to identify who is responsible for the post-incident investigation to discover quickly the exact circumstances and cause of the release. Critiques of real incidents, if handled tactfully, allow improvements to be made based on actual experience. The documentation described above should help this investigation determine if response operations were effective, whether the emergency plan should be amended, and what follow-up responder and public training programs are needed.

Planning Element F: Procedures for Testing and Updating Plan

Planning Element F.1:   Testing the Plan

• Provision for regular tabletop, functional, and full-scale exercises

Comment:

Exercises or drills are important tools in keeping a plan functionally up-to-date. These are simulated accidental releases where emergency response personnel act out their duties. The exercises can be tabletop and/or they can be realistic enough so that equipment is deployed, communication gear is tested, and "victims" are sent to hospitals with simulated injuries. Planners should work with local industry and the private medical community when conducting simulation exercises, and they should provide for drills that comply with state and local legal requirements concerning the content and frequency of drills. After the plan is tested, it should be revised and retested until the planning team is confident that the plan is ready. The public should be involved in or at least informed of these exercises. FEMA, EPA, and CMA provide guidance on simulation exercises through their training programs complementing this guide.

This section should specify:

(1)   The organization in charge of the exercise;

(2)   The types of exercises;

(3)   The frequency of exercises; and

(4)   A procedure for evaluating performance, making changes to plans, and correcting identified deficiencies in response capabilities as necessary. (See Chapter 6 of this guide.)

Planning Element F.2:   Updating the Plan

• Title and organization of responsible person(s)

• Change notification procedures

• How often the plan should be audited and what mechanisms will be used to change the plan

Comment:

Responsibility should be delegated to someone to make sure that the plan is updated frequently and that all plan holders are informed of the changes. Notification of changes should be by written memorandum or letter; the changes should be recorded in the RECORD OF AMENDMENTS page at the front of the completed plan. Changes should be consecutively numbered for ease of tracking and accounting.

Following are examples of information that must regularly be checked for accuracy:

(1)   Identity and phone numbers of response personnel

(2)   Name, quantity, properties, and location of hazardous materials in the community. (If new hazardous materials are made, used, stored, or transported in the community, revise the plan as needed.)

(3)   Facility maps

(4)   Transportation routes

(5)   Emergency services available

(6)   Resource availability

This topic is considered in greater detail in Chapter 6 of this guidance.

Planning Element G: Hazards Analysis (Summary)

- Identification of hazards

- Analysis of vulnerability

- Analysis of risk

Comment:

This analysis is a crucial aspect of the planning process. It consists of determining where hazards are likely to exist, what places would most likely be adversely affected, what hazardous materials could be involved, and what conditions might exist during a spill or release. To prepare a hazards analysis, consult Chapter 3 of this guide, EPA's CEPP technical guidance, and DOT's *Community Teamwork* and *Lessons Learned*. Ask federal offices (listed in Appendix F) for information about available computer programs to assist in a hazards analysis.

Individual data sheets and maps for each facility and transportation routes of interest could be included in this section. Similar data could be included for recurrent shipments of hazardous materials through the area. This section will also assess the probability of damage and/or injury. In communities with a great deal of hazardous materials activity, the hazards analysis will be too massive to include in the emergency plan. In that case, all significant details should be summarized here.

Planning Element H: References

Planning Element H.1:   Laboratory, Consultant, and Other Technical Support Resources

- Telephone director of technical support services

- Laboratories (environmental and public health)

- Private consultants

- Colleges or universities (chemistry departments and special courses)

- Local chemical plants

Comment:

This section should identify the various groups capable of providing technical support and the specific person to be contacted. Medical and environmental laboratory resources to assess the impact of the most probable materials that could be released should be identified. Note should be made about the ability of these laboratories to provide rapid analysis. These technical experts can provide advice during a disaster and also be of great service during the development of this plan. For this reason, one of the first planning steps should be gathering information for this section.

Planning Element H.2:   Technical Library

- List of references, their location, and their availability

  - General planning references

  - Specific references for hazardous materials

  - Technical references and methods for using national data bases

  - Maps

Comment:

Industry sources can provide many specific publications dealing with hazardous materials. This section of the plan will list those published resources that are actually available in the community. Also list any maps (e.g., of facilities, transportation routes) that will aid in the response of an accidental spill or release.

The list of technical references in Appendix E could be helpful. Regional federal offices can also be contacted (see Appendix F).

It is important for planners to acquire, understand, and be able to use available hazardous materials data bases, including electronic data bases available from commercial and government sources. Planning guides such as DOT's *Community Teamwork*, CMA's CAER program, EPA's CEPP technical guidance, and this guide should also be available locally.

6 Plan Appraisal and Continuing Planning

6.1 Introduction

Any emergency plan must be evaluated and kept up-to-date through the review of actual responses, simulation exercises, and regular collection of new data. Effective emergency preparedness requires periodic review and evaluation, and the necessary effort must be sustained at the community level. Plans should reflect any recent changes in: the economy, land use, permit waivers, available technology, response capabilities, hazardous materials present, federal and state laws, local laws and ordinances, road configurations, population change, emergency telephone numbers, and facility location. This chapter describes key aspects of appraisal and provides specific guidance for maintaining an updated hazardous materials emergency plan.

6.2 Plan Review and Approval

Plan review and approval are critically important responsibilities of the planning team. This section discusses the various means by which a plan can be reviewed thoroughly and systematically.

6.2.1 Internal Review

The planning team, after drafting the plan, should conduct an internal review of the plan. It is not sufficient merely to read over the plan for clarity or to search for errors. The plan should also be assessed for adequacy and completeness. Appendix D is an adaptation of criteria developed by the National Response Team that includes questions useful in appraising emergency plans. Individual planning team members can use these questions to conduct self review of their own work and the team can assign a committee to review the total plan. In the case of a hazardous materials appendix (or appendices) to a multi-hazard EOP, the team will have to review the basic EOP as well as the functional annexes to obtain an overall assessment of content. Once the team accomplishes this internal review the plan should be revised in preparation for external review.

6.2.2 External Review

External review legitimizes the authority and fosters community acceptance of the plan. The review process should involve elements of peer review, upper level review, and community input. The planning team must devise a process to receive, review, and respond to comments from external reviewers.

A. *Peer Review*

Peer review entails finding qualified individuals who can provide objective reviews of the plan. Individuals with qualifications similar to those considered for inclusion on the planning team should be selected as peer reviewers. Examples of appropriate individuals include:

• The safety or environmental engineer in a local industry;

• Responsible authorities from other political jurisdictions (e.g., fire chief, police, environmental and/or health officers);

• A local college professor familiar with hazardous materials response operations; and

• A concerned citizen's group, such as the League of Women Voters, that provides a high level of objectivity along with the appropriate environmental awareness.

Exhibit 2 (Chapter 2) presents a comprehensive list of potential peer reviewers. Those selected as peer reviewers should use the criteria contained in Appendix D to develop their assessments of the plan.

B. *Upper Level Review*

Upper level review involves submitting the plan to an individual or group with oversight authority or responsibility for the plan. Upper level review should take place after peer review and modification of the plan.

C. *Community Input*

Community involvement is vital to success throughout the planning process. At the plan appraisal stage, such involvement greatly facilitates formal acceptance of the plan by the community. Approaches that can be used include:

• Community workshops with short presentations by planning team members followed by a question-and-answer period;

• Publication of notice "for comment" in local newspapers, offering interested individuals and groups an opportunity to express their views in writing;

• Public meetings at which citizens can submit oral and written comments;

• Invited reviews by key interest groups that provide an opportunity for direct participation for such groups that are not represented on the planning team; and

• Advisory councils composed of a relatively large number of interested parties that can independently review and comment on the planning team's efforts.

These activities do more than encourage community consensus building. Community outreach at this stage in the process also improves the soundness of the plan by increased public input and expands public understanding of the plan and thus the effectiveness of the emergency response to a hazardous materials incident.

D. *State/Federal Review*

After local review and testing through exercises, a community may want to request review of the plan by state and/or federal officials. Such a review will depend upon the availability of staff resources. Planning committees set up in accordance with Title III of SARA are to submit a copy of the emergency plan to the state emergency response commission for review to ensure coordination of the plan with emergency plans of other planning districts. Federal Regional Response Teams may review and comment upon an emergency plan, at the request of a local emergency planning committee. FEMA regional offices review FEMA-funded multi-hazard EOPs using criteria in CPG 1-8A.

6.2.3 Plan Approval

The planning team should identify and comply with any local or state requirements for formal plan approval. It may be necessary for local officials to enact legislation that gives legal recognition to the emergency plan.

6.3 Keeping the Plan Up-to-Date

All emergency plans become outdated because of social, economic, and environmental changes. Keeping the plan current is a difficult task, but can be achieved by scheduling reviews regularly. As noted in Chapter 5, the plan itself should indicate who is responsible for keeping it up-to-date. Outdated information should be replaced, and the results of appraisal exercises should be incorporated into the plan. The following techniques will aid in keeping abreast of relevant changes:

• Establish a regular review period, preferably every six months, but at least annually. (Title III of SARA requires an annual review.)

• Test the plan through regularly scheduled exercises (at least annually). This testing should include debriefing after the exercises whenever gaps in preparedness and response capabilities are identified.

• Publish a notice and announce a comment period for plan review and revisions.

• Maintain a list of individuals, agencies, and organizations that will be interested in participating in the review process.

- Make one reliable organization responsible for coordination of the review and overall stewardship of the plan. Use of the planning team in this role is recommended, but may not be a viable option due to time availability constraints of team members.

- Require immediate reporting by any facility of an increase in quantities of hazardous materials dealt with in the emergency plan, and require review and revision of plan if needed in response to such new information.

- Include a "Record of Amendments and Changes" sheet in the front section of the plan to help users of the plan stay abreast of all plan modifications.

- Include a "When and Where to Report Changes" notice in the plan and a request for holders of the plan to report any changes or suggested revisions to the responsible organization at the appropriate time.

- Make any sections of the plan that are subject to frequent changes either easily replaceable (e.g., looseleaf, separate appendix), or provide blank space (double- or triple-spaced typing) so that old material may be crossed out and new data easily written in. This applies particularly to telephone rosters and resource and equipment listings.

The organization responsible for review should do the following:

- Maintain a list of plan holders, based on the original distribution list, plus any new copies made or distributed. It is advisable to send out a periodic request to departments/branches showing who is on the distribution list and asking for any additions or corrections.

- Check all telephone numbers, persons named with particular responsibilities, and equipment locations and availability. In addition, ask departments and agencies to review sections of the plan defining their responsibilities and actions.

- Distribute changes. Changes should be consecutively numbered for ease of tracking. Be specific, e.g., "Replace page _____ with the attached new page _____." or "Cross out _____ on page _____ and write in the following" (new phone number, name, location, etc.). Any key change (new emergency phone number, change in equipment availability, etc.) should be distributed as soon as it occurs. Do not wait for the regular review period to notify plan holders.

- If possible, the use of electronic word processing is recommended because it facilitates changing the plan. After a significant number of individual changes, the entire plan should be redistributed to ensure completeness.

- If practical, request an acknowledgement of changes from those who have received changes. The best way to do this is to include a self-addressed postcard to be returned with acknowledgement (e.g., "I have received and entered changes _____. Signed _____").

- Attend any plan critique meetings and issue changes as may be required.

- Integrate changes with other related plans.

6.4   Continuing Planning

In addition to the periodic updates described above, exercises, incident reviews, and training are necessary to ensure current and effective planning.

6.4.1   Exercises

The plan should also be evaluated through exercises to see if its required activities are effective in practice and if the evaluation would reveal more efficient ways of responding to a real emergency. As noted in Chapter 5, the plan itself should indicate who is responsible for conducting exercises. Simulations can be full-scale, functional, or tabletop exercises.

A full-scale exercise is a mock emergency in which the response organizations that would be involved in an actual emergency perform the actions they would take in the emergency. These simulations may focus on limited objectives (e.g., testing the capability of local hospitals to handle relocation problems). The responsible environmental, public safety, and health agencies simulate, as realistically as possible, notification, hazards identification and analysis, command structure, command post staging, communications, health care, containment, evacuation of affected areas, cleanup, and documentation. Responders use the protective gear, radios, and response equipment and act as they would in a real incident. These multi-agency exercises provide a clearer understanding of the roles and resources of each responder.

A functional exercise involves testing or evaluating the capability of individual or multiple functions, or activities within a function.

A low-cost, valuable version of an exercise is the staging of a tabletop exercise. In this exercise, each agency representative describes and acts out what he or she would do at each step of the response under the circumstances given.

Exercises are most beneficial when followed by a meeting of all participants to critique the performance of those involved and the strengths and weaknesses of the plan's operation. The use of an outside reviewer, free of local biases, is desirable. The emergency plan should be amended according to the lessons learned. Provisions should be made to follow up exercises to see that identified deficiencies are corrected.

Communities that want to help in preparing and conducting exercises should consult FEMA's four-volume "Exercise Design Course," which includes sample hazardous materials exercises. CMA's *Community Emergency Response Exercise Handbook* is also helpful. CMA describes four types of exercises: tabletop, emergency operations simulation, drill, and field exercise.

6.4.2 Incident Review

When a hazardous materials incident does occur, a review or critique of the incident is a means of evaluating the plan's effectiveness. Recommendations for conducting an incident review are:

• Assign responsibility for incident review to the same organization that is responsible for plan update, for example, the planning team.

• Conduct the review only after the emergency is under control and sufficient time has passed to allow emergency respondents to be objective about the incident.

• Use questionnaires, telephone interviews, or personal interviews to obtain comments and suggestions from emergency respondents. Follow-up on non-respondents.

• Identify plan and response deficiencies: items that were overlooked, improperly identified, or were not effective.

• Convene the planning team to review comments and make appropriate plan changes.

• Revise the plan as necessary. Communicate personal or departmental deficiencies informally to the appropriate person or department. Follow up to see that deficiencies are corrected.

6.4.3 Training

Training courses can help with continuing planning by sharpening response personnel skills, presenting up-to-date ideas/techniques, and promoting contact with other people involved in emergency response. Everyone who occupies a position that is identified in the plan must have appropriate training. This applies to persons at all levels who serve to coordinate or have responsibilities under the plan, both those directly and indirectly involved at the scene of an incident. One should not assume that a physician in the emergency room or a professional environmentalist is specifically trained to perform his/her assigned mission during an emergency.

The training could be a short briefing on specific roles and responsibilities, or a seminar on the plan or on emergency planning and response in general. However the training is conducted, it should convey a full appreciation of the importance of each role and the effect that each person has on implementing an effective emergency response.

Training is available from a variety of sources in the public and private sectors. At the federal level, EPA, FEMA, OSHA, DOT/RSPA and the USCG offer hazardous materials training. (In some cases, there are limits on attendance in these courses.) FEMA, EPA, and other NRT agencies cooperatively offer the inter-agency "train-the-trainer" course, Hazardous Materials Contingency Planning, at Emmitsburg, MD and in the field.

Title III of SARA authorizes federal funding for training. Communities seeking training assistance should consult appropriate state agencies. States may consult with the RRT and the various federal regional and district offices. (See Appendix F.)

In addition to government agencies, consult universities or community colleges (especially any fire science curriculum courses), industry associations, special interest groups, and the private sector (fixed facilities, shippers, and carriers). Many training films and slide presentations can be borrowed or rented at little cost. Many chemical companies and carriers provide some level of training free.

The Chemical Manufacturers Association has a lending library of audio-visual training aids for use by personnel who respond to emergencies involving chemicals. The training aids are available on a loan basis at no charge to emergency response personnel and the public sector.

Training aids can also be purchased from:

National Chemical Response and Information Center
Chemical Manufacturers Association
2501 M Street, N.W.
Washington, DC 20037

In addition to classroom training, response personnel will need hands-on experience with equipment to be used during an emergency.

Communities should provide for refresher training of response personnel. It is not sufficient to attend training only once. Training must be carried out on a continuing basis to ensure currency and capability. Some communities have found it effective to hold this refresher training in conjunction with an exercise.

The NRT, through its member agencies, is developing a strategy to address issues related to emergency preparedness and response for hazardous materials incidents. The training strategy includes: (1) improved coordination of available federal training programs and courses; (2) sharing information about available training, and lessons learned from responses to recent hazardous materials incidents; (3) the increased use of exercises as a training method; (4) the revision of existing core courses, and the development of any needed new core courses that prepare responders to do the actual tasks expected in their own communities; and (5) decentralizing the delivery of training so that it is more easily available to responders. Further information about this training strategy can be obtained from EPA or FEMA offices in Washington, DC.

Appendix A

IMPLEMENTING TITLE III: EMERGENCY PLANNING AND

COMMUNITY RIGHT-TO-KNOW

SUPERFUND AMENDMENTS AND REAUTHORIZATION

ACT OF 1986

On October 17, 1986, the President signed the "Superfund Amendments and Reauthorization Act of 1986" (SARA) into law. One part of the new SARA provisions is Title III: the "Emergency Planning and Community Right-to-Know Act of 1986." Title III establishes requirements for Federal, State, and local governments and industry regarding emergency planning and community right-to-know reporting on hazardous chemicals. This legislation builds upon the Environmental Protection Agency's (EPA's) Chemical Emergency Preparedness Program (CEPP) and numerous State and local programs aimed at helping communities to meet their responsibilities in regard to potential chemical emergencies.

Title III has four major sections: emergency planning (§301-303), emergency notification (§304), community right-to-know reporting requirements (§311, 312), and toxic chemical release reporting—emissions inventory (§313). The sections are interrelated in a way that unifies the emergency planning and community right-to-know provisions of Title III. (See Exhibit 6.)

In addition to increasing the public's knowledge and access to information on the presence of hazardous chemicals in their communities and releases of these chemicals into the environment, the community right-to-know provisions of Title III will be important in preparing emergency plans.

This appendix includes a summary of these four major sections, followed by a discussion of other Title III topics of interest to emergency planners.

Sections 301-303:   Emergency Planning

The emergency planning sections are designed to develop state and local government emergency preparedness and response capabilities through better coordination and planning, especially at the local level.

403

Title III requires that the governor of each state designate a state emergency response commission (SERC) by April 17, 1987. While existing state organizations can be designated as the SERC, the commission should have broad-based representation. Public agencies and departments concerned with issues relating to the environment, natural resources, emergency management, public health, occupational safety, and transportation all have important roles in Title III activities.

Various public and private sector groups and associations with interest and experience in Title III issues can also be included on the SERC.

The SERC must designate local emergency planning districts by July 17, 1987, and appoint local emergency planning committees (LEPCs) within one month after a district is designated. The SERC is responsible for supervising and coordinating the activities of the LEPCs, for establishing procedures for receiving and processing public requests for information collected under other sections of Title III, and for reviewing local emergency plans.

The LEPC must include elected state and local officials, police, fire, civil defense, public health professionals, environmental, hospital, and transportation officials as well as representatives of facilities, community groups, and the media. Interested persons may petition the SERC to modify the membership of an LEPC.

No later than September 17, 1987, facilities subject to the emergency planning requirements must notify the LEPC of a representative who will participate in the planning process as a facility emergency coordinator.

Facility emergency coordinators will be of great service to LEPCs. For example, they can provide technical assistance, an understanding of facility response procedures, information about chemicals and their potential effects on nearby persons and the environment, and response training opportunities. CEPP experience revealed that, as a result of CMA's CAER initiative, there already exist a large number of plant managers and other facility personnel who want to cooperate with local community planners.

The LEPC must establish rules, give public notice of its activities, and establish procedures for handling public requests for information.

The LEPC's primary responsibility will be to develop an emergency response plan by October 17, 1988. In developing this plan, the local committee will evaluate available resources for preparing for and responding to a potential chemical accident. The plan must include:

• Identification of facilities and extremely hazardous substances transportation routes;

• Emergency response procedures, on site and off site;

- Designation of a community coordinator and facility coordinator(s) to implement the plan;

- Emergency notification procedures;

- Methods for determining the occurrence of a release and the probable affected area and population;

- Description of community and industry emergency equipment and facilities, and the identity of persons responsible for them;

- Evacuation plans;

- Description and schedules of a training program for emergency response to chemical emergencies; and

- Methods and schedules for exercising emergency response plans.

To assist the LEPC in preparing and reviewing plans, Congress required the National Response Team (NRT), composed of 14 federal agencies with emergency preparedness and response responsibilities, to publish guidance on emergency planning. This Hazardous Materials Emergency Planning Guide is being published by the NRT to fulfill this requirement.

The emergency plan must be reviewed by the SERC upon completion and reviewed annually by the LEPC. The Regional Response Teams (RRTs), composed of federal regional officials and state representatives, may review the plans and provide assistance if the LEPC so requests.

The emergency planning activities of the LEPC and facilities should initially be focused on, but not limited to, the extremely hazardous substances published as an interim final rule in the November 17, 1986, *Federal Register*. The list included the threshold planning quantity (TPQ) for each substance. EPA can revise the list and TPQs but must take into account the toxicity, reactivity, volatility, dispersability, combustibility, or flammability of a substance. Consult EPA regional offices for a copy of the Title III (Section 302) list of extremely hazardous substances.

Any facility that produces, uses, or stores any of the listed chemicals in a quantity greater than the TPQ must meet all emergency planning requirements. In addition, the SERC or the governor can designate additional facilities, after public comment, to be subject to these requirements. By May 17, 1987, facilities must notify the SERC that they are subject to these requirements. If, after that time, a facility first begins to produce, use, or store an extremely hazardous substance in an amount exceeding the threshold planning quantity, it must notify the SERC and LEPC within 60 days.

Each SERC must notify EPA regional offices of all facilities subject to Title III planning requirements.

In order to complete information on many sections of the emergency plan, the LEPC will require data from the facilities covered under the plan. Title III provides authority for the LEPC to secure from a facility information that it needs for emergency planning and response. This is provided by Section 303(d)(3), which states that:

"Upon request from the emergency planning committee, the owner or operator of the facility shall promptly provide information to such committee necessary for developing and implementing the emergency plan."

Within the trade secret restrictions contained in Section 322, LEPCs should be able to use this authority to secure from any facility subject to the planning provisions of the law information needed for such mandatory plan contents as: facility equipment and emergency response capabilities, facility emergency response personnel, and facility evacuation plans.

Some of the facilities subject to Section 302 planning requirements may not be subject to Sections 311-12 reporting requirements, which are currently limited to manufacturers and importers in SIC codes 20-39. LEPCs may use Section 303(d)(3) authority to gain information such as name(s), MSDSs, and quantity and location of chemicals present at facilities subject to Section 302.

Section 304:   Emergency Notification

If a facility produces, uses, or stores one or more hazardous chemical, it must immediately notify the LEPC and the SERC if there is a release of a listed hazardous substance that exceeds the reportable quantity for that substance. Substances subject to this notification requirement include substances on the list of extremely hazardous substances published in the *Federal Register* on November 17, 1986, and substances subject to the emergency notification requirements of CERCLA Section 103(a).

Information included in this initial notification (as well as the additional information in the follow-up written notice described below) can be used by the LEPC to prepare and/or revise the emergency plan. This information should be especially helpful in meeting the requirement to list methods for determining if a release has occurred and identifying the area and population most likely to be affected.

The initial notification of a release can be by telephone, radio, or in person. Emergency notification requirements involving transportation incidents may be satisfied by dialing 911 or, in the absence of a 911 emergency number, calling the operator.

This emergency notification needs to include: the chemical name; an indication of whether the substance is an extremely hazardous substance; an estimate of the quantity released into the environment; the time and duration of the release; the medium into which the release occurred; any known or anticipated acute or chronic health risks associated with the emergency and, where appropriate, advice regarding medical attention necessary for exposed individuals; proper precautions, such as evacuation; and the name and telephone number of a contact person.

Section 304 also requires a follow-up written emergency notice after the release. The follow-up notice or notices shall update information included in the initial notice and provide additional information on actual response actions taken, any known or anticipated data on chronic health risks associated with the release, and advice regarding medical attention necessary for exposed individuals.

The requirement for emergency notification comes into effect with the establishment of the SERC and LEPC. If no SERC is established by April 17, 1987, the governor becomes the SERC and notification should be made to him/her. If no LEPC is established by August 17, 1987, local notification must be made to the appropriate local emergency response personnel, such as the fire department.

Sections 311-312:   Community Right-to-Know Reporting Requirements

As noted above, Section 303(d)(3) gives LEPCs access to information from facilities subject to Title III planning requirements. Sections 311-12 provide information about the nature, quantity, and location of chemicals at many facilities not subject to the Section 303(d)(3) requirement. For this reason, LEPCs will find Sections 311-12 information especially helpful when preparing a comprehensive plan for the entire planning district.

There are two community right-to-know reporting requirements. Section 311 requires a facility which must prepare or have available material safety data sheets (MSDSs) under the Occupational Safety and Health Administration (OSHA) hazard communications regulations to submit either copies of its MSDSs or a list of MSDS chemicals to the LEPC, the SERC, and the local fire department. Currently, only facilities in Standard Industrial Classification (SIC) Codes 20-39 (manufacturers and importers) are subject to these OSHA regulations.

The initial submission of the MSDSs or list is required no later than October 17, 1987, or 3 months after the facility is required to prepare or have available an MSDS under OSHA regulations. A revised MSDS must be provided to update an MSDS which was originally submitted if significant new information regarding a chemical is discovered.

EPA encourages LEPCs and fire departments seriously to consider contacting facilities prior to the deadline of October 17, 1987 to request the submission of lists rather than MSDS forms. In communities with a large number of facilities, handling large numbers of chemicals, and in communities with limited capabilities to store and manage the MSDSs, the list of MSDS chemicals from the facility would be more useful than the forms themselves, and likely to be more easily produced.

LEPCs also have the option of using the chemical names provided to develop additional data on each of the chemicals, using a variety of data sources, including several on-line data bases maintained by agencies of the federal government.

Specific MSDSs could be requested on chemicals that are of particular concern. In general every MSDS will provide the LEPC and the fire departments in each community with the following information on each of the chemicals covered:

* *The chemical name;*

* *Its basic characteristics, for example:*

  * *toxicity, corrosivity, reactivity*

  * *known health effects, including chronic effects from exposure,*

  * *basic precautions in handling, storage, and use,*

  * *basic countermeasures to take in the event of a fire, explosion, leak, and*

  * *basic protective equipment to minimize exposure.*

In any case, these data should be useful for the planning to be accompanied by the LEPC and first responders, especially fire departments and hazmat teams. Both hazards analysis and the development of emergency countermeasures should be facilitated by the availability of MSDS information.

If the facility owner or operator chooses to submit a list of MSDS chemicals, the list must include the chemical name or common name of each substance and any hazardous component as provided on the MSDS. This list must be organized in categories of health and physical hazards as set forth in OSHA regulations or as modified by EPA.

If a list is submitted, the facility must provide the MSDS for any chemical on the list upon the request of the LEPC. Under Section 311, EPA may establish threshold quantities for hazardous chemicals below which no facility must report.

The reporting requirement of Section 312 requires facilities to submit an emergency and hazardous chemical inventory form to the LEPC, the SERC, and the local fire department. The hazardous chemicals covered by Section 312 are the same chemicals for which facilities are required to submit MSDS forms or the list for Section 311.

Under Sections 311-12, EPA may establish threshold quantities for hazardous chemicals below which no facility is subject to this requirement. See the proposed rule in the January 27, 1987 *Federal Register*. The Final Rule will be published before October 1987.

The inventory form incorporates a two-tier approach. Under Tier I, facilities must submit the following aggregate information for each applicable OSHA category of health and physical hazard:

• An estimate (in ranges) of the maximum amount of chemicals for each category present at the facility at any time during the preceding calendar year;

• An estimate (in ranges) of the average daily amount of chemicals in each category; and

• The general location of hazardous chemicals in each category.

Tier I information shall be submitted on or before March 1, 1988 and annually thereafter on March 1.

The public may also request additional information for specific facilities from the SERC and LEPC. Upon the request of the LEPC, the SERC, or the local fire department, the facility must provide the following Tier II information for each covered substance to the organization making the request:

• The chemical name or the common name as indicated on the MSDS;

• An estimate (in ranges) of the maximum amount of the chemical present at any time during the preceding calendar year;

• A brief description of the manner of storage of the chemical;

• The location of the chemical at the facility; and

• An indication of whether the owner elects to withhold information from disclosure to the public.

The information submitted by facilities under Sections 311 and 312 must generally be made available to the public by local and state governments during normal working hours.

As in the case of the MSDS data, this Section 312 information may be useful for LEPCs interested in extending the scope of their planning beyond the facilities covered by Section 302, and for reviewing and updating existing plans. Section 312 information about the quantity and location of chemicals can be of use to fire departments in the development of pre-fire plans. Section 312 data may be of limited use in the initial planning process, given the fact that initial emergency plans are to be completed by October 17, 1988, but they will be useful for the subsequent review and update of plans. Facility owners or operators, at the request of the fire department, must allow the fire department to conduct an on-site inspection and provide specific information about the location of hazardous chemicals.

Section 313:   Toxic Chemical Release Reporting

Section 313 of Title III requires EPA to establish an inventory of toxic chemical emissions from certain facilities. Facilities subject to this reporting requirement must complete a toxic chemical release form (to be prepared by EPA by June 1987) for specified chemicals. The form must be submitted to EPA and those state officials designated by the governor on or before July 1, 1988, and annually thereafter on July 1, reflecting releases during each preceding calendar year.

The purpose of this reporting requirement is to inform government officials and the public about releases of toxic chemicals into the environment. It will also assist in research and the development of regulations, guidelines, and standards.

The reporting requirement applies to owners and operators of facilities that have 10 or more full-time employees, that are in Standard Industrial Classification (SIC) Codes 20 through 39, and that manufactured, processed, or otherwise used a listed toxic chemical in excess of specified threshold quantities. The SIC Codes mentioned cover basically all manufacturing industries.

Facilities using listed toxic chemicals in quantities over 10,000 pounds in a calendar year are required to submit toxic chemical release forms by July 1 of the following year. Facilities manufacturing or processing any of these chemicals in excess of 75,000 pounds in 1987 must report by July 1, 1988. Facilities manufacturing or processing in excess of 50,000 pounds in 1988 must report by July 1, 1989. Thereafter, facilities manufacturing or processing more than 25,000 pounds in a year are required to submit the form. EPA can revise these threshold quantities and the SIC categories involved.

The list of toxic chemicals subject to reporting consists initially of chemicals listed for similar reporting purposes by the states of New Jersey and Maryland. There are over 300 chemicals and categories on these lists. EPA can modify this combined list. In adding a chemical to the combined Maryland and New Jersey lists, EPA must consider the following factors:

(1)   Is the substance known to cause cancer or serious reproductive or neurological disorders, genetic mutations, or other chronic health effects?

(2)   Can the substance cause significant adverse acute health effects as a result of continuous or frequently recurring releases?

(3)   Can the substance cause an adverse effect on the environment because of its toxicity, persistence, or tendency to bioaccumulate?

Chemicals can be deleted if there is not sufficient evidence to establish any of these factors. State governors or any other person may petition the EPA administrator to add or delete a chemical from the list for any of the above reasons. EPA must either publish its reasons for denying the petition, or initiate action to implement the petition within 180 days.

Through early consultation with states or EPA regions, petitioners can avoid duplicating previous petitions and be assisted in locating sources of data already collected on the problem of concern and data sources to support their petitions. EPA will conduct information searches on chemicals contained in a petition, focusing on the effects the petitioners believe warrant addition or deletion.

The toxic chemical release form includes the following information for released chemicals:

• The name, location, and type of business;

• Whether the chemical is manufactured, processed, or otherwise used and the general categories of use of the chemical;

• An estimate (in ranges) of the maximum amounts of the toxic chemical present at the facility at any time during the preceding year;

• Waste treatment and disposal methods and the efficiency of methods for each wastestream;

• The quantity of the chemical entering each environmental medium annually; and

• A certification by a senior official that the report is complete and accurate.

EPA must establish and maintain a national toxic chemical inventory based on the data submitted. This information must be computer accessible on a national database.

In general these Section 313 reports appear to be of limited value in emergency planning. Over time, however, they may contain information that can be used by local planners in developing a more complete understanding of the total spectrum of hazards that a given facility may pose to a community. These reports will not be available to states until July 1, 1988. These reports do not go to the LEPCs directly but they are likely to become available if the LEPCs request them from the states.

Other Title III Provisions

In addition to these four major sections of Title III, there are other provisions of interest to local communities.

Preemption

Section 321 stipulates that (with the exception of the MSDS format and content required by Section 311) Title III does not preempt any state and local laws. In effect, Title III imposes minimum planning and reporting standards where no such standards (or less stringent standards) exist, while permitting states and localities to pursue more stringent requirements as they deem appropriate.

Trade Secrets

Section 322 of Title III addresses trade secrets and applies to Section 303 emergency planning and Sections 311, 312, 313 regarding planning information, community right-to-know reporting requirements, and toxic chemical release reporting. Any person may withhold the specific chemical identity of an extremely hazardous substance or toxic chemical for specific reasons. Even if the chemical identity is withheld, the generic class or category of the chemical must be provided. Such information may be withheld if the facility submits the withheld information to EPA along with an explanation of why the information is a trade secret. The information may not be withheld as a trade secret unless the facility shows each of the following:

• The information has not been disclosed to any other person other than a member of the LEPC, a government official, an employee of such person, or someone bound by a confidentiality agreement, and that measures have been taken to protect the confidentiality;

• The information is not required to be disclosed to the public under any other federal or state law;

• The information is likely to cause substantial harm to the competitive position of the person; and

• The chemical identity could not reasonably be discovered by anyone in the absence of disclosure.

Even if information can be legally withheld from the public, Section 323 requires it not to be withheld from health professionals who require the information for diagnostic purposes or from local health officials who require the information for assessment activities. In these cases, the person receiving the information must be willing to sign a confidentiality agreement with the facility.

Information claimed as trade secret and substantiation for that claim must be submitted to EPA. People may challenge trade secret claims by petitioning EPA, which must then review the claim and rule on its validity.

EPA will publish regulations governing trade secret claims. The regulations will cover the process for submission of claims, petitions for disclosure, and a review process for these petitions.

Enforcement

Section 325 identifies the following enforcement procedures:

- Civil penalties for facility owners or operators who fail to comply with emergency planning requirements;

- Civil, administrative, and criminal penalties for owners or operators who fail to comply with the emergency notification requirements of Section 304;

- Civil and administrative penalties for owners or operators who fail to comply with the reporting requirements in Sections 311-313;

- Civil and administrative penalties for frivolous trade secret claims; and

- Criminal penalties for the disclosure of trade secret information.

In addition to the federal government, state and local governments and individual citizens may enforce the provisions of Title III through the citizen suit authority provided in Section 326.

Training

Section 305 mandates that federal emergency training programs must emphasize hazardous chemicals. It also authorizes the Federal Emergency Management Agency (FEMA) to provide $5 million for each of fiscal years 1987, 1988, 1989, and 1990 for training grants to support state and local governments. These training grants are designed to improve emergency planning, preparedness, mitigation, response, and recovery capabilities. Such programs must give special emphasis to hazardous chemical emergencies. The training grants may not exceed 80 percent of the cost of any such programs. The remaining 20 percent must come from non-federal sources. Consult FEMA and/or EPA Regional offices for a list of training courses.

Review of Emergency Systems

Under Section 305, EPA has initiated a review of emergency systems for monitoring, detecting, and preventing releases of extremely hazardous substances at representative facilities that produce, use, or store these substances. It also is examining public alert systems. EPA will report interim findings to the Congress no later than May 17, 1987 and issue a final report of findings and recommendations to the Congress by April 17, 1988.

The report must include EPA's findings regarding each of the following:

- Status of current technological capabilities to 1) monitor, detect, and prevent significant releases of extremely hazardous substances; 2) determine the magnitude and direction of the hazard posed by each release; 3) identify specific substances; 4) provide data on the specific chemical composition of such releases; and 5) determine relative concentrations of the constituent substances;

• Status of public emergency alert devices or systems for effective public warning of accidental releases of extremely hazardous substances into any media; and

• The technical and economic feasibility of establishing, maintaining, and operating alert systems for detecting releases.

The report must also include EPA's recommendations for the following:

• Initiatives to support development of new or improved technologies or systems that would assist the timely monitoring, detection, and prevention of releases of extremely hazardous substances; and

• Improving devices or systems for effectively alerting the public in the event of an accidental release.

Exhibit 5   Key Title III Dates

The following is a list of some key dates relative to the implementation of the "Emergency Planning and Community Right-to-Know Act of 1986."

November 17, 1986	• EPA publishes interim final List of Extremely Hazardous Substances and their Threshold Planning Quantities in *Federal Register* [§302(a) (2–3)]
November 17, 1986	• EPA initiates comprehensive review of emergency systems [§ 305(b)]
January 27, 1987	• EPA publishes proposed formats for emergency inventory forms and reporting requirements in *Federal Register* (§ 311–12)
March 17, 1987	• National Response Team publishes guidance for preparation and implementation of emergency plans [§ 303 (f)]
April 17, 1987	• State Governors appoint SERCs [§ 301(a)]
May 17, 1987	• Facilities subject to Section 302 planning requirements notify SERC [§ 302(c)]
June 1, 1987	• EPA publishes toxic chemicals release (i.e., emissions inventory) form [§ 302(c)]
July 17, 1987	• SERC designates emergency planning districts [§ 301(b)]
August 17, 1987 (or 30 days after designation of districts, whichever is sooner)	• SERC appoints members of LEPCs [§ 301(c)]
September 17, 1987 (or 30 days after local committee is formed, whichever is earlier)	• Facility notifies LEPC of selection of a facility representative to serve as facility emergency coordinator [§ 303(d) (1)]
October 17, 1987	• MSDSs or list of MSDS chemicals submitted to SERC, LEPC, and local fire department [§ 311(d)]
March 1, 1988	• Facilities submit their initial emergency inventory forms to SERC, LEPC, and local fire department [§ 312(a) (2)]
April 17, 1988	• Final report on emergency systems study due to Congress [§ 305(b)]
July 1, 1988 (and anually hereafter)	• Facilities to submit initial toxic chemical release forms to EPA and designated State officials [§ 313(a)]
October 17, 1988	• LEPCs complete preparation of an emergency plan [§ 303(a)]

Exhibit 6
Title III — Major Information Flow/Requirements

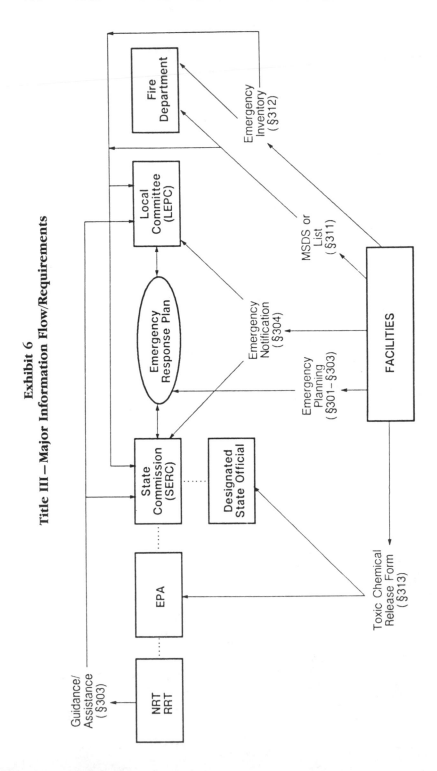

Exhibit 7 Information from Facilities Provided by Title III in Support of LEPC Plan Development

Information Generated by Title III Compliance	Authority	How LEPC Can Use the Information
Facilities subject to Title III planning requirements (including those designated by the Governor or SERC)	Section 302; Notice from Governor/SERC	Hazards analysis — Hazards identification
Additional facilities near subject facilities (such as hospitals, natural gas facilities, etc.)	Sections 302(b)(2); 303(c)(1)	Hazards analysis — Vulnerability analysis
Transportation routes	Sections 303(c)(1); 303(d)(3)	Hazards analysis — Hazards identification
Major chemical hazards (chemical name, properties, location, and quantity)	Section 303(d)(3) for extremely hazardous substances used, produced, stored	Hazards analysis — Hazards identification
	Section 311 MSDSs for chemicals manufactured or imported	
	Section 312 inventories for chemicals manufactured or imported	
Facility and community response methods, procedures, and personnel	Sections 303(c)(2); 303(d)(3)	Response functions
Facility and community emergency coordinators	Sections 303(c)(3); 303(d)(1)	Assistance in preparing and implementing the plan
Release detection and notification procedures	Sections 303(c)(4); 303(d)(3)	Initial notification Warning systems
Methods for determining release occurrence and population affected	Sections 303(c)(5); 303(d)(3)	Hazards analysis — Vulnerability analysis and risk analysis
Facility equipment and emergency facilities; persons responsible for such equipment and facilities	Sections 303(c)(6); 303(d)(3)	Resource management
Evacuation plans	Sections 303(c)(7); 303(d)(3)	Evacuation planning
Training programs	Sections 303(c)(8); 303(d)(3)	Resource management
Exercise methods and schedules	Sections 303(c)(9); 303(d)(3)	Testing and updating

Exhibit 8   Title III Chemical Lists and Their Purposes

List	Required in Section	Purpose
Extremely Hazardous Substances (*Federal Register* 11/17/86 — initially 402 chemicals listed in CEPP Interim Guidance)	Section 302: Emergency Planning	• Facilities with more than established planning quantities of these substances must notify the SERC. • Initial focus for preparation of emergency plans by LEPCs
Substance requiring notification under Section 103(a) of CERCLA (717 chemicals)	Section 304: Emergency Notification	• Certain releases of these chemicals trigger Section 304 notification to SERC and LEPC.
	Section 304: Emergency Notification	• Certain releases of these chemicals trigger Section 304 notification to SERC and LEPC as well as CERCLA Section 103(a) requirement to notify National Response Center.
Hazardous Chemicals considered physical or health hazards under OSHA's Hazard Communication Standard (This is a performance standard, there is no specific list of chemicals.)	Section 304: Emergency Notification	• Identifies facilities subject to emergency notification requirements
	Section 311: Material Safety Data Sheets	• MSDS or list of MSDS chemicals provided by facilities to SERC, LEPC, and local fire department
	Section 312 Emergency and Hazardous Chemical Inventory	• Covered facilities provide site-specific information on the quality and location of chemicals to SERC, LEPC, and local fire departments to inform the community and assist in plan preparation.
Toxic Chemicals identified as chemicals of concern by States of New Jersey and Maryland (329 chemicals/chemical categories)	Section 313: Toxic Chemical Release Reporting	• These chemicals are reported on an emissions inventory to inform government officials and the public about releases of toxic chemicals in the environment.

Appendix B

LIST OF ACRONYMS AND RECOGNIZED ABBREVIATIONS

AAR/BOE	Association of American Railroads/Bureau of Explosives
AIChE	American Institute of Chemical Engineers
ASCS	Agricultural Stabilization and Conservation Service
ASME	American Society of Mechanical Engineers
ASSE	American Society of Safety Engineers
ATSDR	Agency for Toxic Substances and Disease Registry (HHS)
CAER	Community Awareness and Emergency Response (CMA)
CDC	Centers for Disease Control (HHS)
CEPP	Chemical Emergency Preparedness Program
CERCLA	Comprehensive Environmental Response, Compensation, and Liability Act of 1980 (PL 96-510)
CFR	Code of Federal Regulations
CHEMNET	A mutual aid network of chemical shippers and contractors
CHEMTREC	Chemical Transportation Emergency Center
CHLOREP	A mutual aid group comprised of shippers and carriers of chlorine.
CHRIS/HACS	Chemical Hazards Response Information System/Hazard Assessment Computer System
CMA	Chemical Manufacturers Association
CPG 1-3	Federal Assistance Handbook: Emergency Management, Direction and Control Programs
CPG 1-8	Guide for Development of State and Local Emergency Operations Plans
CPG 1-8A	Guide for the Review of State and Local Emergency Operations Plans
CWA	Clean Water Act
DOC	U.S. Department of Commerce
DOD	U.S. Department of Defense
DOE	U.S. Department of Energy
DOI	U.S. Department of the Interior
DOJ	U.S. Department of Justice
DOL	U.S. Department of Labor
DOS	U.S. Department of State
DOT	U.S. Department of Transportation
EENET	Emergency Education Network (FEMA)
EMA	Emergency Management Agency
EMI	Emergency Management Institute

EOC	Emergency Operating Center
EOP	Emergency Operations Plan
EPA	U.S. Environmental Protection Agency
ERD	Emergency Response Division (EPA)
FEMA	Federal Emergency Management Agency
FEMA-REP-5	Guidance for Developing State and Local Radiological Emergency Response Plans and Preparedness for Transportation Accidents
FWPCA	Federal Water Pollution Control Act
HAZMAT	Hazardous Materials
HAZOP	Hazard and Operability Study
HHS	U.S. Department of Health and Human Services
ICS	Incident Command System
IEMS	Integrated Emergency Management System
LEPC	Local Emergency Planning Committee
MSDS	Material Safety Data Sheet
NACA	National Agricultural Chemicals Association
NCP	National Contingency Plan
NCRIC	National Chemical Response and Information Center (CMA)
NETC	National Emergency Training Center
NFA	National Fire Academy
NFPA	National Fire Protection Association
NIOSH	National Institute of Occupational Safety and Health
NOAA	National Oceanic and Atmospheric Administration
NRC U.S.	Nuclear Regulatory Commission; National Response Center
NRT	National Response Team
NUREG 0654/	Criteria for Preparation and Evaluation of Radiological
FEMA-REP-1	Emergency Response Plans and Preparedness in Support of Nuclear Power Plants
OHMTADS	Oil and Hazardous Materials Technical Assistance Data System
OSC	On-Scene Coordinator
OSHA	Occupational Safety and Health Administration (DOL)
PSTN	Pesticide Safety Team Network
RCRA	Resource Conservation and Recovery Act
RQs	Reportable Quantities
RRT	Regional Response Team
RSPA	Research and Special Programs Administration (DOT)
SARA	Superfund Amendments and Reauthorization Act of 1986 (PL 99-499)
SCBA	Self-Contained Breathing Apparatus
SERC	State Emergency Response Commission
SPCC	Spill Prevention Control and Countermeasures
TSD	Treatment, Storage, and Disposal Facilities
USCG	U.S. Coast Guard (DOT)
USDA	U.S. Department of Agriculture
USGS	U.S. Geological Survey
USNRC	U.S. Nuclear Regulatory Commission

Appendix C

CAER — Community Awareness and Emergency Response program developed by the Chemical Manufacturers Association. Guidance for chemical plant managers to assist them in taking the initiative in cooperating with local communities to develop integrated (community/industry) hazardous materials response plans.

CEPP — Chemical Emergency Preparedness Program developed by EPA to address accidental releases of acutely toxic chemicals.

CERCLA — Comprehensive Environmental Response, Compensation, and Liability Act regarding hazardous substance releases into the environment and the cleanup of inactive hazardous waste disposal sites.

CHEMNET — A mutual aid network of chemical shippers and contractors. CHEMNET has more than fifty participating companies with emergency teams, twenty-three subscribers (who receive services in an incident from a participant and then reimburse response and cleanup costs), and several emergency response contractors. CHEMNET is activated when a member shipper cannot respond promptly to an incident involving that company's product(s) and requiring the presence of a chemical expert. If a member company cannot go to the scene of the incident, the shipper will authorize a CHEMNET-contracted emergency response company to go. Communications for the network are provided by CHEMTREC, with the shipper receiving notification and details about the incident from the CHEMTREC communicator.

CHEMTREC — Chemical Transportation Emergency Center operated by the Chemical Manufacturers Association. Provides information and/or assistance to emergency responders. CHEMTREC contacts the shipper or producer of the material for more detailed information, including on-scene assistance when feasible. Can be reached 24 hours a day by calling 800-424-9300. (Also see *HIT.*)

CHLOREP — Chlorine Emergency Plan operated by the Chlorine Institute. 24-hour mutual aid program. Response is activated by a CHEMTREC call to the designated CHLOREP contact, who notifies the appropriate team leader, based upon CHLOREP's geographical sector assignments for teams. The team leader in turn calls the emergency caller at the incident scene and determines what advice and assistance are needed. The team leader then decides whether or not to dispatch his team to the scene.

CHRIS/HACS — Chemical Hazards Response Information System/Hazard Assessment Computer System developed by the U.S. Coast Guard. HACS is a computerized model of the four CHRIS manuals that contain chemical-specific data. Federal OSCs use HACS to find answers to specific questions during a chemical spill/response. State and local officials and industry representatives may ask an OSC to request a HACS run for contingency planning purposes.

CPG 1-3 — Federal Assistance Handbook: Emergency Management, Direction and Control Programs, prepared by FEMA. Provides states with guidance on administrative and programmatic requirements associated with FEMA funds.

CPG 1-5 — Objectives for Local Emergency Management, prepared by FEMA. Describes and explains functional objectives that represent a comprehensive and integrated emergency management program. Includes recommended activities for each objective.

CPG 1-8 — Guide for Development of State and Local Emergency Operations Plans, prepared by FEMA (see *EOP* below).

CPG 1-8A — Guide for the Review of State and Local Emergency Operations Plans, prepared by FEMA. Provides FEMA staff with a standard instrument for assessing EOPs that are developed to satisfy the eligibility requirement to receive Emergency Management Assistance funding.

CPG 1-35 — Hazard Identification, Capability Assessment, and Multi-Year Development Plan for Local Governments, prepared by FEMA. As a planning tool, it can guide local jurisdictions through a logical sequence for identifying hazards, assessing capabilities, setting priorities, and scheduling activities to improve capability over time.

EBS
— Emergency Broadcasting System to be used to inform the public about the nature of a hazardous materials incident and what safety steps they should take.

EMI
— The Emergency Management Institute is a component of FEMA's National Emergency Training Center located in Emmitsburg, Maryland. It conducts resident and nonresident training activities for federal, state, and local government officials, managers in the private economic sector, and members of professional and volunteer organizations on subjects that range from civil nuclear preparedness systems to domestic emergencies caused by natural and technological hazards. Nonresident training activities are also conducted by State Emergency Management Training Offices under cooperative agreements that offer financial and technical assistance to establish annual training programs that fulfill emergency management training requirements in communities throughout the nation.

ERT
— Environmental Response Team, a group of highly specialized experts available through EPA 24 hours a day.

EOP
— Emergency Operations Plan developed in accord with the guidance in CPG 1-8. EOPs are multi-hazard, functional plans that treat emergency management activities generically. EOPs provide for as much generally applicable capability as possible without reference to any particular hazard; then they address the unique aspects of individual disasters in hazard-specific appendices.

FAULT-TREE ANALYSIS
— A means of analyzing hazards. Hazardous events are first identified by other techniques such as HAZOP. Then all combinations of individual failures that can lead to that hazardous event are shown in the logical format of the fault tree. By estimating the individual failure probabilities, and then using the appropriate arithmetical expressions, the top-event frequency can be calculated.

FEMA-REP-5
— Guidance for Developing State and Local Radiological Emergency Response Plans and Preparedness for Transportation Accidents, prepared by FEMA. Provides a basis for state and local governments to develop emergency plans and improve emergency preparedness

for transportation accidents involving radioactive materials.

HAZARDOUS MATERIALS — Refers generally to hazardous substances, petroleum, natural gas, synthetic gas, acutely toxic chemicals, and other toxic chemicals.

HAZOP — Hazard and operability study, a systematic technique for identifying hazards or operability problems throughout an entire facility. One examines each segment of a process and lists all possible deviations from normal operating conditions and how they might occur. The consequences on the process are assessed, and the means available to detect and correct the deviations are examined.

HIT — Hazard Information Transmission program provides a digital transmission of the CHEMTREC emergency chemical report to first responders at the scene of a hazardous materials incident. The report advises the responder on the hazards of the materials, the level of protective clothing required, mitigating action to take in the event of a spill, leak, or fire, and first aid for victims. HIT is a free public service provided by the Chemical Manufacturers Association. Reports are sent in emergency situations only to organizations that have preregistered with HIT. Brochures and registration forms may be obtained by writing: Manager, CHEMTREC/ CHEMNET, 2501 M Street, N.W., Washington, DC 20037.

ICS — Incident Command System, the combination of facilities, equipment, personnel, procedures, and communications operating within a common organizational structure with responsibility for management of assigned resources to effec- tively accomplish stated objectives at the scene of an incident.

IEMS — Integrated Emergency Management System, developed by FEMA in recognition of the economies realized in planning for all hazards on a generic functional basis as opposed to developing independent structures and resources to deal with each type of hazard.

NCP — National Oil and Hazardous Substances Pollution Contingency Plan (40 CFR Part 300), prepared by EPA to put into effect the response powers and responsibilities created by CERCLA and the authorities established by

Section 311 of the Clean Water Act.

NFA   — The National Fire Academy is a component of FEMA's National Emergency Training Center located in Emmitsburg, Maryland. It provides fire prevention and control training for the fire service and allied services. Courses on campus are offered in technical, management, and prevention subject areas. A growing off-campus course delivery system is operated in conjunction with state fire training program offices.

NHMIE — National Hazardous Materials Information Exchange, provides information on hazmat training courses, planning techniques, events and conferences, and emergency response experiences and lessons learned. Call toll-free 1-800-752-6367 (in Illinois, 1-800-367-9592). Planners with personal computer capabilities can access NHMIE by calling FTS 972-3275 or (312) 972-3275.

NRC   — National Response Center, a communications center for activities related to response actions, is located at Coast Guard headquarters in Washington, DC. The NRC receives and relays notices of discharges or releases to the appropriate OSC, disseminates OSC and RRT reports to the NRT when appropriate, and provides facilities for the NRT to use in coordinating a national response action when required. The toll-free number (800-424-8802, or 202-426-2675 or 202-267-2675 in the Washington, DC area) can be reached 24 hours a day for reporting actual or potential pollution incidents.

NRT   — National Response Team, consisting of representatives of 14 government agencies (DOD, DOI, DOT/RSPA, DOT/USCG, EPA, DOC, FEMA, DOS, USDA, DOJ, HHS, DOL, Nuclear Regulatory Commission, and DOE), is the principal organization for implementing the NCP. When the NRT is not activated for a response action, it serves as a standing committee to develop and maintain preparedness, to evaluate methods of responding to discharges or releases, to recommend needed changes in the response organization, and to recommend revisions to the NCP. The NRT may consider and make recommendations to appropriate agencies on the training, equipping, and protection of response teams; and necessary research, development, demonstration, and evaluation to improve response capabilities.

NSF — National Strike Force, made up of three Strike Teams. The USCG counterpart to the EPA ERTs.

NUREG 0654/ — Criteria for Preparation and Evaluation of Radiological
FEMA-REP-1 Emergency Response Plans and Preparedness in Support of Nuclear Power Plants, prepared by NRC and FEMA. Provides a basis for state and local government and nuclear facility operators to develop radiological emergency plans and improve emergency preparedness. The criteria also will be used by federal agency reviewers in determining the adequacy of State, local, and nuclear facility emergency plans and preparedness.

OHMTADS — Oil and Hazardous Materials Technical Assistance Data System, a computerized data base containing chemical, biological, and toxicological information about hazardous substances. OSCs use OHMTADS to identify unknown chemicals and to learn how to best handle known chemicals.

OSC — On-Scene Coordinator, the federal official predesignated by EPA or USCG to coordinate and direct federal responses and removals under the NCP; or the DOD official designated to coordinate and direct the removal actions from releases of hazardous substances, pollutants, or contaminants from DOD vessels and facilities. When the NRC receives notification of a pollution incident, the NRC Duty Officer notifies the appropriate OSC, depending on the location of an incident. Based on this initial report and any other information that can be obtained, the OSC makes a preliminary assessment of the need for a federal response. If an on-scene response is required, the OSC will go to the scene and monitor the response of the responsible party or state or local government. If the responsible party is unknown or not taking appropriate action, and the response is beyond the capability of state and local governments, the OSC may initiate federal actions, using funding from the FWPCA Pollution Fund for oil discharges and the CERCLA Trust Fund (Superfund) for hazardous substance releases.

PSTN — Pesticide Safety Team Network operated by the National Agricultural Chemicals Association to minimize environmental damage and injury arising from accidental pesticide spills or leaks. PSTN area coordinators in ten regions nationwide are available 24 hours a day to receive pesticide incident notifications from CHEMTREC.

RCRA   — Resource Conservation and Recovery Act (of 1976) established a framework for the proper management and disposal of all wastes. RCRA directed EPA to identify hazardous wastes, both generically and by listing specific wastes and industrial process waste streams. Generators and transporters are required to use good management practices and to track the movement of wastes with a manifest system. Owners and operators of treatment, storage, and disposal facilities also must comply with standards, which are generally implemented through permits issued by EPA or authorized states.

RRT   — Regional Response Teams composed of representatives of federal agencies and a representative from each state in the federal region. During a response to a major hazardous materials incident involving transportation or a fixed facility, the OSC may request that the RRT be convened to provide advice or recommendations in specific issues requiring resolution. Under the NCP, RRTs may be convened by the chairman when a hazardous materials discharge or release exceeds the response capability available to the OSC in the place where it occurs; crosses regional boundaries; or may pose a substantial threat to the public health, welfare, or environment, or to regionally significant amounts of property. Regional contingency plans specify detailed criteria for activation of RRTs. RRTs may review plans developed in compliance with Title III, if the local emergency planning committee so requests.

SARA   — The "Superfund Amendments and Reauthorization Act of 1986." Title III of SARA includes detailed provisions for community planning.

Superfund — The trust fund established under CERCLA to provide money the OSC can use during a cleanup.

Title III — The "Emergency Planning and Community Right-to-Know Act of 1986." Specifies requirements for organizing the planning process at the state and local levels for specified extremely hazardous substances; minimum plan content; requirements for fixed facility owners and operators to inform officials about extremely hazardous substances present at the facilities; and mechanisms for making information about extremely hazardous substances available to citizens. (See Appendix A.)

Appendix D

CRITERIA FOR ASSESSING STATE AND LOCAL PREPAREDNESS

C.1   Introduction

The criteria in this appendix, an adaptation of criteria developed by the Preparedness Committee of the NRT in August 1985, represent a basis for assessing a state or local hazardous materials emergency response preparedness program. These criteria reflect the basic elements judged to be important for a successful emergency preparedness program.

The criteria are separated into six categories, all of which are closely interrelated. These categories are hazards analysis, authority, organizational structure, communications, resources, and emergency planning.

These criteria may be used for assessing the emergency plan as well as the emergency preparedness program in general. It must be recognized, however, that few state or local governments will have the need and/or capability to address all these issues and meet all these criteria to the fullest extent. Resource limitations and the results of the hazards analysis will strongly influence the necessary degree of planning and preparedness. Those governmental units that do not have adequate resources are encouraged to seek assistance and take advantage of all resources that are available.

Other criteria exist that could be used for assessing a community's preparedness and emergency planning. These include FEMA's CPG 1-35 (Hazard Identification, Capability Assessment and Multi-Year Development Plan for Local Governments) and CPG 1-8A. Additionally, states may have issued criteria for assessing capability.

C.2   The Criteria

C.2.1   Hazards Analysis

"Hazards Analysis" includes the procedures for determining the susceptibility or vulnerability of a geographical area to a hazardous materials release, for identifying potential sources of a hazardous materials release from fixed facilities that manufacture, process, or otherwise use, store, or dispose of materials that are generally considered hazardous in an unprotected environment. This also includes an analysis of the potential or probable hazard of transporting hazardous materials through a particular area.

A hazards analysis is generally considered to consist of identification of potential hazards, determination of the vulnerability of an area as a result of the existing hazards, and an assessment of the risk of a hazardous materials release or spill.

The following criteria may assist in assessing a hazards analysis:

• Has a hazards analysis been completed for the area? If one exists, when was it last updated?

• Does the hazards analysis include the location, quantity, and types of hazardous materials that are manufactured, processed, used, disposed, or stored within the appropriate area?

• Was it done in accordance with community right-to-know laws and prefire plans?

• Does it include the routes by which the hazardous materials are transported?

• Have areas of public health concern been identified?

• Have sensitive environmental areas been identified?

• Have historical data on spill incidents been collected and evaluated?

• Have the levels of vulnerability and probable locations of hazardous materials incidents been identified?

• Are environmentally sensitive areas and population centers considered in analyzing the hazards of the transportation routes and fixed facilities?

C.2.2   Authority

"Authority" refers to those statutory authorities or other legal authorities vested in any personnel, organizations, agencies, or other entities in responding to or being prepared for responding to hazardous materials emergencies resulting from releases or spills.

The following criteria may be used to assess the existing legal authorities for response actions:

• Do clear legal authorities exist to establish a comprehensive hazardous materials response mechanism (federal, state, county, and local laws, ordinances, and policies)?

• Do these authorities delegate command and control responsibilities between the different organizations within the same level of government (horizontal), and/or provide coordination procedures to be followed?

• Do they specify what agency(ies) has (have) overall responsibility for directing or coordinating a hazardous materials response?

• Do they specify what agency(ies) has (have) responsibility for providing assistance or support for hazardous materials response and what comprises that assistance or support?

• Have the agency(ies) with authority to order evacuation of the community been identified?

• Have any limitations in the legal authorities been identified?

C.2.3   Organizational Structure

"Organizational" refers to the organizational structure in place for responding to emergencies. This structure will, of course, vary considerably from state to state and from locality to locality.

There are two basic types of organizations involved in emergency response operations. The first is involved in the planning and policy decision process similar to the NRT and RRT. The second is the operational response group that functions within the precepts set forth in the state or local plan. Realizing that situations vary from state to state and locality to locality and that emergency planning for the state and local level may involve the preparation of multiple situation plans or development of a single comprehensive plan, the criteria should be broadly based and designed to detect a potential flaw that would then precipitate a more detailed review.

• Are the following organizations included in the overall hazardous materials emergency preparedness activities?

  • Health organizations (including mental health organizations)

  • Public safety

    • fire

    • police

    • health and safety (including occupational safety and health)

    • other responders

  • Transportation

  • Emergency management/response planning

  • Environmental organizations

  • Natural resources agencies (including trustee agencies)

- Environmental agencies with responsibilities for:

  - fire

  - health

  - water quality

  - air quality

  - consumer safety

- Education system (in general)

  - public education

  - public information

- Private sector interface

  - trade organizations

  - industry officials

- Labor organizations

- Have each organization's authorities, responsibilities, and capabilities been determined for pre-response (planning and prevention), response (implementing the plan during an incident), and post-response (cleanup and restoration) activities?

- Has one organization been given the command and control responsibility for these three phases of emergency response?

- Has a "chain of command" been established for response control through all levels of operation?

- Are the roles, relationships, and coordination procedures between government and non-government (private entities) delineated? Are they understood by all affected parties? How are they instituted (written, verbal)?

- Are clear interrelationships, and coordination procedures between government and non-government (private entities) delineated? Are they understood by all affected parties? How are they instituted (written, verbal)?

- Are the agencies or departments that provide technical guidance during a response the same agencies or departments that provide technical guidance in non-emergency situations? In other words, does the organizational structure vary with the type of situation to be addressed?

- Does the organizational structure provide a mechanism to meet regularly for planning and coordination?

- Does the organizational structure provide a mechanism to regularly exercise the response organization?

- Has a simulation exercise been conducted within the last year to test the organizational structure?

- Does the organizational structure provide a mechanism to review the activities conducted during a response or exercise to correct shortfalls?

- Have any limitations within the organizational structure been identified?

- Is the organizational structure compatible with the federal response organization in the NCP?

- Have trained and equipped incident commanders been identified?

- Has the authority for site decisions been vested in the incident commanders?

- Have the funding sources for a response been identified?

- How quickly can the response system be activated?

C.2.4 Communication

"Communication" means any form or forms of exchanging information or ideas for emergency response with other entities, either internal or external to the existing organizational structure.

Coordination:

- Have procedures been established for coordination of information during a response?

- Has one organization been designated to coordinate communications activities?

- Have radio frequencies been established to facilitate coordination between different organizations?

Information Exchange:

- Does a formal system exist for information sharing among agencies, organizations, and the private sector?

- Has a system been established to ensure that "lessons learned" are passed to the applicable organizations?

Information Dissemination:

• Has a system been identified to carry out public information/community relations activities?

• Has one organization or individual been designated to coordinate with or speak to the media concerning the release?

• Is there a communication link with an Emergency Broadcast System (EBS) point of entry (CPCS-1) station?

• Does a communications system/method exist to disseminate information to responders, affected public, etc.?

• Is this system available 24 hours a day?

• Have alternate systems/methods of communications been identified for use if the primary method fails?

• Does a mechanism exist to keep telephone rosters up-to-date?

• Are communications networks tested on a regular basis?

Information Sources and Data Base Sharing:

• Is a system available to provide responders with rapid information on the hazards of chemicals involved in an incident?

• Is this information available on a 24-hour basis? Is it available in computer software?

• Is a system in place to update the available information sources?

Notification Procedures:

• Have specific procedures for notification of a hazardous materials incident been developed?

• Are multiple notifications required by overlapping requirements (e.g., state, county, local each have specific notification requirements)?

• Does the initial notification system have a standardized list of information that is collected for each incident?

• Does a network exist for notifying and activating necessary response personnel?

• Does a network exist for notifying or warning the public of potential hazards resulting from a release? Does this network have provisions for informing the public of what hazards to expect, what precautions to take, whether evacuation is required, etc.?

- Has a central location or phone number been established for initial notification of an incident?

- Is the central location or phone number accessible on a 24-hour basis?

- Does the central location phone system have the ability to expand to a multiple line system during an emergency?

Clearinghouse Functions:

- Has a central clearinghouse for hazardous materials information been established with access by the public and private sector?

C.2.5   Resources

"Resource" means the personnel, training, equipment, facilities, and other sources available for use in responding to hazardous materials emergencies. To the extent that the hazards analysis has identified the appropriate level of preparedness for the area, these criteria may be used in evaluating available resources of the jurisdiction undergoing review.

Personnel:

- Have the numbers of trained personnel available for hazardous materials been determined?

- Has the location of trained personnel available for hazardous materials been determined? Are these personnel located in areas identified in the hazards analysis as:

  - heavily populated;

  - high hazard areas — i.e., numbers of chemical (or other hazardous materials) production facilities in well-defined areas;

  - hazardous materials storage, disposal, and/or treatment facilities; and

  - transit routes?

- Are sufficient personnel available to maintain a given level of response capability identified as being required for the area?

- Has the availability of special technical expertise (chemists, industrial hygienists, toxicologists, occupational health physicians, etc.) necessary for response been identified?

- Have limitations on the use of above personnel resources been identified?

- Do mutual aid agreements exist to facilitate interagency support between organizations?

Training:

- Have the training needs for the state/local area been identified?

- Are centralized response training facilities available?

- Are specialized courses available covering topics such as:

  - organizational structures for response actions (i.e., authorities and coordination);

  - response actions;

  - equipment selection, use, and maintenance; and

  - safety and first aid?

- Does the organizational structure provide training and cross training for or between organizations in the response mechanism?

- Does an organized training program for all involved response personnel exist? Has one agency been designated to coordinate this training?

- Have training standards or criteria been established for a given level of response capability? Is any certification provided upon completion of the training?

- Has the level of training available been matched to the responsibilities or capabilities of the personnel being trained?

- Does a system exist for evaluating the effectiveness of training?

- Does the training program provide for "refresher courses" or some other method to ensure that personnel remain up-to-date in their level of expertise?

- Have resources and organizations available to provide training been identified?

- Have standardized curricula been established to facilitate consistent State-wide training?

Equipment:

- Have response equipment requirements been identified for a given level of response capability?

- Are the following types of equipment available?

  - personal protective equipment

  - first aid and other medical emergency equipment

- emergency vehicles available for hazardous materials response

- sampling equipment (air, water, soil, etc.) and other monitoring devices (e.g., explosivity meters, oxygen meters)

- analytical equipment or facilities available for sample analyses

- fire fighting equipment/other equipment and material (bulldozers, boats, helicopters, vacuum trucks, tank trucks, chemical retardants, foam)

- Are sufficient quantities of each type of equipment available on a sustained basis?

- Is all available equipment capable of operating in the local environmental conditions?

- Are up-to-date equipment lists maintained? Are they computerized?

- Are equipment lists available to all responders?

- Are these lists broken down into the various types of equipment (e.g., protective clothing, monitoring instruments, medical supplies, transportation equipment)?

- Is there a mechanism to ensure that the lists are kept up to date?

- Have procedures necessary to obtain equipment on a 24-hour basis been identified?

- Does a program exist to carry out required maintenance of equipment?

- Are there maintenance and repair records for each piece of equipment?

- Have mutual aid agreements been established for the use of specialized response equipment?

- Is sufficient communications equipment available for notifying personnel or to transmit information? Is the equipment of various participating agencies compatible?

- Is transportation equipment available for moving equipment rapidly to the scene of an incident, and its state of readiness assured?

Facilities:

- Have facilities capable of performing rapid chemical analyses been identified?

- Do adequate facilities exist for storage and cleaning/reconditioning of response equipment?

• Have locations or facilities been identified for the storage, treatment, recycling, and disposal of wastes resulting from a release?

• Do adequate facilities exist for carrying out training programs?

• Do facilities exist that are capable of providing medical treatment to persons injured by chemical exposure?

• Have facilities and procedures been identified for housing persons requiring evacuation or temporary relocation as a result of an incident?

• Have facilities been identified that are suitable for command centers?

C.2.6   Emergency Plan

The emergency plan, while it relates to many of the above criteria, also stands alone as a means to assess preparedness at the state and local level of government, and in the private sector. The following questions are directed more toward evaluating the plan rather than determining the preparedness level of the entity that has developed the plan. It is not sufficient to ask if there is a plan, but rather to determine if the plan that does exist adequately addresses the needs of the community or entity for which the plan was developed.

• Have the levels of vulnerability and probable locations of hazardous materials incidents been identified in the plan?

• Have areas of public health concern been identified in the plan?

• Have sensitive environmental areas been identified in the plan?

• For the hazardous materials identified in the area, does the plan include information on the chemical and physical properties of the materials, safety and emergency response information, and hazard mitigation techniques? (NOTE: It is not necessary that all this information be included in the emergency plan, the plan should, however, at least explain where such information is available.)

• Have all appropriate agencies, departments, or organizations been involved in the process of developing or reviewing the plan?

• Have all the appropriate agencies, departments, or organizations approved the plan?

• Has the organizational structure and notification list defined in the plan been reviewed in the last six months?

• Is the organizational structure identified in the plan compatible with the federal response organization in the NCP?

• Has one organization been identified in the plan as having command and control responsibility for the pre-response, response, and post-response phases?

• Does the plan define the organizational responsibilities and relationships among city, county, district, state, and federal response agencies?

• Are all organizations that have a role in hazardous materials response identified in the plan (public safety and health, occupational safety and health, transportation, natural resources, environmental, enforcement, educational, planning, and private sector)?

• Are the procedures and contacts necessary to activate or deactivate the organization clearly given in the plan for the pre-response, response, and post-response phases?

• Does the organizational structure outlined in the plan provide a mechanism to review the activities conducted during a response or exercise to correct shortfalls?

• Does the plan include a communications system/method to disseminate information to responders, affected public, etc.?

• Has a system been identified in the plan to carry out public information/community relations activities?

• Has a central location or phone number been included in the plan for initial notification of an incident?

• Have trained and equipped incident commanders been identified in the plan?

• Does the plan include the authority for vesting site decisions in the incident commander?

• Have government agency personnel that may be involved in response activities been involved in the planning process?

• Have local private response organizations (e.g., chemical manufacturers, commercial cleanup contractors) that are available to assist during a response been identified in the plan?

• Does the plan provide for frequent training exercises to train personnel or to test the local contingency plans?

• Are lists/systems that identify emergency equipment available to response personnel included in the plan?

• Have locations of materials most likely to be used in mitigating the effects of a release (e.g., foam, sand, lime) been identified in the plan?

• Does the plan address the potential needs for evacuation, what agency is authorized to order or recommend an evacuation, how it will be carried out, and where people will be moved?

• Has an emergency operating center, command center, or other central location with the necessary communications capabilities been identified in the plan for coordination of emergency response activities?

• Are there follow-up response activities scheduled in the plan?

• Are there procedures for updating the plan?

• Are there addenda provided with the plan, such as: laws and ordinances, statutory responsibilities, evacuation plans, community relations plan, health plan, and resource inventories (personnel, equipment, maps [not restricted to road maps], and mutual aid agreements)?

• Does the plan address the probable simultaneous occurrence of different types of emergencies (e.g., power outage and hazardous materials releases) and the presence of multiple hazards (e.g., flammable and corrosive) during hazardous materials emergencies?

Appendix E

BIBLIOGRAPHY

General Emergency Planning for Hazardous Materials

American Institute of Chemical Engineers, Center for Chemical Plant Safety. *Guidelines for Hazard Evaluation Procedures*. Washington, DC: A.I.Ch.E., 1985.

American Society of Testing & Materials. *Toxic and Hazardous Industrial Chemicals Safety Manual*. 1983.

Association of Bay Area Governments. *San Francisco Bay Area: Hazardous Spill Prevention and Response Plan*. Volumes I & II. Berkeley, CA: 1983.

Avoiding and Managing Environmental Damage from Major Industrial Accidents. Proc. of Conference of the Air Pollution Control Association. 1985.

Bretherick, L. *Handbook of Reactive Chemical Hazards*. 2nd ed. Butterworth, 1979.

Brinsko, George A. et al. *Hazardous Material Spills and Responses for Municipalities*. (EPA-600/2-80-108, NTIS PB80-214141). 1980.

Cashman, John R. *Hazardous Materials Emergencies: Response and Control*. 1983.

Chemical Manufacturers Association. *Community Awareness and Emergency Response Program Handbook*. Washington, DC: CMA, 1985.

Chemical Manufacturers Association. *Community Emergency Response Exercise Program*. Washington, DC: CMA, 1986.

Chemical Manufacturers Association. *Risk Analysis in the Chemical Industry–Proceedings of a Symposium*. Rockville, MD: Government Institutes, Inc., 1985.

Chemical Manufacturers Association. *Site Emergency Response Planning*. Washington, DC: CMA, 1986.

Copies of the CMA guides can be obtained by writing to:

Publications Fulfillment
Chemical Manufacturers Association
2501 M Street, N.W.
Washington, DC 20037

Emergency Management and Civil Defense Division, Consolidated City of Indianapolis. *Final Report: Demonstration Project to Develop a Hazardous Materials Accident Prevention and Emergency Response Program, Phases I, II, III, IV.* Indianapolis: 1983.

Energy Resources Co., Inc.; Cambridge Systematics, Inc.; Massachusetts Department of Environmental Quality Engineering. *Demonstration Project to Develop a Hazardous Materials Accident Prevention and Emergency Response Program for the Commonwealth of Massachusetts.* Volumes I & II. Cambridge and Boston, MA: 1983.

Environmental and Safety Design, Inc. *Development of a Hazardous Materials Accident Prevention and an Emergency Response Program.* Memphis, TN: 1983.

Federal Emergency Management Agency. *Disaster Operations: A Handbook for Local Governments.* Washington, DC: 1981.

Federal Emergency Management Agency. *Hazard Identification, Capability Assessment, and Multi-Year Development Plan for Local Governments.* CPG 1-35, Washington, DC: 1985.

Federal Emergency Management Agency. *Objectives for Local Emergency Management.* CPG 1-5, Washington, DC: 1984.

Federal Emergency Management Agency. *Professional Development Series: Emergency Planning—Student Manual.* Washington, DC.

Federal Emergency Management Agency. *Professional Development Series: Introduction to Emergency Management—Student Manual.* Washington, DC.

Gabor, T. and T.K., Griffith. *The Assessment of Community Vulnerability to Acute Hazardous Materials Incidents.* Newark, DE: University of Delaware, 1985.

Government Institutes, Inc. Md. *R.C.R.A. Hazardous Waste Handbook.* Volumes 1 & 2. 1981.

Green, Don W., ed. *Perry's Chemical Engineers' Handbook.* 6th ed. McGraw-Hill, 1984.

Hawley, Gessner G., ed. *Condensed Chemical Dictionary.* 10th ed. New York: Van Nostrand Reinhold, 1981.

Hildebrand, Michael S. *Disaster Planning Guidelines for Fire Chiefs.* Washington, DC: International Association of Fire Chiefs, 1980.

Multnomah County Office of Emergency Management. *Hazardous Materials Management System: A Guide for Local Emergency Managers.* Portland, OR: 1983.

National Fire Protection Association. *Fire Protection Guide on Hazardous Materials.* Boston: NFPA, 1986.

National Institute of Occupational Safety and Health. *Pocket Guide to Chemical Hazards.* Washington, DC: DHEW (NIOSH) 78-210, 1985. (GPO Stock No. 017-033-00342-4)

New Orleans, City of. *Demonstration Project to Develop a Hazardous Materials Accident Prevention and Emergency Response Program for the City of New Orleans, Phases I, II, III, IV.* New Orleans: 1983.

Portland Office of Emergency Management. *Hazardous Materials Hazard Analysis.* Portland, OR: 1981.

Puget Sound Council of Governments. *Hazardous Materials Demonstration Project Report: Puget Sound Region.* Seattle, WA: 1981.

Sax, N. Irving. *Dangerous Properties of Industrial Materials.* 6th ed. New York: Van Nostrand Reinhold, 1984.

Sittig, Marshall. *Handbook of Toxic and Hazardous Chemicals and Carcinogens.* Noyes, 1985.

Smith, Al J. *Managing Hazardous Substances Accidents.* 1981.

U.S. Department of Transportation. *CHRIS: Manual I, A Condensed Guide to Chemical Hazards.* U.S. Coast Guard, 1984.

U.S. Department of Transportation. *CHRIS: Manual II, Hazardous Chemical Data.* U.S. Coast Guard, 1984.

U.S. Department of Transportation. *Emergency Response Guidebook.* Washington, DC: 1984.

U.S. Environmental Protection Agency. *Community Relations in Superfund: A Handbook.* Washington, DC.

U.S. Environmental Protection Agency. *The National Oil and Hazardous Substances Pollution Contingency Plan.* 40 CFR 300.

Verschuaren, Karel. *Handbook of Environmental Data on Organic Chemicals.* 2nd ed. New York: Van Nostrand Reinhold, 1983.

Waste Resource Associates, Inc. *Hazmat—Phases I, II, III, IV: Demonstration Project to Develop a Hazardous Materials Accident Prevention and Emergency Response Program.* Niagara Falls, NY: 1983.

Zajic, J.E. and W.A. Himmelman. *Highly Hazardous Material Spills and Emergency Planning.* Dekker, 1978.

Transportation Emergency Planning

American Trucking Association. *Handling Hazardous Materials.* Washington, DC: 1980.

Association of American Railroads. *Emergency Action Guides.* Washington, DC: 1984.

Association of American Railroads. *Emergency Handling of Hazardous Materials in Surface Transportation.* Washington, DC: 1981.

Battelle Pacific Northwest Laboratories. *Hazardous Material Transportation Risks in the Puget Sound Region.* Seattle, WA: 1981.

Portland Office of Emergency Management. *Establishing Routes for Trucks Hauling Hazardous Materials: The Experience in Portland, Oregon.* Portland, Oregon: 1984.

Portland Office of Emergency Management. *Hazardous Materials Highway Routing Study: Final Report.* Portland, OR: 1984.

Russell, E.R., J.J. Smaltz, et al. *A Community Model for Handling Hazardous Materials Transportation Emergencies: Executive Summaries.* Washington, DC: U.S. Department of Transportation, January 1986.

Russell, E.R., J.J. Smaltz, et al. *Risk Assessment/Vulnerability Users Manual for Small Communities and Rural Areas.* Washington, DC: U.S. Department of Transportation, March 1986.

Russell, E.R., W. Brumgardt, et al. *Risk Assessment/Vulnerability Validation Study Volume 2: 11 Individual Studies.* Washington, DC: U.S. Department of Transportation, June 1983.

Urban Consortium Transportation Task Force. *Transportation of Hazardous Materials.* Washington, DC: U.S. Department of Transportation, September 1980.

Urban Systems Associates, Inc., St. Bernard Parish Planning Commission. *St. Bernard Parish: Hazardous Materials Transportation and Storage Study.* New Orleans, LA: 1981.

Urganek, G. and E. Barber. *Development of Criteria to Designate Routes for Transporting Hazardous Materials.* Springfield, VA: National Technical Information Service, 1980.

U.S. Department of Transportation. *Community Teamwork: Working Together to Promote Hazardous Materials Transportation Safety.* Washington, DC: 1983.

U.S. Department of Transportation. *A Guide for Emergency Highway Traffic Regulation.* Washington, DC: 1985.

U.S. Department of Transportation. *A Guide to the Federal Hazardous Transportation Regulatory Program.* Washington, DC: 1983.

U.S. Department of Transportation. *Guidelines for Selecting Preferred Highway Routes for Highway Route Controlled Quantity Shipments of Radioactive Materials.* Washington, DC: 1984.

U.S. Department of Transportation and U.S. Environmental Protection Agency. *Lessons Learned from State and Local Experiences in Accident Prevention and Response Planning for Hazardous Materials Transportation.* Washington, DC, December 1985.

U.S. Department of Transportation. Three-Phase/Four-Volume Report: Volume I, *A Community Model for Handling Hazardous Materials Transportation Emergencies*; Volume II, *Risk Assessment Users Manual for Small Communities and Rural Areas*; Volume III, *Risk Assessment/Vulnerability Model Validation*; and, Volume IV, *Manual for Small Towns and Rural Areas to Develop a Hazardous Materials Emergency Plan.* 7/81–12/85. Document is available to the U.S. Public through the National Technical Information Service, Springfield, VA 22161.

Transportation Research Board. *Transportation of Hazardous Materials: Toward a National Strategy.* Volumes 1 & 2. Washington, DC: 1983.

Spill Containment and Cleanup

Guswa, J.H. *Groundwater Contamination and Emergency Response Guide.* Noyes, 1984.

U.S. Environmental Protection Agency. *State Participation in the Superfund Remedial Program.* Washington, DC: 1984.

Personal Protection

International Association of Fire Chiefs. *Fire Service Emergency Management Handbook.* Washington, DC: 1985.

National Institute of Occupational Safety and Health. *Occupational Safety and Health Guidance Manual for Hazardous Waste Site Activities.* Washington, DC: DHHS Publication No. 85-115, 1985.

U.S. Environmental Protection Agency. *Standard Operating Safety Guides.* Washington, DC: 1984.

Videotapes

The following videotapes are available from the Chemical Manufacturers Association:

- CAER: "Reaching Out"

- CAER: "How a Coordinating Group Works"

- CAER: "Working with the Media"

- CAER: "Planning and Conducting Emergency Exercises"

- NCRIC: "First on the Scene"

The following videotapes are available from FEMA's National Emergency Training Center/Emergency Management Information Center:

- "Livingston, LA, Hazardous Materials Spills" (September 28, 1982)

- "Waverly, TN, Hazardous Materials Blast" (February 22, 1978)

Also available for purchase from FEMA's National Emergency Training Center (see Appendix F for address and telephone number) are videotapes of teleconferences produced by FEMA's Emergency Education Network (EENET). One available teleconference is:

- "Emergency Exercises—Getting Involved in Community Preparedness," originally seen on December 11, 1986, and co-sponsored by FEMA, EPA, DOT/RSPA, USCG, and CMA.

The following documentary videotape (produced by the League of Women Voters of California and available from Bullfrog Films, Oley, PA 19547) provides public education on the nature and need for local emergency planning and hazardous materials data bases from a citizen's perspective.

- "Toxic Chemicals: Information Is The Best Defense"

Appendix F

1. National Offices

Federal Emergency Management Agency
Technological Hazards Division
Federal Center Plaza
500 C Street, S.W.
Washington, DC 20472
(202) 646-2861

FEMA National Emergency Training Center
Emmitsburg, MD 21727
(301) 447-6771

U.S. Environmental Protection Agency
OSWER Preparedness Staff
401 M Street, S.W.
Washington, DC 20460
(202) 475-8600
CEPP Hotline: 1-800-535-0202
(479-2449 in Washington, DC area)

U.S. Environmental Protection Agency
OERR Emergency Response Division
401 M Street, S.W.
Washington, DC 20460
(202) 475-8720

Agency for Toxic Substances
and Disease Registry
Department of Health & Human Services
Chamblee Building 30S
ytlanta, GA 30333
(404) 452-4100

U.S. Department of Energy
1000 Independence Avenue, S.W.
Washington, DC 20585
(202) 252-5000

447

Department of Agriculture
Forest Service
P.O. Box 96090
Washington, DC 20013-6090
(703) 235-8019

Department of Labor
Occupational Safety & Health Admin.
Directories of Field Operations
200 Constitution Avenue, N.W.
Washington, DC 20210
(202) 523-7741

U.S. Coast Guard (G-MER)
Marine Environmental Response Division
2100 2nd Street, S.W.
Washington, DC 20593
(202) 267-2010 (Info.)

NATIONAL RESPONSE CENTER
1-800-424-8802
(202-426-2675 or 202-267-2675 in Washington, DC area)

U.S. Department of Transportation
Research and Special Programs Admin.
Office of Hazardous Materials
Transportation (Attention: DHM-50)
400 7th Street, S.W.
Washington, DC 20590
(202) 366-4000

Department of Justice
Environmental Enforcement Section
Room 7313
10th and Constitution, N.W.
Washington, DC 20530
(202) 633-3646

Department of the Interior
18th and C St., N.W.
Washington, DC 20240
(202) 343-3891

Department of Commerce
NOAA—Superfund Program Coordinator
11400 Rockville Pike
Rockville, MD 20852
(301) 443-8465

Department of Defense
OASD (A + L)E
Room 3D 833
The Pentagon
Washington, DC 20301-8000
(202) 695-7820

Department of State
Office of Oceans and Polar Affairs
Room 5801
2201 C St., N.W.
Washington, DC 20520
(202) 647-3263

Nuclear Regulatory Commission
Washington, DC 20555
(202) 492-7000

2. Regional Offices
EPA, FEMA, HHS, ATSDR, OSHD

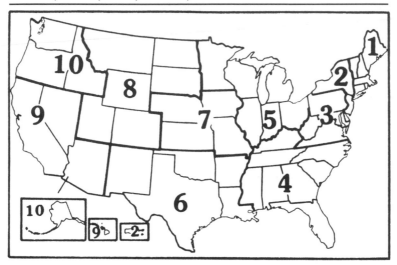

U.S. COAST GUARD DISTRICTS

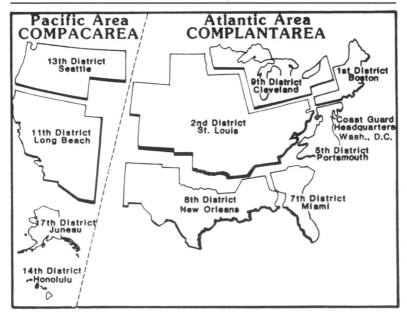

Department of Energy Regional Coordinating Offices for Radiological Assistance and Geographical Areas of Responsibility

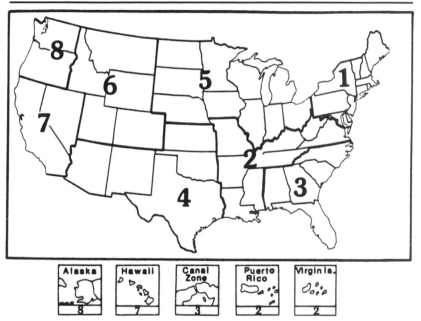

United States Nuclear Regulatory Commission

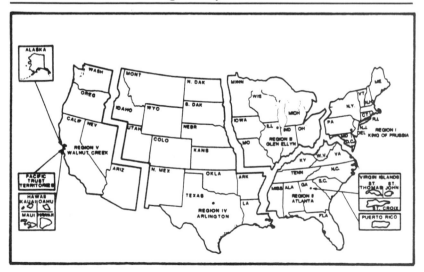

2. Regional Offices

A. EPA Regional Offices

(Note: Direct all requests to the "EPA Regional Preparedness Coordinator" (RPC) of the appropriate EPA Regional office.)

Region I

(Connecticut, Maine, Massachusetts, New Hampshire, Rhode Island, Vermont)

John F. Kennedy Building, Rm. 2203
Boston, MA 02203
(617) 565-3715
RPC: (617) 861-6700

Region II

(New Jersey, New York, Puerto Rico, Virgin Islands)

26 Federal Plaza, Room 900
New York, NY 10278
(212) 264-2525
RPC: (201) 321-6657

Region III

(Delaware, Washington DC, Maryland, Pennsylvania, Virginia, West Virginia)

841 Chestnut Street
Philadelphia, PA 19107
(215) 597-9800
RPC: (215) 597-8907

Region IV

(Alabama, Florida, Georgia, Kentucky, Mississippi, North Carolina, South Carolina, Tennessee)

345 Courtland Street, N.E.
Atlanta, GA 30365
(404) 347-4727
RPC: (404) 347-3931

Region V

(Illinois, Indiana, Michigan, Minnesota, Ohio, Wisconsin)

230 S. Dearborn Street
Chicago, IL 60604
(312) 353-2000
RPC: (312) 886-1964

Region VI

(Arkansas, Louisiana, New Mexico, Oklahoma, Texas)

1445 Ross Avenue, 12th Floor
Dallas, TX 75202-2733
(214) 655-6444
RPC: (214) 655-2270

Region VII

(Iowa, Kansas, Missouri, Nebraska)

726 Minnesota Avenue
Kansas City, KS 66101
(913) 236-2800
RPC: (913) 236-2806

Region VIII

(Colorado, Montana, North Dakota, South Dakota, Utah, Wyoming)

One Denver Place
999 18th Street, Suite 1300
Denver, CO 80202-2413
(303) 293-1603
RPC: (303) 293-1723

Region IX

(Arizona, California, Hawaii, Nevada, American Samoa, Guam)

215 Fremont Street
San Francisco, CA 94105
(415) 974-8071
RPC: (415) 974-7460

Region X

(Alaska, Idaho, Oregon, Washington)

1200 6th Avenue
Seattle, WA 98101
(206) 442-5810
RPC: (206) 442-1263

B.   FEMA Regional Offices

(Note: Direct all requests to the "Hazmat Program Staff" of the appropriate FEMA Regional office.)

Region I

(Connecticut, Maine, Massachusetts, New Hampshire, Rhode Island, Vermont)

442 J.W. McCormack POCH
Boston, MA 02109
(617) 223-9540

Region II

(New Jersey, New York, Puerto Rico, Virgin Islands)

Room 1337
26 Federal Plaza
New York, NY 10278
(212) 264-8980

Region III

(Delaware, Washington DC, Maryland, Pennsylvania, Virginia, West Virginia)

Liberty Square Building
105 S. 7th Street
Philadelphia, PA 19106
(215) 597-9416

Region IV

(Alabama, Florida, Georgia, Kentucky, Mississippi, North Carolina, South Carolina, Tennessee)

Suite 700
1371 Peachtree Street, N.E.
Atlanta, GA 30309
(404) 347-2400

Region V

(Illinois, Indiana, Michigan, Minnesota, Ohio, Wisconsin)

24th Floor
300 S. Wacker Drive
Chicago, IL 60606
(312) 353-8661

Region VI

(Arkansas, Louisiana, New Mexico, Oklahoma, Texas)

Federal Regional Center, Room 206
800 N. Loop 288
Denton, TX 76201-3698
(817) 387-5811

Region VII

(Iowa, Kansas, Missouri, Nebraska)

911 Walnut Street, Room 300
Kansas City, MO 64106
(816) 374-5912

Region VIII

(Colorado, Montana, North Dakota, South Dakota, Utah, Wyoming)

Denver Federal Center, Building 710
Box 25267
Denver, CO 80225-0267
(303) 235-4811

Region IX

(Arizona, California, Hawaii, Nevada, American Samoa, Guam)

Building 105
Presidio of San Francisco, CA 94129
(415) 923-7000

Region X

(Alaska, Idaho, Oregon, Washington)

Federal Regional Center
130 228th St., S.W.
Bothell, WA 98021-9796
(206) 481-8800

C. HHS Regional Offices

(Note: Consult the map on Page 450 to determine which States are assigned to each Region.)

Region I

Division of Preventive Health Services
John Fitzgerald Kennedy Building
Boston, Massachusetts 02203
(617) 223-4045

Region II

Division of Preventive Health Services
Federal Building
26 Federal Plaza, Room 3337
New York, New York 10278
(212) 264-2485

Region III

Division of Preventive Health Services
Gateway Building #1
Post Office Box 13716
Philadelphia, Pennsylvania 19101
(215) 596-6650

Region IV

Division of Preventive Health Services
101 Marietta Tower
Atlanta, Georgia 30323
(404) 331-2313

Region V

Division of Preventive Health Services
300 South Wacker Drive
Chicago, Illinois 60606
(312) 353-3652

Region VI

Division of Preventive Health Services
1200 Main Tower Building, Room 1835
Dallas, Texas 75202
(214) 767-3916

Region VII

Division of Preventive Health Services
601 East 12th Street
Kansas City, Missouri 64106
(816) 374-3491

Region VIII

Division of Preventive Health Services
1185 Federal Building
1961 Stout Street
Denver, Colorado 80294
(303) 844-6166, ext. 28

Region IX

Division of Preventive Health Services
50 United Nations Plaza
San Francisco, California 94102
(415) 556-2219

Region X

Division of Preventive Health Services
2901 Third Avenue, M.S. 402
Seattle, Washington 98121
(206) 442-0502

D. ATSDR Public Health Advisors Assigned to EPA Regional Offices

(Note: Consult the map on Page 450 to determine which States are assigned to each Region.)

Region I

ATSDR Public Health Advisor
EPA Superfund Office
Room 1903
John F. Kennedy Building
Boston, MA 02203
(617) 861-6700

Region II

ATSDR Public Health Advisor
Emergency & Remedial Response
Room 737
26 Federal Plaza
New York, New York 10007
(212) 264-8676

Region III

ATSDR Public Health Advisor
EPA Superfund Office
841 Chestnut Street, 6th Floor
Philadelphia, PA 19106
(215) 597-7291

Region IV

ATSDR Public Health Advisor
Air & Waste Management Division
345 Courtland Street, N.E.
Atlanta, GA 30365
(404) 347-3931/2

Region V

ATSDR Public Health Advisor
Emergency & Remedial Branch (5HR)
230 S. Dearborn
Chicago, IL 60604
(312) 886-9293

Region VI

ATSDR Public Health Advisor
EPA Superfund Office
1201 Elm Street
Dallas, TX 75270
(214) 767-9872

Region VII

ATSDR Public Health Advisor
Waste Management Branch
726 Minnesota Avenue
Kansas City, KS 66101
(913) 236-2856

Region VIII

ATSDR Public Health Advisor
Waste Management Division
1860 Lincoln Street
Denver, CO 80295
(303) 293-1526

Region IX

ATSDR Public Health Advisor
Toxics & Waste Management Division
215 Freemont Street
San Francisco, CA 94105
(415) 974-7742
Mailing address: P.O. Box 2453
Daly City, CA 94017

Region X

ATSDR Public Health Advisor
Hazardous Waste (M/S 525)
1200 6th Avenue
Seattle, WA 98101
(206) 442-2711

E.   OSHA Regional Offices

(Note: Consult the map on Page 450 to determine which States are assigned to each Region.)

Region I

16–18 North Street – 4th Floor
1 Dock Square Building
Boston, Massachusetts 02109
(617) 223-6710

Region II

1515 Broadway (1 Astor Plaza)
Room 3445
New York, New York 10036
(212) 944-3432

Region III

Gateway Building—Suite 2100
3535 Market Street
Philadelphia, Pennsylvania 19104
(215) 596-1201

Region IV

1375 Peachtree Street, N.E.
Suite 587
Atlanta, Georgia 30367
(404) 347-3573

Region V

32nd Floor—Room 3244
230 Dearborn Street
Chicago, Illinois 60604
(312) 353-2220

Region VI

525 Griffin Street
Room 602
Dallas, Texas 75202
(214) 767-4731

Region VII

911 Walnut Street
Room 406
Kansas City, Missouri 64106
(816) 374-5861

Region VIII

Federal Building—Room 1576
1961 Stout Street
Denver, Colorado 80294
(303) 844-3061

Region IX

11349 Federal Building
450 Golden Gate Avenue
P.O. Box 36017
San Francisco, California 94102
(415) 556-7260

Region X

Federal Office Building
Room 6003
909 First Avenue
Seattle, Washington 98174
(206) 442-5930

F.   U.S. Coast Guard District Offices

1st District

(Maine, Massachusetts, New York, New Hampshire, Connecticut, Rhode Island, Vermont, Northern Pennsylvania, Northern New Jersey)

Commander (mep)
408 Atlantic Avenue
Boston, MA 02110-2209
(617) 223-8444

2nd District

(Alabama, Arkansas, Colorado, Illinois, Indiana, Iowa, Kansas, Kentucky, Minnesota, Mississippi, Missouri, Nebraska, North Dakota, Ohio, Western Pennsylvania, South Dakota, Tennessee, West Virginia, Wyoming)

Commander (meps)
1430 Olive Street
St. Louis, MO 63103
(314) 425-4655

5th District

(Maryland, Delaware, North Carolina, Southern Pennsylvania, Southern New Jersey, Virginia)

Commander (mep)
Federal Building
431 Crawford Street
Portsmouth, VA 23705
(804) 398-6638

7th District

(Georgia, Florida, South Carolina, Puerto Rico, Virgin Islands)

Commander (mep)
Federal Building
51 S.W. 1st Avenue
Miami, FL 33130
(305) 350-5276

8th District

(Alabama, Florida, Georgia, Louisiana, Mississippi, New Mexico, Texas)

Commander (mpes)
Hale Boggs Federal Building
500 Camp Street,
New Orleans, LA 70130
(504) 589-6296

9th District

(Indiana, Illinois, Michigan, Minnesota, Ohio, Pennsylvania, New York, Wisconsin)

Commander (mep)
1240 East 9th Street
Cleveland, OH 44199
(216) 522-3918

11th District

(Arizona, California, Nevada, Utah)

Commander (mep)
Union Bank Building
400 Oceangate
Long Beach, CA 90822
(213) 590-2301

13th District

(Idaho, Montana, Oregon, Washington)

Commander (mep)
Federal Building
915 Second Avenue
Seattle, WA 98174
(206) 442-5850

14th District

(Hawaii, Guam, American Samoa, Trust Territory of the Pacific Island, Commonwealth of Northern Mariana Islands)

Commander (mep)
Prince Kalanianaole Federal Building
300 Ala Moana Boulevard, 9th Floor
Honolulu, HI 96850
(808) 541-2114

17th District

(Alaska)

Commander (mep)
P.O. Box 3–5000
Juneau, AK 99802
(907) 586-7195

G. Department of Energy (DOE) Regional Coordinating Offices for Radiological Emergency Assistance Only

Region 1

(Connecticut, Delaware, District of Columbia, Maine, Maryland, Massachusetts, New Hampshire, New Jersey, New York, Pennsylvania, Rhode Island, Vermont)

Brookhaven Area Office:
Upton, NY 11973
(516) 282-2200
FTS – 666-2200
(312) 972-5731 (off hours)
(Use same 7-digit number for FTS)

Region 2

(Arkansas, Kentucky, Louisiana, Mississippi, Missouri, Puerto Rico, Tennessee, Virgin Islands, Virginia, West Virginia)

Oak Ridge Operations Office:
P.O. Box E
Oak Ridge, TN 37830
(615) 576-1005
FTS 626-1005

Region 3

(Alabama, Canal Zone, Florida, Georgia, North Carolina, South Carolina)

Savannah River Operations Office:
P.O. Box A
Aiken, SC 29801
(803) 725-3333
FTS — 239-3333

Region 4

(Arizona, Kansas, New Mexico, Oklahoma, Texas)

Albuquerque Operations Office:
P.O. Box 5400
Albuquerque, NM 87115
(505) 844-4667
(Use same 7-digit number for FTS)

Region 5

(Illinois, Indiana, Iowa, Michigan, Minnesota, Nebraska, North Dakota, Ohio, South Dakota, Wisconsin)

Chicago Operations Office:
9800 South Cass Avenue
Argonne, IL 60439
(312) 972-4800 (duty hours)
(Use same 7-digit number for FTS)
(312) 972-5731 (off hours)

Region 6

(Colorado, Idaho, Montana, Utah, Wyoming)

Idaho Operations Office:
550 Second Street
Idaho Falls, ID 83401
(208) 526-1515
FTS 582-1515

Region 7

(California, Hawaii, Nevada)

San Francisco Operations Office:
1333 Broadway
Oakland, CA 94612
(415) 273-4237
FTS 537-4237

Region 8

(Alaska, Oregon, Washington)

Richland Operations Office:
P.O. Box 550
Richland, WA 99352
(509) 373-3800
FTS—440-3800

H.   Department of Transportation, Regional Pipeline Offices

Office of Pipeline Safety
Eastern Region, DPS-4, Room 8321
400 7th Street, S.W.
Washington, D.C. 20590
(202) 366-4585

(Connecticut, Delaware, District of Columbia, Maine, Maryland, Vermont, Massachusetts, New Hampshire, New Jersey, New York, Pennsylvania, Rhode Island, Virginia, West Virginia, Puerto Rico)

Office of Pipeline Safety
Southern Region, DPS-5, Ste. 504N.
1720 Peachtree Road, N.W.
Atlanta, Georgia 30309
(404) 347-2632

(Alabama, Florida, Georgia, Kentucky, North Carolina, South Carolina, Tennessee)

Office of Pipeline Safety
Central Region, DPS-6
911 Walnut Street, Room 1811
Kansas City, Missouri 64106
(816) 374-2653

(Iowa, Illinois, Indiana, Kansas, Michigan, Minnesota, Ohio, Missouri, Nebraska, Wisconsin)

Office of Pipeline Safety
Southeast Region, DPS-7
2320 La Branch, Room 2116
Houston, Texas 77704
(713) 750-1746

(Arkansas, Louisiana, New Mexico, Oklahoma, Texas)

Office of Pipeline Safety
Western Region, DPS-8
555 Zang Street, 2nd Floor
Lakewood, Colorado 80228
(303) 235-3424

(Arizona, California, Colorado, Idaho, Montana, Nevada, North Dakota, Oregon, South Dakota, Utah, Washington, Wyoming, Alaska, Hawaii)

I.   U.S. Nuclear Regulatory Commission Regional Offices

Region 1

(Connecticut, Delaware, District of Columbia, Maine, Maryland, Massachusetts, New Hampshire, New Jersey, New York, Pennsylvania, Rhode Island, Vermont)

USNRC
631 Park Avenue
King of Prussia, PA 19406
(215) 337-5000

Region 2

(Alabama, Florida, Georgia, Kentucky, Mississippi, North Carolina, Puerto Rico, South Carolina, Tennessee, Virginia, Virgin Islands, West Virginia)

USNRC
Suite 2900
101 Marietta Street, NW
Atlanta, GA 30323
(404) 331-4503

Region 3

(Illinois, Indiana, Iowa, Michigan, Minnesota, Missouri, Ohio, Wisconsin)

USNRC
799 Roosevelt Road
Glen Ellyn, IL 60137
(312) 790-5500

Region 4

(Arkansas, Colorado, Idaho, Kansas, Louisiana, Montana, Nebraska, New Mexico, North Dakota, Oklahoma, South Dakota, Texas, Utah, Wyoming)

USNRC
Suite 1000
611 Ryan Plaza Drive
Arlington, TX 76011
(817) 860-8100

Region 5

(Alaska, Arizona, California, Hawaii, Nevada, Oregon, Pacific Trust Territories, Washington)

USNRC
Suite 210
1450 Maria Lane
Walnut Creek, CA 94596
(415) 943-3700

VI Criteria for Review of Hazardous Materials Emergency Plans

National Response Team
National Oil and Hazardous Substances Contingency Plan
Washington, DC 20593

This supplement is a reprint of the US government document and describes the criteria a regional response team can use to review local emergency plans under the provisions of Section 303(g) of the Emergency Planning and Community Right to Know Act of 1986 (SARA Title III).

Criteria for Review of Hazardous Materials Emergency Plans

Introduction

This document contains a set of criteria which may be used by the Regional Response Teams (RRTs) in the review of local plans under the provisions of Section 303(g) of the Superfund Amendments and Reauthorization Act of 1986 (SARA). These criteria also may be used by local emergency planning committees (LEPCs) for preparing plans as required under Section 303(a) and by state emergency response commissions (SERCs) for reviewing plans as required under Section 303(e) of the Act. This review guide is intended as a companion document to the *Hazardous Materials Emergency Planning Guide* (NRT-1), and can be viewed as a supplement to the planning process as implemented by local emergency planning committees.

Background

Section 303(a) of the Superfund Amendments and Reauthorization Act of 1986 requires each local emergency planning committee to prepare comprehensive hazardous substances emergency response plans by October 1988. The local emergency planning committee is required to review the plan once a year, or more frequently as changed circumstances in the community or at any facility may require.

Section 303(b) requires each local emergency planning committee to evaluate the need for resources necessary to develop, implement, and exercise the emergency plan, and to make recommendations with respect to additional resources that may be required and the means for providing these additional resources.

Section 303(c) specifically states that "Each emergency management plan shall include (but is not limited to) each of the following:

(1)   Identification of facilities subject to the requirements of this subtitle that are within the emergency planning district, identification of routes likely to be used for the transportation of substances on the list of extremely hazardous substances referred to in Section 303(a), and identification of additional facilities contributing or subjected to additional risk due to their proximity to facilities subject to the requirements of this subtitle, such as hospitals or natural gas facilities.

471

(2)   Methods and procedures to be followed by facility owners and operators and local emergency and medical personnel to respond to any release of such substances.

(3)   Designation of a community emergency coordinator and facility emergency coordinators, who shall make determinations necessary to implement the plan.

(4)   Procedures providing reliable, effective, and timely notification by the facility emergency coordinators and the community emergency coordinator to persons designated in the emergency plan, and to the public, that a release has occurred (consistent with the emergency notification requirements of Section 304).

(5)   Methods for determining the occurrence of a release, and the area or population likely to be affected by such release.

(6)   A description of emergency equipment and facilities in the community and at each facility in the community subject to the requirements of this subtitle, and an identification of the persons responsible for such equipment and facilities.

(7)   Evacuation plans, including provisions for a precautionary evacuation and alternative traffic routes.

(8)   Training programs, including schedules for training of local emergency response and medical personnel.

(9)   Methods and schedules for exercising the emergency plan.''

Under Section 303(e) of the Act, state emergency response commissions are required to review and make recommendations on each plan to ensure "coordination" with the plans of other local emergency planning districts.

Under Section 303(g) of the Act, the Regional Response Teams "may review and comment upon an emergency plan or other issues related to preparation, implementation, or exercise of such a plan upon request of a local emergency planning committee." This review is viewed by the National Response Team to be a form of technical assistance to the local emergency planning committees and the state emergency response commissions, and is not to be considered as an approval of these plans.

Finally, under Section 303(f), the National Response Team is required to issue guidance documents for the preparation and implementation of emergency plans. In March 1987 the National Response Team published and distributed the first such guidance document by issuing NRT-1, the *Hazardous Materials Emergency Planning Guide*. NRT-1 contains extensive discussion of both the planning process and the elements or contents required for an effective hazardous materials emergency response plan. The following plan review criteria are issued as supplemental technical guidance to NRT-1.

Basis for the Criteria

The review criteria are based on the guidelines for plans as contained in Section 303(c) of the Act, NRT-1, and CPG 1-8, *Guide for Development of State and Local Emergency Operations Plans*, published by the Federal Emergency Management Agency. Section 303(c) outlines the minimum requirements for local emergency response plans. The criteria which address these minimum requirements are introduced throughout the document by the phrase "the plan shall." The criteria based upon NRT-1, which expand upon the above minimum requirements, include all the elements that the NRT considers essential for an effective hazardous materials emergency response plan. While local emergency planning committee plans are not required to contain all these elements, the NRT believes that they should. Accordingly, the criteria based on the planning elements in NRT-1 (and many of the elements in CPG 1-8) are introduced by the phrase "the plan should." In those cases where a plan may be improved by including other considerations, the criteria are introduced by the phrase "the plan might." In these cases, the criteria are not recommended either by Title III, NRT-1, or CPG 1-8.

There also are criteria included in this document that are considered to be of such merit that they are placed under the category "the plan should," but cannot be specifically cited from Title III, NRT-1, or CPG 1-8. These criteria will be included in subsequent revisions to NRT-1 and are highlighted in this document with an asterisk (*).

CPG 1-8 (used as one of the sources for these criteria) is used by local governments to develop emergency operations plans which are required by the Comprehensive Cooperative Agreements between the Federal Emergency Management Agency and the states. CPG 1-8a, *Guide for the Review of State and Local Emergency Operations Plans*, also was used as a resource for developing these review criteria. The planning elements in CPG 1-8 have been incorporated into NRT-1, and most of the review criteria in CPG 1-8a are included in the attached RRT Plan Review Document. The relevant sections of Title III, NRT-1, and CPG 1-8 are indexed in the review document for informational purposes.

The plan criteria outlined below are structured to correspond to the sequence of plan elements suggested by Chapters 4 through 6 of NRT-1 and Chapters 2 and 3 of CPG 1-8.

Use of Criteria

The NRT expects that the primary use of these criteria will be in plan review by Regional Response Teams. Through the use of these criteria and the development of comments related thereto, the RRTs can both conduct organized and systematic reviews of local plans and ensure that plan elements of particular interest to RRTs are covered. The RRTs should also use the criteria as a basis for ensuring coordination between federal plans developed under the National Contingency Plan (e.g., Regional Contingency Plans and OSC Plans) and plans developed at the local level.

As mentioned above, the local emergency planning committees may find the criteria useful in the development of plans required under Section 303(c) of the Act. These criteria are concise statements of the contents of plan elements covered in NRT-1 and CPG 1-8, and all of the plan elements required in Section 303(c). It is essential, however, that the criteria be used by local emergency planning committees only in concert with the full range of available guidance.

State emergency response commissions may find the plan criteria useful in the coordination of local emergency planning committees and in the review of each local plan. The criteria offer a useful guide for all of the planning elements which may require coordinated and consistent treatment among the local emergency planning committees within a state. They also provide the basis for a more general review of plans.

RRT Consideration of the LEPC Planning Process

One of the major themes of NRT-1 is that the way in which a local hazardous materials emergency plan is developed is as important as the actual contents of such a plan. Thus, the Regional Response Teams may find it useful to secure the following information pertaining to the local emergency planning committee under review:

1. A list of the names and affiliations of the members of the LEPC;

2. A description of the activities and accomplishments (with completion dates) of the committee in compliance with Section 301, including:

Appointment of a chairperson;

Establishment of rules for committee operations;

Development of methods for public notification of committee activities;

Conduct of public meetings on the emergency plan;

Receiving and responding to public comments;

Public notice of availability of emergency response plan, Material Safety Data Sheets, and inventory forms under Section 324;

Dealing with public requests for information under Sections 311, 312 and 313; and

Securing information from facilities covered by the plan.

3. A description of the major activities of the committee in completing the tasks for the hazard analysis and capability assessment.

4. A summary of the data produced by these tasks, if not already described in the plan.

5. A summary of the resources expended in developing the plan, including local funds, staff effort and technical expertise, plus a summary of resources required for maintaining and revising the plan.

6. A description of any findings on ways to fund hazardous materials emergency planning within the district.

CRITERIA FOR PLANS		DOCUMENTATION		
		TITLE III	NRT-1	CPG 1-8
1.0	INCIDENT INFORMATION SUMMARY			
	The Plan should[1] contain:			
1.1	Detailed description of the essential information that is to be developed and recorded by the local response system in an actual incident, e.g., date, time, location, type of release, and material released;		A.1	
2.0	PROMULGATION DOCUMENT			
	The Plan should contain:			
2.1	A document signed by the chairperson of the LEPC, promulgating the plan for the district;		A.2	2.3(a)(1)
2.2	Documents signed by the chief executives of all local jurisdictions within the district[2]; and		*	
2.3	Letters from affected facilities endorsing the plan.		*	
	The Plan might contain:			
2.4	Letters of agreement between the affected facilities and local jurisdictions for emergency response and notification responsibilities.		A.2	

---

[1](a) All criteria with a Title III reference are required by section 303(c) of the Act and are introduced by the phrase "The Plan shall."

(b) All criteria with a NRT-1 reference are not required by Title III, but are regarded as essential by the NRT for an effective hazardous materials emergency response plan. They can be found in NRT-1, Chapter 5, "Planning Elements," and are introduced by the phrase "The Plan should."

(c) All CPG 1-8 references include those criteria that address requirements for emergency operations plans prepared under the provisions of the Comprehensive Cooperative Agreements with the Federal Emergency Management Agency.

[2]These criteria are considered to be of such merit that they are included under the heading "the plan should," but cannot specifically be cited from Title III, NRT-1, or CPG 1-8. They are designated in the documentation section by an asterisk (*). They will be included in subsequent revisions to NRT-1.

	CRITERIA FOR PLANS	DOCUMENTATION		
		TITLE III	NRT-1	CPG 1-8
3.0	LEGAL AUTHORITY AND RESPONSIBILITY FOR RESPONSE			
	The Plan should:			
3.1	Describe, reference, or include legal authorities of the jurisdictions whose emergency response roles are described in the plan, including authorities of the emergency planning district and the local jurisdictions within the district; and		A.3	2.3(h)
3.2	List all other authorities the LEPC regards as essential for response within the district, including state and federal authorities		"	
4.	TABLE OF CONTENTS			
	The Plan should:			
4.1	List all elements of the plan, provide tabs for each and provide a cross-reference for all of the nine required elements in Section 303 of the Act. Plans that are prepared in the context of requirements of CPG 1-8 should contain an index to the location of both NRT-1 and Section 303 elements.		A.4	2.3(a)(3)
5.	ABBREVIATIONS AND DEFINITIONS			
	The Plan should:			
5.1	Explain all abbreviations and define all essential terms included in the plan text.		A.5	2.3(i)
6.0	PLANNING FACTORS			
	Assumptions: Assumptions are the advance judgments concerning what might happen in the case of an accidental spill or release.			
	The Plan should:			
6.1	List all of the assumptions about conditions that might develop in the district in the event of accidents from any of the affected facilities or along any of the transportation routes.		A.6	2.3(c)
	Planning Factors			
	The planning factors consist of all the local conditions that make an emergency plan necessary.			
	The Plan shall:			
6.2	Identify and describe the facilities in the district that possess extremely hazardous substances and the transportation routes along which such substances may move within the district;	303(c)(1)	A.6	2.3(c)
6.3	Identify and describe other facilities that may contribute to additional risk by virtue of their proximity to the above mentioned facilities;	"	"	"

	CRITERIA FOR PLANS	DOCUMENTATION		
		TITLE III	NRT-1	CPG 1-8
6.4	Identify and describe additional facilities included in the plan that are subject to additional risks due to their proximity to facilities with extremely hazardous substances; and	303(c)(1)	A.6	2.3(c)
6.5	Include methods for determining that a release of extremely hazardous substances has occurred, and the area of population likely to be affected by such release.	303(c)(5)	A.6	2.3(c)
	The Plan should:			
6.6	Include the major findings from the hazard analysis (date of analysis should be provided), which should consist of:		A.6	2.3(a)(4)
6.6.1	Major characteristics of affected facilities/transportation routes impacting on the types and levels of hazards posed, including the types, identities, characteristics, and quantities of hazardous materials related to facilities and transportation routes;		"	"
6.6.2	Potential release situations with possible consequences beyond the boundaries of facilities, or adjacent to transportation routes. Use may be made of historical data on spills and any data secured from facilities under Section 303(d)(3) of the Act;		"	"
6.6.3	Maps showing locations of facilities, transportation routes, and special features of district, including vulnerable areas;		"	"
6.7	Geographical features of the district, including sensitive environmental areas, land use patterns, water supplies, and public transportation;		"	"
6.8	Major demographic features of the district, including those features that impact most on emergency response, e.g., population density, special populations, and particularly sensitive institutions;		"	"
6.9	The district's climate and weather as they affect airborne distribution of chemicals; and		"	"
6.10	Critical time variables impacting on emergencies, e.g., time of day and month of year in which they would be most likely to occur		"	"
7.0	CONCEPT OF OPERATIONS			
	The Plan shall:			
7.1	Designate a community emergency coordinator and facility emergency coordinators, who shall make determinations necessary to implement the plan.	303(c)(3)		
	The Plan should:			
7.2	Identify, by title, the individual designated as the community emergency coordinator and each of the facility emergency coordinators;		A.7b	2.3(d)

	CRITERIA FOR PLANS	DOCUMENTATION		
		TITLE III	**NRT-1**	**CPG 1-8**
7.3	Explain the relationships between these coordinators, their organizations, and the other local governmental response authorities within the district, e.g., the county emergency management authority;		A.7b	2.3(d)
7.4	Describe the relationship between this plan and other response plans within the district which deal in whole or in part with hazardous materials emergency response, e.g., the county Emergency Operations Plan and plans developed by fire departments under OSHA Regulation CFR 29 Part 1910-120;		A.7c	
7.5	List all the facility emergency plans within the district that apply to hazardous materials emergency response, including all plans developed under OSHA Regulation on Hazardous Waste Operations and Emergency Response (CFR 29 Part 1910.120);		"	
7.6	Describe the way in which the above plans are integrated with local response plans;		"	
7.7	Describe the functions and responsibilities of all the local response organizations within the district, including public and private sector as well as volunteer and charitable organizations;		A.7b	2.3(e)
7.8	List mutual aid agreements or other arrangements for sharing data and response resources;		A.3	
7.9	Describe conditions under which the local government will coordinate its response with other districts and the means or sequence of activities to be followed by districts in interacting with other districts;		A.7b	
7.10	Describe the relationship between plans of the district and related state plans;		A.7c	
7.11	Describe the relationship between local and state emergency response authorities; and		A.7b	
7.12	Describe the relationships between emergency response plans and activities in the district and response plans and activities by federal agencies, including all plans and responses outlined in the National Contingency Plan.		A.7b; A.7c	
	[Emphasis should be given to the allocation of responsibilities among federal emergency response agencies, including listing the names and the responsibilities of federal response agencies and entities such as the RRT, describing the means for their notification, the types of resources to be sought from them, the means for obtaining them, the conditions under which assistance is to be provided, and the methods of coordination during a response.]			

	CRITERIA FOR PLANS	DOCUMENTATION		
		TITLE III	NRT-1	CPG 1-8
8.0	INSTRUCTIONS FOR PLAN USE			
	The Plan should:			
8.1	Contain a discussion of the purpose of the plan; and		A.8a	2.3(a)(5)(b)
8.2	Contain a list of organizations and persons receiving the plan or plan amendments and the date that the plan was transmitted as well as a specific identification number for each plan.		A.8b	2.3(a)(5)(c)
9.0	RECORD OF AMENDMENTS			
	The Plan should:			
9.1	Contain a section that describes methods for maintaining and revising the plan and recording all changes in the plan, including a method for controlling distribution.		A.9	2.3(a)(6)
10.0	EMERGENCY NOTIFICATION PROCEDURES			
	The Plan shall:			
10.1	Include procedures for providing reliable, effective, and timely notification by the facility emergency coordinators and the community emergency coordinator to persons designated in the emergency plan, and to the public, that a release has occurred.	303(c)(4)	C.4	3.3(c)
	The Plan should:			
10.2	Include procedures for immediately notifying the appropriate 24-hour hotline first, and should locate these procedures in a prominent place in the plan;		C.1	
10.3	List the 24-hour emergency hotline number(s) for the local emergency response organization(s) within the district;		"	
10.4	Contain an accurate and up-to-date Emergency Assistance Telephone Roster that includes numbers for the:		B	
10.4.1	Technical and response personnel;		"	
10.4.2	Community emergency coordinator, and all facility emergency coordinators;		*	
10.4.3	CHEMTREC;		B	
10.4.4	National Response Center;[3]		"	
10.4.5	Other participating agencies;		"	
10.4.6	Community emergency coordinators in neighboring emergency planning districts;		*	
10.4.7	Public and private sector support groups; and the		B	

[3] Spills exceeding the CERCLA reportable quantities are required to be reported to the National Response Center.

	CRITERIA FOR PLANS	DOCUMENTATION		
		TITLE III	NRT-1	CPG 1-8
10.4.8	Points of contact for all major carriers on transportation routes within the district;		*	
10.5	List all local organizations to be notified of a release, and the order of their notification, and list names and telephone numbers of primary and alternate points of contact;		*	
10.6	List all local institutions to be notified of the occurrence of a release and the order of their notification, and the names and telephone numbers of contacts;		*	
10.7	List all state organizations to be notified, and list the names and telephone numbers of contacts; and		*	
10.8	List all federal response organizations to be notified, and the names and telephone numbers of the contacts.		*	
11.0	INITIAL NOTIFICATION OF RESPONSE AGENCIES			
	The Plan should:			
11.1	Describe methods or means to be used by facility emergency coordinators (FECs) within the district to notify community emergency coordinators (CECs) of any potentially affected districts, and SERCs of any potentially affected states, and any other persons to whom the facility is to give notification of any release, in compliance with Section 304 of Title III;		C.1	
11.2	Describe methods by which the CECs and local response organizations will be notified of releases from transportation accidents, following notification through 911 systems or specified alternative means;		"	
11.3	Describe methods by which the CEC, or his designated agent, will ensure that contents of notification match the requirements of Section 304, including the regulations contained in 40 CFR Part 355 (Notification Requirements, Final Rule);		"	
11.4	List procedures by which the CEC will assure that both the immediate and follow-on notifications from facility operators are made within the time frames specified by Notification of Final Rule in 40 CFR Part 355; and		"	
11.5	Identify the person or office responsible for receiving the notification for the community emergency coordinator or his/her designated agent and list the telephone number;		"	
12.0	DIRECTION AND CONTROL			

	CRITERIA FOR PLANS	DOCUMENTATION		
		TITLE III	NRT-1	CPG 1-8
	The Plan shall:			
12.1	Include methods and procedures to be followed by facility owners and operators and local emergency and medical personnel to respond to a release of extremely hazardous substances.	303(c)(2)		
	The Plan should:			
12.2	Identify the organization within the district responsible for providing direction and control to the overall emergency response system described in the Concept of Operations;		C.2	3.3(a)
12.3	Identify persons or offices within each response organization who provide direction and control to each of the organizations;		"	"
12.4	Identify persons or offices providing direction and control within each of the emergency response functions;		"	"
12.5	Describe persons or offices responsible for the performance of incident command functions and the way in which the incident command system is used in hazardous substances incidents;		"	"
12.6	Describe the chain of command for the total response system, for each of the major response functions and for the organization controlled by the incident commander; and		"	"
12.7	Identify persons responsible for the activation and operations of the emergency operations center, the on-scene command post, and the methods by which they will coordinate their activities;		"	"
12.8	List three levels of incident severity and associated response levels;		"	"
12.9	Identify the conditions for each level; and		"	"
12.10	Indicate the responsible organizations at each level.		"	"
13.0	COMMUNICATION AMONG RESPONDERS			
	The Plan should:			
13.1	Describe all the methods by which identified responders will exchange information and communicate with each other during a response, including the communications networks and common frequencies to be used;		C.3	3.3(b)
	[At a minimum, these methods should be described for each function. Both communications among local response units and between these units and facilities where incidents occur should be described.]			

	CRITERIA FOR PLANS		DOCUMENTATION	
		TITLE III	**NRT-1**	**CPG 1-8**
13.2	Describe the methods by which emergency responders can receive information on chemical and related response measures; and		C.3	
	[May include a description of computer systems with on line data bases.]			
13.3	Describe primary and back-up systems for all communication channels and systems.		*	3.3(b)
	The Plan might:			
13.4	Contain a diagram or matrix showing the flows of information within the response system.		*	
14.0	WARNING SYSTEMS AND EMERGENCY PUBLIC NOTIFICATION			
	The Plan should:			
14.1	Identify responsible officials within the district and describe the methods by which they will notify the public of a release from any facility or along any transportation route, including sirens or other signals, and use of the broadcast media and the Emergency Broadcast System. This should include a description of:		C.4	3.3(c)
14.1.1	The sirens and other signals to be employed, their meaning, their methods of coordination, and their geographical coverage;		"	"
14.1.2	Other methods, such as door-to-door alerting, that may be employed to reach segments of the population that may not be reached by sirens or other signals; and		"	"
14.1.3	Time frames within which notification to the public can be accomplished;		"	"
14.2	Describe methods for the coordination of emergency public notification during a response; and		"	"
14.3	Describe any responsibilities or activities of facilities covered by the Act for emergency public notification during a response.		"	"
15.	PUBLIC INFORMATION AND COMMUNITY RELATIONS			
	The Plan should:			
15.1	Describe the methods used by local governments, prior to emergencies, for educating the public about possible emergencies and planned protective measures;		C.5	3.3(d)
15.2	Describe the role and organizational position of the public information officer during emergencies;		"	
15.3	Designate a spokesperson and describe the methods for keeping the public informed during an emergency situation, including a list of all radio, TV, and press contacts; and		"	

CRITERIA FOR PLANS	DOCUMENTATION		
	TITLE III	NRT-1	CPG 1-8
15.4 Describe any related public information activities of affected facilities, both prior to an emergency and during an emergency.		*	
16.0 RESOURCE MANAGEMENT			
The Plan shall:			
16.1 Include a description of emergency equipment and facilities in the community and at each facility in the community subject to the requirements of this subtitle and an identification of the persons responsible for such equipment and facilities.	303(c)(6)		
The Plan should:			
16.2 List personnel resources available for emergency response by major categories, including governmental, volunteer, and the private sector;		C.6	3.3(n)
16.3 Describe the types, quantities, capabilities and locations of emergency response equipment available to the local emergency response units, including fire, police and emergency medical response units.		"	"
[Categories of equipment should include transportation, communications, monitoring and detection, containment, decontamination, removal, and cleanup.]			
16.4 List the emergency response equipment available to each of the affected facilities and describe them in the same way as community equipment is described;		"	"
16.5 Describe the emergency operating centers or other facilities available to the local community and the facility emergency coordinators and other response coordinators, such as incident commanders;		*	
16.6 Describe emergency response equipment and facilities available to each affected facility and the conditions under which they are to be used in support of local responders;		"	"
16.7 Describe significant resource shortfalls and mutual support agreements with other jurisdictions whereby the district might increase its capabilities in an emergency;		"	"
[This may be discussed under the Concept of Operations.]		"	"
16.8 Describe procedures for securing assistance from federal and state agencies and their emergency support contractors;		"	"
[This may be discussed under the Concept of Operations.]			

	CRITERIA FOR PLANS	DOCUMENTATION		
		TITLE III	NRT-1	CPG 1-8
16.9	Describe emergency response capabilities and the expertise in the private sector that might be available to assist local responders, facility managers, and transportation companies during emergencies.		C.6	3.3(n)
17.0	HEALTH AND MEDICAL			
	The Plan shall:			
17.1	Include methods and procedures to be followed by facility owners and operators and local emergency and medical personnel to respond to a release of extremely hazardous substances.	303(c)(2)		
	The Plan should:			
17.2	Describe the procedures for summoning emergency medical and health department personnel;		C.7	3.3(h)
17.3	Describe the procedures for the major types of emergency medical services, including first aid, triage, ambulance service, and emergency medical care, using both the resources available within the district and those that can be secured in neighboring districts;		"	"
17.4	Describe the procedures to be followed for decontamination of exposed people;		"	"
17.5	Describe the procedures for providing sanitation, food, water supplies, and safe re-entry of persons to the accident area;		"	"
17.6	Describe procedures for conducting health assessments upon which to base protective action decisions;		*	
17.7	Describe the level and types of emergency medical capabilities in the district to deal with exposure of people to extremely hazardous substances;		C.7	
17.8	Describe the provisions for emergency mental health care; and		"	
17.9	Indicate mutual aid agreements with other communities to provide backup emergency medical and health department personnel, and equipment.		*	
18.0	RESPONSE PERSONNEL SAFETY			
	The Plan should:			
18.1	Describe initial and follow-up procedures for entering and leaving incident sites, including personnel accountability, personnel safety precautions, and medical monitoring;		C.8	
18.2	Describe personnel and equipment decontamination procedures; and		"	
18.3	List sampling, monitoring and personnel protective equipment appropriate to various degrees of hazards based on EPA levels of protection (A, B, C, & D).		"	

	CRITERIA FOR PLANS	DOCUMENTATION		
		TITLE III	NRT-1	CPG 1-8
	[Just prior to publication of NRT-1, the Occupational Safety and Health Administration (OSHA) published proposed rules (29 CFR Part 1910.120) to provide more definitive requirements to plan for emergency response personnel safety. If the LEPC plans include a section on this function, the plan elements listed in the OSHA regulation should be used.]			
19.0	PERSONAL PROTECTION OF CITIZENS/INDOOR PROTECTION			
	The Plan shall:			
19.1	Describe methods in place in the community and in each of the affected facilities for determining the areas likely to be affected by a release.	303(c)(5)		
	The Plan should:			
19.2	Include methods to predict the speed, direction, and concentration of plumes resulting from airborne releases, and methods for modeling vapor cloud dispersion as well as methods to monitor the release and concentration in real time;		C.9a	3.3(g)
	[19.1 and 19.2 may be considered in the hazard analysis, included in Section 6, Planning Factors.]			
19.3	Identify the decision making process, including the decision making authority for indoor protection;		"	"
19.4	Describe the roles and activities of affected facilities in the decision making for indoor protection decisions, including the determination that indoor sheltering is no longer required;		*	
19.5	Indicate the conditions under which indoor protection would be recommended, including the decision making criteria;		C.9a	3.3(g)
19.6	Describe the methods for indoor protection that would be recommended for citizens, including provisions for shutting off ventilation systems; and		*	
19.7	Describe the methods for educating the public on indoor protective measures;		C.9a	
	[May be discussed in the section on public information.]			
20.0	PERSONAL PROTECTIVE MEASURES/EVACUATION PROCEDURES			
	The Plan shall:			
20.1	Describe evacuation plans, including those for precautionary evacuations and alternative traffic routes;	303(c)(7)	C.9b	3.3(e)

	CRITERIA FOR PLANS	DOCUMENTATION		
		TITLE III	NRT-1	CPG 1-8
	The Plan should:			
20.2	Describe the authority for ordering or recommending evacuation, including the personnel authorized to recommend evacuation;		C.9b	
20.3	Describe the authority and responsibility of various governmental agencies and supporting private sector organizations, such as the Red Cross, and the chain of command among them;		"	
20.4	Describe the role of the affected facilities in the evacuation decision-making;		*	
20.5	Describe methods to be used in evacuation, including methods for assisting the movement of mobility impaired persons and in the evacuation of schools, hospitals, prisons and other facilities;		C.9b	
20.6	Describe the relationship of evacuation procedures to other protective measures.		*	
20.7	Describe potential conditions requiring evacuation, i.e., the types of accidental release and spills that may require evacuation;		C.9b	
20.8	Describe evacuation routes, including primary and alternative routes;		"	3.3(e)
	[These may be either established routes for the community or special routes appropriate to the location of facilities.]			
20.9	Describe evacuation zones and distances and the basis for their determination;		"	"
	[These should be related to the location of facilities and transportation routes and the potential pathways to exposure.]			
20.10	Describe procedures for precautionary evacuations of special populations;		"	
20.11	List the mass care facilities for providing food, shelter and medical care to relocated populations;		"	3.3(f)
	[This may be discussed under the human services section.]			
20.12	Describe procedures for providing security for the evacuation, for evacuees and of the evacuated areas;		C.11	3.3(i)
	[May be covered under the law enforcement discussions.]			
20.13	Describe methods for managing the flow of traffic along evacuation routes and for keeping the general public from entering threatened areas, including maps with traffic and other control points; and		C.9b	3.3(e)
	[May be covered in the law enforcement section.]			

	CRITERIA FOR PLANS	DOCUMENTATION		
		TITLE III	NRT-1	CPG 1-8
20.14	Describe the procedures for managing an orderly return of people to the evacuated area;		C.9b	3.3(e)
21.0	FIRE AND RESCUE			
	The Plan should:			
21.1	List the major tasks to be performed by firefighters in coping with releases of extremely hazardous substances;		C.10	3.3(k)
21.2	Identify the public and private sector fire protection organizations with a response capability and responsibility for hazardous materials incidents;		"	"
21.3	Describe the command structure of multi-agency, multi-jurisdictional incident management systems in place, and identify applicable mutual aid agreements and good samaritan provisions in place;		"	"
21.4	List available support systems, e.g., protective equipment and emergency response guides, DOT Emergency Response Guidebook, mutual aid agreements, and good samaritan provisions; and		"	"
	[May be covered under resource management.]			
21.5	List and describe any HAZMAT teams in the district.		*	
	[May be covered in Section 21.2 above.]			
22.0	LAW ENFORCEMENT			
	The Plan should:			
22.1	Describe the command structure of multi-agency, multi-jurisdictional incident management systems in place, and identify applicable mutual aid agreements and good samaritan provisions in place;		C.11	3.3(i)
22.2	List the major law enforcement tasks related to responding to releases of extremely hazardous materials, including those related to security for the accident site and for evacuation activities; and		"	"
22.3	List the locations of control points for the performance of tasks, with appropriate maps.		*	
23.0	ON-GOING INCIDENT ASSESSMENT			
	The Plan should:			
23.1	Describe methods in place in the community and/or each of the affected facilities for determining the areas likely to be affected by an ongoing release.		C.12	3.3(p)
23.2	Describe methods for determining the private and public property that may be in the affected areas and the nature of the impact of the release on this property;		"	"

CRITERIA FOR PLANS	DOCUMENTATION		
	TITLE III	NRT-1	CPG 1-8
23.3 Describe methods and capabilities of both local response organizations and facilities for monitoring the size, concentration, and migration of leaks, spills, and releases, including sampling around the site; and		C.12	
23.4 Describe provisions for environmental assessments, biological monitoring, and contamination surveys;		"	
24.0 HUMAN SERVICES			
The Plan should:			
24.1 List the agencies responsible for providing emergency human services, e.g., food, shelter, clothing, continuity of medical care, and crisis counseling; and		C.13	3.3(m)
24.2 Describe the major human services activities and the means for their accomplishment.		"	"
25. PUBLIC WORKS			
The Plan should:			
25.1 Describe the chain of command for the performance of public works actions in an emergency; and		C.14	3.3(j)
25.2 List all major tasks to be performed by the public works department in a hazardous materials incident.		"	"
26.0 TECHNIQUES FOR SPILL CONTAINMENT AND CLEANUP			
The Plan should:			
26.1 Explain the allocation of responsibilities among local authorities and affected facilities and responsible parties for these activities;		D.1	
26.2 Describe the major containment and mitigation activities for all major types of HAZMAT incidents;		"	
26.3 Describe cleanup and disposal services to be provided by the responsible parties and/or local community;		"	
26.4 Describe major methods for cleanup;		"	
26.5 Describe methods to restore the surrounding environment, including natural resource areas, to pre-emergency conditions;		"	
26.6 Describe the provisions for long term site control;		D.2	
26.7 List the location of approved disposal sites;		"	
26.8 List cleanup material and equipment available within the district;		"	
[May be covered in the resource management section.]			

CRITERIA FOR PLANS		DOCUMENTATION		
		TITLE III	NRT-1	CPG 1-8
26.9	Describe the capabilities of cleanup personnel; and		D.2	
26.10	List the applicable regulations governing disposal of hazardous materials in the district.		"	
27.	DOCUMENTATION AND INVESTIGATIVE FOLLOW-UP			
	The Plan should:			
27.1	List all reports required in the district and all offices and agencies that are responsible for preparing them following a release;		E	
27.2	Describe the methods of evaluating responses and identify persons responsible for evaluations; and		"	
27.3	Describe provisions for cost recovery.		"	
28.0	PROCEDURES FOR TESTING AND UPDATING THE PLAN			
	The Plan shall:			
28.1	Include methods and schedules for exercising the emergency plan.	303(c)(9)	F.1	
	The Plan should:			
28.2	Describe the nature of the exercises for testing the adequacy of the plan;		"	
28.3	List the frequency of such exercises, by type;		"	
28.4	Include an exercise schedule for the current year and for future years;		"	
28.5	Describe the role of affected facilities or transportation companies in these exercises; and		"	
28.6	Describe the procedures by which performance will be evaluated in the exercise, revisions will be made to plans, and deficiencies in response capabilities will be corrected.		F.6	
29.0	TRAINING			
	The Plan shall:			
29.1	Include the training programs, including schedules, for training of local emergency response and medical personnel.	303(c)(8)	6.4.3	
	The Plan should:			
29.2	Describe training requirements for LEPC members and all emergency planners within the district;		*	
29.3	Describe training requirements for all major categories of hazardous materials emergency response personnel, including the types of courses and the number of hours;		*	

	CRITERIA FOR PLANS	DOCUMENTATION		
		TITLE III	NRT-1	CPG 1-8
29.4	List and describe the training programs to support these requirements, including all training to be provided by the community, state and federal agencies, and the private sector; and		*	
29.5	Contain a schedule of training activities for the current year and for the following three years.		*	

VII

Quick Reference Chemical Compatibility Chart for Protective Clothing

Compiled by Gerald L. Grey
Captain, Hazardous Materials Team
San Francisco Fire Department

The following supplement provides an easy reference source for determining the personal protective clothing suitable for dealing with a release of a given compound.

Our special thanks to Captain Gerald L. Grey for providing this valuable reference and allowing its inclusion as a supplement in the *Hazardous Materials Response Handbook*.

Quick Reference
Chemical Compatibility Chart
for Protective Clothing

Introduction

This chemical compatibility chart was compiled by Captain Gerald Grey of the San Francisco Fire Department for use by their hazardous materials team.

The need for a quick reference source of chemical compatibility became apparent after Captain Grey's chemical suit suffered severe damage to the facepiece when exposed to anhydrous dimethylamine at an incident in Benicia in 1983. The use of butyl rubber suits was recommended by a reference source, but no consideration was given to the compatibility of the polycarbonate facepiece that was installed as an integral part of the chemical suit.

Several sources provide information concerning the applicability of products and materials in hazardous environments. As certain materials are more suited for use in particular environments, it is important for the user to select the most desirable material for that environment. Likewise, certain product designs are more or less suited for particular environments.

Unfortunately, no agency or manufacturer has tested all the materials used in the construction of personal protective equipment with all the different environments that are likely to be encountered. (And it is unlikely that such an undertaking will be done.)

There have, however, been tests done by manufacturers for agencies such as the Coast Guard on the compatibility of various chemicals with their products.

Because each agency or manufacturer acted independently, a standard compilation of the information does not necessarily provide the information needed in the field. For example, one manufacturer may have tested butyl rubber with a particular chemical and found the butyl rubber totally unaffected by that chemical, whereas another manufacturer may have conducted the tests using different criteria or after the garment was constructed, and, due to the design or components of the garment, found that butyl rubber clothing was unacceptable.

It is critical to get as much information on a particular product as possible. This chemical compatibility chart is a composite of several charts currently available from various sources. When using this data, keep in mind that the format is intended as a quick reference, and the user should consult the original information as provided by the manufacturer or agency. No evaluation of the information has been made, only the transferring of the information onto one chart for a quick reference.

As additional charts are received, the intention is to incorporate the information onto this chart. Also, as the charts or studies used are replaced or revised, those changes will be made.

Any comments or suggestions for improving the usefulness of this chemical compatibility chart for emergency responders should be directed to:

Captain Gerald L. Grey
Hazardous Materials Team
San Francisco Fire Department
260 Golden Gate Ave.
San Francisco, California 94102
(415) 861-8000 x236

Explanation

The information for compiling this chart came from the following sources:

Coast Guard (Study, August 1980)

Materials Tested:
  Butyl rubber
  Polycarbonate

The compatibility of butyl rubber and polycarbonate with the chemicals was determined from a literature search.

Symbols used:

Y   Yes (compatible)
N   No (non-compatible)

Eastwind (Specification #200)

Materials tested:
  Polyvinyl chloride
  Neoprene
  Butyl

Symbols used:

R   recommended
L   limited resistance
N   not recommended (N used instead of NR)
*   info not available

ILC Dover Chemical Compatibility Chart

Materials tested:
  CPE (chlorinated polyethylene)
  PVC (polyvinyl chloride)
  Butyl
Symbols used:

A   recommended (little or no effect)
B   minor to moderate effect
C   conditional (varies from moderate to severe under different conditions)
X   not recommended (severe effect)
I   insufficient data to rate

Service temperature 140 degrees Fahrenheit maximum

Andover LL Series Positive Pressure Suits, Care and Use Manual (Schedule A, chemical resistance tables)

Materials tested:
  Butyl
  Nitrile
  Neoprene
  Viton

Symbols used:

A   recommended — no effect to garment
B   minor effect — garment still useful
C   moderate effect — continuous use will damage garment
U   not recommended — will damage garment

MSA (Data Sheet 13-01-07 & Bulletin No. 0105-64)

Materials tested:
  Butyl — butadiene on nylon
  Viton — fluorocarbon on dacron
  Nitrile — 80% acrylonitrile on nylon

Symbols used:

E   excellent   — resists permeation for minimum of 8 hours
G   good   — resists permeation for minimum of 2 hours
F   fair   — resists permeation for minimum of 1 hour
P   poor   — resists permeation for minimum of 1/2 hour
N   not recommended — challenge substance may permeate fabric and
                         expose wearer to skin contact within a few min-
                         utes after exposure to the challenge substance.
                         (N used instead of NR)

Edmont (Chemical Degradation Guide for Gloves, 1984. Becton, Dickson and Company)

Materials tested:
  Nitrile (NBR)
  Neoprene - unsupported
  Polyvinyl chloride (PVC)

Symbols used:

e	excellent	—fluid has little degrading effect
g	good	—fluid has minor degrading effect
f	fair	—fluid has moderate degrading effect
p	poor	—fluid has pronounced degrading effect
n	not recommended	—fluid is not recommended with this material (n used instead of nr)

Several chemicals have a bracketed [C] following the name of the chemical. These chemicals were identified in the sixth edition of Sax's Dangerous Properties of Industrial Materials as suggested carcinogens, experimental carcinogens, and other materials that pose a lesser risk of cancer.

Letters appearing in { } following the name of the chemical are common synonyms of the chemical.

Letters and numbers appearing in ( ) following the name of the chemical indicate the structure of the chemical, (n) normal, (o) ortho, (p) para.

Some general characteristics of chemical-protective clothing with groups or "families" of chemicals are listed below.

Butyl rubber *Resistant to:* Most acids and bases, inorganic salts, animal and vegetable fats and oils, esters, ketones, and ozone.
Not Resistant to: Aromatic hydrocarbons, coal tar, petroleum solvents, and solvents.

CPE (chlorinated polyethylene) *Resistant to:* Acids, bases, salts, aliphatic hydrocarbons, esters, ketones, and phenols.

Neoprene *Resistant to:* Aliphatic hydroxy compounds (methyl and ethyl alcohols and ethylene glycol), animal and vegetable fats and oils, fluorinated hydrocarbons, straight chain hydrocarbons, moderate acids, concentrated caustics, ozone, and some solvents.
Not Resistant to: Chlorinated, aromatic, and nitro hydrocarbons, esters, ketones, phenols, and strong oxidizers including nitric and sulfuric acids.

PVC (polyvinyl chloride) *Resistant to:* Amines, aromatic hydrocarbons, inorganic acids, bases, and salts.
Not Resistant to: Esters, halogenated hydrocarbons, and ketones.

Viton *Resistant to:* Aliphatic, aromatic, and halogenated hydrocarbons, animal and vegetable oils, fuels, hydraulic fluids, lubricants, oils, caustics, and most mineral acids.

Not Resistant to: Aldehydes, ketones, esters, and nitro containing compounds.

The following shows the order and source of data as entered in the various columns of the "chemical compatibility chart."

Butyl Rubber (Butyl)

Coast Guard	ILC Dover	Eastwind	Andover	MSA
Y/N	A/B/C/X/I	R/L/N/*	A/B/C/U	E/G/F/P/N

Polycarbonate (PC)

Coast Guard
Y/N

Chlorinated Polyethylene (CPE)

ILC Dover
A/B/C/X/I

Nitrile (NBR)

Andover	MSA	Edmont	ILC Dover
A/B/C/U	E/G/F/P/N	e/g/f/p/n	A/B/C/X/I

Neoprene (NEO)

Andover	Eastwind	Edmont	ILC Dover
A/B/C/U	R/L/N/*	e/g/f/p/n	A/B/C/X/I

Polyvinyl chloride (PVC)

Eastwind	ILC Dover	Edmont
R/L/N/*	A/B/C/X/I	e/g/f/p/n

Viton (VIT)

Andover	MSA	ILC Dover
A/B/C/U	E/G/F/P/N	A/B/C/X/I

Chemical Compatibility Chart

CHEMICAL NAME	BUTYL	PC	CPE	NBR	NEO	PVC	VIT
Acetaldehyde	YARA	N	B	U pX	CReC	NXn	U X
Acetamide	ALA		B	A A	BL B	*B	B B
Acetate	L				L	*	
Acetic acid	Y	Y					
Acetic acid, glacial	B B		A	C gC	C eX	Cf	C C
Acetic acid (30%)	B		A	B B	A A	B	B B
Acetic acid (80%)	R				R	R	
Acetic anhydride	YB B	Y	A	C C	A A	X	U X
Acetone	YARA	N	A	U nX	BLgB	NXn	U X
Acetone cyanohydrin	Y	Y					
Acetonitrile	Y E	Y		F			N
Acetophenone	YA- A	Y	X	U X	UN X	*X	U X
Acetyl bromide	Y	Y					
Acetyl chloride	YC--	N	A	– X	UN X	*X	A A
Acetyl peroxide solution	Y	Y					
Acetyl acetone	Y	N					
Acetylene	YARA	Y	A	B B	BR B	*A	A A
Acetylene tetrachloride	N				N	*	
Acridene	Y	Y					
Acrolein	Y	Y					
Acrylamide	YA	N	A	A	A	B	I
Acrylic acid	YI	Y	A			X	
Acrylonitrile	YXLU	Y	I	U X	CL C	NX	U X
Adipic acid	YA –	Y	A	A A	– A	A	– A
Adiponitrile	Y	Y					
Aldrin	YB	Y	C			C	
Alkanes	X		A			A	
Alkylbenzenesulfonic acid							
Allyl alcohol	YA E	Y	A	E A	A	A	EA
Allyl bromide	Y	Y					
Allyl chloride	YB	N	C			X	
Allyl chloroformate	Y	N					
Allyl trichlorosilane	Y	N					
Alum	A		A			A	
Aluminum acetate	A A		A	B B	B B	A	– A
Aluminum chloride	YA*A	Y	A	A A	AR A	RB	A A
Aluminum fluoride	YA A	Y	A	A A	A A	B	A A
Aluminum hydroxide	B		A	A	A	B	A
Aluminum nitrate	YA*A	Y	A	A A	AR A	RA	– A
Aluminum phosphate	A		A	A	A	A	A
Aluminum sulfate	YA*A	Y	A	A A	AR A	RA	A A
Amidol	R				L	*	
Aminoethylethanolamine(n)	Y	N					
Ammonia, anhydrous	YARA	Y	A	A B	AL A	*A	U X
Ammonia gas (cold)	YA F		A	G A	A	A	PA
Ammonia, liquid	*				R	N	
Ammonium acetate	Y	Y					

CHEMICAL NAME	BUTYL	PC	CPE	NBR	NEO	PVC	VIT
Ammonium benzoate	Y	Y					
Ammonium bicarbonate	Y	Y					
Ammonium bifluoride	Y	Y					
Ammonium bisulfite							
Ammonium bromide							
Ammonium carbamate							
Ammonium carbonate	YA A	Y	A	U X	A A	A	– A
Ammonium chloride	YA A	Y	A	A A	A A	B	– A
Ammonium chromate							
Ammonium citrate	Y	Y					
Ammonium dichromate	Y	Y					
Ammonium fluoborate							
Ammonium fluoride	Y	Y					
Ammonium fluoride (40%)				e	e	e	
Ammonium formate	Y	Y					
Ammonium gluconate	Y	Y					
Ammonium hydroxide	ARA		A	U C	AR A	RA	B B
Ammonium hydroxide (29%)	Y	N		e	e	e	
Ammonium hypophosphate							
Ammonium iodide	Y	Y					
Ammonium lactate	Y	Y					
Ammonium lauryl sulfate	Y	Y					
Ammonium molybdate	Y	Y					
Ammonium nitrate	YA A	Y	A	A A	B B	A	– A
Ammonium nitrate-phos/mix	Y	Y					
Ammonium nitrate-sulfate	Y	Y					
Ammonium nitrate-urea sol	Y	Y					
Ammonium oleate	Y	Y					
Ammonium oxalate	Y	Y					
Ammonium pentaborate	Y	Y					
Ammonium perchlorate	Y	Y					
Ammonium persulfate	YA A	Y	A	U X	A A	A	– A
Ammonium phosphate	YA A	Y	A	A A	A A	A	– A
Ammonium salts, general	A		A	A	A	A	A
Ammonium silicofluoride	Y	Y					
Ammonium stearate	Y	Y					
Ammonium sulfamate	Y	Y					
Ammonium sulfate	YA A	Y	A	A A	A A	A	– A
Ammonium sulfide	Y	Y					
Ammonium sulfite	Y	Y					
Ammonium tartrate	Y	Y					
Ammonium thiocyanate	Y	Y					
Ammonium thiosulfate	Y	Y					
Amyl acetate (n)	YANA	N	C	U eX	UNnX	NXp	U X
Amyl alcohol (n)	YARA	Y	A	B eB	ARe A	* A	B B
Amyl borate	X U		I	A A	A A	I	A A

CHEMICAL NAME	BUTYL	PC	CPE	NBR	NEO	PVC	VIT
Amyl chloride (n)	Y	N					
Amyl chloronapthalene	X U		I	– X	U X	I	A A
Amyl mercaptan (n)							
Amyl methyl ketone (n)	Y	N					
Amyl nitrite (iso)	Y	N					
Amyl nitrate (n)	Y	Y					
Amyltrichlorosilane (n)	Y	N					
Aniline	YBNBE	N	B	UGnX	CNgX	NXf	CNC
Aniline hydrochloride	BNB		A	B B	UN X	NA	B B
Anisoyl chloride	Y	N					
Anthracene [C]	Y N	Y			N	*	
Antimony	I		A			I	
Antimony pentachloride	Y	N					
Antimony pentafluoride	Y	Y					
Antimony potassium tartrate	Y	Y					
Antimony tribromide	Y	Y					
Antimony trichloride	Y	Y					
Antimony trifluoride	Y	Y					
Antimony trioxide	Y	Y					
Aqua regia	X U		C	– fX	U gX	Cg	B B
Aromatic fuels	*				R	R	
Arsenic acid	YARA	Y	A	A A	AR A	A	A A
Arsenic, inorganic	A		A			A	
Arsenic pentoxide	Y	N				Y	
Arsenic sulfide	Y	Y					
Arsenic trichloride	YC –	Y	B	A A	A A	B	– A
Arsenic trioxide	Y	Y					
Arsenic trisulfide							
Asbestos	AR		A		R	*A	
Askarel	X U		C	B B	U X	B	A A
Asphalt	YXNU	Y	A	B B	CL C	*B	A A
Atrazine	Y	Y					
Azinphosmethyl	Y	Y					
Banana oil	N				N	*	
Barium carbonate	Y	Y					
Barium chlorate	Y	Y					
Barium chloride	A*A		A	A A	AR A	RA	A A
Barium cyanide							
Barium nitrate	Y	Y					
Barium perchlorate	Y	Y					
Barium permanganate	Y	Y					
Barium peroxide	Y	Y					
Barium salts & hydroxide	A		A			A	
Benzaldehyde	YA*A	N	C	U nX	UNnX	NXn	U X
Benzaldehyde (10%)	L				N	R	
Benzene [C]	NXNUN		C	UNpX	UNnX	NCn	AGA
Benzene benzol	*				N	N	

CHEMICAL NAME	BUTYL	PC	CPE	NBR	NEO	PVC	VIT
Benzene hexachloride	Y	Y					
Benzene sulfonic acid	X –		A	– X	A A	A	A A
Benzene sulfonic acid (10%)	*				R	R	
Benzoic acid	YX –	Y	A	– X	– X	B	A A
Benzonitrile	Y	Y					
Benzophenone	Y	Y					
Benzoyl chloride	YI	N	X			X	
Benzoyl peroxide	I		B			I	
Benzyl acetate	I		B			B	
Benzyl alcohol	YBRB	Y	A	U X	AR B	*X	A A
Benzylamine	Y	N					
Benzyl benzoate	BNB		A	– X	–L X	*A	A A
Benzyl bromine	Y	Y					
Benzyl chloride	NXL–	Y	X	U X	UN X	*X	A A
Benzyl chloroformate	N						
Benzyl n–butyl phthalate	N						
Beryllium	A		A	A	A	A	A
Beryllium chloride	Y	Y					
Beryllium fluoride	Y	Y					
Beryllium, metallic	Y	Y					
Beryllium nitrate	Y	Y					
Beryllium oxide	Y	Y					
Beryllium sulfate	Y	Y					
Bismuth oxychloride	Y	Y					
Bisphenol A	Y	Y					
Bleach solutions	A A		A	– X	C C	A	A A
Boiler compound - liquid	Y	Y					
Boric acid	YA*A	Y	A	A A	AR A	RA	A A
Boron tribromide	Y N	Y			L	*	
Boron trichloride	Y	N					
Boron trifluoride	C		C	C	C	B	C
Brine	A A		A	A A	A A	A	– A
Bromine	YXN–	N	A	– X	UL X	*A	A
Bromine pentafluoride	N						
Bromine trifluoride	N U			U	U		U
Bromobenzene	NX U		X	U	U	X	A
Bromobutane	X		C			X	
Bromo chloromethane	X		C			X	
Bromoform	N				L	*	
Bromotoluene	X		X			X	
Brucine	Y	N					
Butadiene, inhibited	YX C	Y	B	U X	B B	B	B B
Butane	YXNU	Y	B	A A	AR A	RB	A A
Butanediol (1,3)	Y	Y					
Butenediol (1,4)	Y	Y					
Butric acid	YI*	Y	A		R	RA	
Butyl acetate (n)	YBNB	N	X	– fX	UNnX	LXn	U X
Butyl acetate (sec)	Y	N					
Butyl acrylate (iso)	YX U	Y	C	– X	– X	X	U X
Butyl alcohol (n)	YBRB	Y	A	A eA	AReA	LBg	A A

CHEMICAL NAME	BUTYL	PC	CPE	NBR	NEO	PVC	VIT
Butylaldehyde	L				L	*	
Butylamine	YX U	N	B	C C	U X	X	U X
Butyl bromide	C		C			X	
Butyl cellosolve	A A		B	C eC	B eC	B–	U X
Butylene	Y LU	Y		B	CR	N	A
Butylene oxide (1,2)	Y	Y					
Butyl hydroperoxide	Y	Y					
Butyl mercaptan (n)	N						
Butyl methacrylate (n)	Y	Y					
Butylphenol (p-tert)	Y	Y					
Butyltrichlorosilane	Y	N					
Butynediol (1,4)	Y	N					
Butyraldehyde	YB B	N	B	C X	C C	X	U X
Butyric acid	B		A	X	X	A	A
Butyric acid (50%)	*				R	R	
Cacodylic acid	Y	Y					
Cadmium	A		A			A	
Cadmium acetate	Y	Y					
Cadmium bromide	Y	Y					
Cadmium chloride	Y	Y					
Cadmium fluoborate							
Cadmium nitrate	Y	Y					
Cadmium oxide	Y R	Y			R	*	
Cadmium sulfate	Y	Y					
Calcium acetate	A A		A	B B	B B	A	U X
Calcium arsenate	Y	Y					
Calcium arsenite							
Calcium bisulfite	X U		A	A A	A A	A	A A
Calcium carbide	Y	Y					
Calcium chlorate	Y	Y					
Calcium chloride	YA A	Y	A	A A	A A	A	A A
Calcium chromate	Y	Y					
Calcium cyanide	Y	Y					
Calcium fluoride	Y	Y					
Calcium hydroxide	YA A	Y	A	A A	A A	A	A A
Calcium hydroxide (50%)	R				R	R	
Calcium hypochlorite	YA A	Y	A	C C	C C	A	A A
Calcium, metallic	Y	Y					
Calcium nitrate	YA A	Y	A	A A	A A	A	A A
Calcium oxide	Y	Y					
Calcium peroxide	Y	Y					
Calcium phosphate	Y	Y					
Calcium phosphide	Y	Y					
Calcium resinate							
Camphene							
Camphor (oil)	Y	Y					
Cane sugar liquors	A A		A	A A	A A	A	A A
Caprolactam (solution)	Y	Y					
Captan	Y	Y					
Carbamate	B B		A	C C	B B	A	A A

CHEMICAL NAME	BUTYL	PC	CPE	NBR	NEO	PVC	VIT
Carbaryl	YI	Y	B			I	
Carbitol®	B B		A	B B	B B	B	B B
Carbolic acid (phenol)	BRB		A	U X	CR C	*A	A A
Carbolic oil	Y	Y					
Carbon dioxide	YBRB	Y	A	A A	BR B	RA	A A
Carbon disulfide	NXNU		X	C gC	UNnX	NXn	A A
Carbonic acid	A A		A	A B	A A	A	A A
Carbon monoxide	YA A	Y	A	A A	A A	A	A A
Carbon tetrachloride [C]	NXNU		C	C gC	UNnX	NXf	A A
Castor oil	BN B		A	A eA	ANeA	RBe	A A
Catechol	Y	Y					
Caustic potash (solution)	Y	N					
Caustic soda (solution)	Y	N					
Cellosolve acetate	BLB		B	U fx	−L fX	NB−	U X
Cellosolve solvent	BLB		B	− gX	−NeX	L B	C C
Chlordane	Y N	Y			L	*	
Chlorinated aliphatics	X		C			X	
Chlorine	YXR−E	Y	A	−E X	CL C	NB	AEA
Chlorine, liquid	CRC		A	− X	UN X	NX	A A
Chlorine dioxide	C C		A	U X	U X	A	A A
Chlorine trifluoride	NX U		B	U X	U X	B	U X
Chloroacetone	BR		X	X	L B	*X	X
Chloroacetophenone	Y	Y					
Chloracetyl chloride	Y	N					
Chloroaniline (p)	Y	Y					
Chloroaromatics	X		X			X	
Chlorobenzene	NXNU		X	U nX	UNnX	NXn	A A
Chlorobromomethane	BNB		X	− X	UN X	*X	B X
Chlorobutyronitrile (4)	N	N					
Chloroform [C]	NXNU		X	U nX	UNnX	NXn	A A
Chlorohydrin (crude)	N						
Chloromethyl methyl ether							
Chloronapthalene	XNU		A	U pX	UNnX	NCn	A A
Chloronitrobenzene (o)							
Chloro-o-toluidine (4)	N						
Chlorophenol (p)	Y	Y					
Chlorophenylene diamine	N				L	*	
Chloropicrin	Y N	Y			N	N	
Chloroprene	NI	N	X			I	
Chlorosulfonic acid	NX U	N	A	U X	U X	A	C C
Chlorothene (VG)	N			f	Nn	N p	
Chlorotoluene (p)	NX U		C	U X	U X	B	A A
Chrome plating solutions	X U		A	U X	U X	A	A A
Chromic acetate							
Chromic acid	CNC		A	U X	UN X	RA	A A
Chromic acid (30%)	C		A			I	
Chromic acid (50%) [C]	*			f	Nn	R g	
Chromic anhydride	Y	Y					
Chromic sulfate	Y	N					
Chromium (VI)	A		A			A	

CHEMICAL NAME	BUTYL	PC	CPE	NBR	NEO	PVC	VIT
Chromous chloride							
Chromyl chloride							
Chrysene	C		C			C	
Citric acid	YARA	Y	A	A A	AR A	RA	A A
Citric acid (10%)				e	e	e	
Coal tar products	YX*	Y	C	A	N B	NC	C
Cobalt acetate (ous)	Y	Y					
Cobalt bromide (ous)							
Cobalt chloride (ous)	YA A	Y	A	A A	A A	A	– A
Cobalt fluoride (ous)							
Cobalt formate (ous)							
Cobalt nitrate (ous)	Y	Y					
Cobalt sulfamate (ous)							
Cobalt sulfate (ous)	Y	Y					
Coconut oil	B A		A	A A	B B	A	– A
Collodion	Y	N					
Copper acetate	A A		A	B B	B B	A	– A
Copper acetic (ic)	Y	Y					
Copper acetoarsenite (ic)	Y	Y					
Copper arsenite (ic)	Y	Y					
Copper bromide (ic)	Y	Y					
Copper bromide (ous)							
Copper chloride (ic)	YA A	Y	A	A A	A A	A	A A
Copper cyanide (ous)	YA A	Y	A	A A	A A	A	A A
Copper fluoborate (ic)	Y	Y					
Copper formate (ic)							
Copper glycinate	Y	Y					
Copper iodide (ous)							
Copper lactate (ic)							
Copper napthenate (ic)	Y	Y					
Copper nitrate (ic)	Y	Y					
Copper oxalate (ic)	Y	Y					
Copper salts (solutions)	A		A			A	
Copper subacetate (ic)	Y						
Copper sulfate (ic)	YB A	Y	A	A A	A A	A	A
Copper tartrate (ic)							
Cottonseed oil	CNC		A	A A	BL B	*A	A A
Coumaphos							
Cresols	YXNU	Y	A	C X	CL C	LC	A A
Cresote	YXNU	Y	C	B B	CL C	LC	A A
Cresyl glycidyl ether							
Cresylic acid (50%)	*				N	R	
Crotonaldehyde	Y	N					
Cumene	YX –	Y	B	– X	U X	B	A A
Cumene hydroperoxide	Y	Y					
Cupric nitrate	R				L	*	
Cupriethylenediamine sol.							
Cyanide	L				L	*	
Cyanide, H$_2$ & CN salts	A		A			A	
Cyanoacetic acid	Y	Y					

CHEMICAL NAME	BUTYL	PC	CPE	NBR	NEO	PVC	VIT
Cyanogen	Y	Y					
Cyanogen bromide	Y	Y					
Cyanogen chloride	Y	Y					
Cyclohexane	NXNU	Y	A	A A	UL X	*B	A A
Cyclohexanol	YXLU	Y	A	B eB	ALeB	NCe	A A
Cyclohexanone	YBLB	N	X	U X	UL X	NX	U X
Cyclohexanone peroxide	Y	Y					
Cyclohexylamine	Y	N					
Cyclopentane	Y	Y					
Cyclopropane	Y	Y					
Cymene (p)	YX –	Y	X	– X	U X	X	A A
DDD	Y	Y					
DDT	YA	Y	B			B	
Decaborane	Y N	Y			N	*	
Decahydronaphthalene	Y	N					
Decaldehyde	Y	N					
Decane	X –		B	B B	U B	C	A A
Decene (1)	Y	Y					
Decyl alcohol	Y	Y					
Degreasing fluids	N				N	*	
Demeton	Y	Y					
Denatured alcohol	A A		A	A A	A A	B	A A
Detergent solutions	A A		A	A A	A B	B	A A
Developing fluids	B B		A	A A	A A	A	A A
Diacetone	A			–	–		U
Diacetone alcohol	YARA	Y	A	U X	AR B	*C	– I
Diazinon	Y	Y					
Dibenzoyl peroxide	Y	Y					
Dibenzyl ether	BLB		C	U X	BL C	*C	– I
Diborane	N				N	*	
Dibromochloropropane	I		X			X	
Dibutylamine	X U		A	U B	U B	B	U X
Dibutyl ether	C C		A	C X	C C	X	C C
Dibutyl phthalate	YCLB	N	C	U gX	UNfX	NXn	B B
Dibutyl sebacate	B B		B	U X	U X	X	B B
Dichlorobenzene (o)	NX*U		X	U X	UN X	NX	A A
Dichlorobenzene (p)	Y	Y					
Dichlorobutene							
Dichlorodifluoromethane	Y	Y					
Dichloroethane (1,1)	N N				N	*	
Dicloroethylene	YI	N	C			X	
Dichloroethyl ether	N						
Dichloromethane	YX	N	C			X	
Dichlolophenol (2,4)	Y	Y					
Dichlolophenoxy acetic acid	Y	Y					
Dichloropropane (1,2)	Y	N					
Dichloropropene	Y N	N			N	*	
Dicyclopentadiene	Y	Y					
Dieldrin	YB	Y	C			C	
Diesel oil	XNU		A	A A	BR B	RA	A A

CHEMICAL NAME	BUTYL	PC	CPE	NBR	NEO	PVC	VIT
Diethanolamine	Y R	N			R	*	
Diethylamine	YBLB	N	B	C fC	CLpC	NBn	U X
Diethylbenzene	Y U	N		U	U		A
Diethyl carbonate	Y	Y					
Diethylene glycol	YA A	Y	A	A A	AI A	B	A A
Diethylenetriamine	Y	Y					
Diethyl ether	X U		A	U X	C C	C	U X
Diethyltriamine	L				L	*	
Difluoroethane (1,1)	Y	Y					
Difluorophosphoric acid	Y	N					
Diisobutylcarbinol	Y	Y					
Diisobutylene	YX –	Y	C	B B	C X	C	A A
Diisobutyl ketone	YCL	N	C	e	Np	NXp	
Diisocyanates	CR		C		L	*C	
Diisodecylphthalate	N						
Diisopropanolamine	Y	N					
Diisopropylamine	Y	N					
Dimethylacetamide	Y	N					
Dimethylamine	Y	N					
Dimethyl benzene (xylene)	X		X			X	
Dimethyldichlorosilane	Y	N					
Dimethyl ether	N						
Dimethyl formamide {DMF}	YBL–	N	C	B nB	CNgC	NXn	U X
Dimethylhydrazine (1,1)	Y	N					
Dimethyl sulfate	Y	Y					
Dimethyl sulfide	Y	Y					
Dimethyl sulfoxide {DMSO}	Y	Y		e	e	–	
Dimethyl terephthalate	Y	Y					
Dinitrobenzene (M–)	Y	Y					
Dinitrocresols	YA	Y	X			I	
Dinitrophenol (2,4)	Y	Y					
Dinitrotoluene (2,4)	YX U	Y	C	U X	U X	C	C C
Dioctyl adipate	Y	N					
Dioctyl phthalate {DOP} [C]	YBNB	N	C	– gC	ULgX	NCn	B B
Dioxane (1,4) [C]	NBLB		B	– n	–Rn	*Xn	–
Dipentene	Y –	Y		B	–		A
Diphenylamine	Y	Y					
Diphenyldichlorosilane							
Diphenylether	N						
Dipropylene glycol	Y	Y					
Diquat							
Distillates, straight run	N						
Disulfton							
Diuron							
Dodecanol	Y	Y					
Dodecene	Y	Y					
Dodecene (1)	Y	Y					
Dodecylbenzene							
Dodecyltrichlorosilane	Y	N					
Dowtherm oil	YX U	Y	C	– X	U X	C	A A

CHEMICAL NAME	BUTYL	PC	CPE	NBR	NEO	PVC	VIT
Drycleaning fluids	X U		B	C C	U X	C	A A
Dursban							
Emulsifying agent	R				L	*	
Endosulfane							
Endrin	Y	Y					
Epichlorohydrin	YBLB	Y	I	– X	–L X	*I	U X
Epoxied vegetable oils	Y	Y					
Esters	N				N	*	
Ethane	YXRU	Y	B	A A	BR B	*B	A A
Ethanol	R				R	L	
Ethanolamine	B B		A	B B	B B	A	U X
Ethers	L				R	*	
Ethion							
Ethoxylated dodecanol	Y	Y					
Ethoxylated monylphenol							
Ethoxylated pentadecanol	Y	Y					
Ethoxylated tetradecanol	Y	Y					
Ethoxylated tridecanol	Y	Y					
Ethoxy triglycol	Y	Y					
Ethyl acetate	YBLB	Y	B	U nX	CNfC	NXn	U X
Ethyl acetoacetate	YB B	N	A	U X	C C	A	U X
Ethyl acrylate	YB BP	N	B	–N X	– X	X	UNX
Ethyl alcohol	YARA	Y	A	A eA	AReA	*Bg	A A
Ethyl aluminum dichloride	Y	N					
Ethylamine	Y	N					
Ethylaniline	L				N	*	
Ethylbenzene	NX U		C	U X	U X	B	A A
Ethyl bromide	X		C	B	X	X	I
Ethyl butanol	Y	Y					
Ethyl butyrate	Y	N					
Ethyl chloride	YA A	N	C	A A	B B	C	A A
Ethyl chloroacetate	Y	N					
Ethylchloroformate	Y –	N		–	C		A
Ethyldichlorosilane	Y	N					
Ethylene	YBR–	Y	A	A A	–R C	*A	A A
Ethylene chloride	C C		I	– X	– X	I	A A
Ethylene chlorohydrin	YB –	N	I	U X	B B	I	A A
Ethylene cyanohydrin	Y	Y					
Ethylenediamine	YALA	N	B	A A	AR A	*B	U X
Ethylene dibromide	YC P	Y	C	F X	X	X	EC
Ethylene dichloride [C]	YCNC	N	C	U nX	UNnX	NXn	A A
Ethylene glycol	YARA	Y	A	A eA	AReA	LAe	A A
Ethylene glycol diacetate							
Ethyleneimene	Y	N					
Ethylene oxide	YC*C	Y	A	U X	UL X	NX	U X
Ethylene thiourea	A		A			A	
Ethylene trichloride	CNC		C	U X	UN X	*C	A A
Ethyl ether	YCRC	N	B	C eC	ULeX	NXn	U X
Ethyl formate	YBLB	N	A	U X	BL B	NB	A A
Ethylhexaldehyde							

CHEMICAL NAME	BUTYL	PC	CPE	NBR	NEO	PVC	VIT
Ethyl hexanol (2)	Y	Y					
Ethylhexyl (2) acrylate	Y	N					
Ethylhexyl tallate	Y	Y					
Ethylidenenorbornene							
Ethyl iodide	X		C			X	
Ethyl lactate	Y	N					
Ethyl mercaptan	YX U	Y	B	U X	– C	B	A A
Ethyl methacrylate	Y	Y					
Ethyl nitrate	Y	Y					
Ethylphenyldichlorosilane	Y	N					
Ethyl silicate	YA A	Y	A	A A	A A	A	A A
Ethyltrichlorosilane	Y	N					
Ferric ammonium citrate	Y	Y					
Ferric ammonium oxalate	Y	Y					
Ferric chloride	YA*A	Y	A	A A	AL A	RA	A A
Ferric fluoride							
Ferric glycerophosphate	Y	Y					
Ferric nitrate	YA*A	Y	A	A A	AR A	RA	A A
Ferric & ferrous salts	A		A			A	
Ferric sulfate	YA*A	Y	A	A A	AR A	RA	A A
Ferrocyanide	R				N	*	
Ferrous ammonium sulfate	Y	Y					
Ferrous chloride	Y	Y					
Ferrous fluoborate	Y	Y					
Ferrous oxalate	Y	Y					
Ferrous sulfate	Y *	Y			R	R	
Fluorine	NCLC		X	– X	–L X	*X	B B
Fluorine gas, wet	L				L	R	
Fluosilicic acid	NB*–		A	A A	AR A	RC	– B
Fluosulfonic acid	N						
Formaldehyde solution	YA A	Y	A	B eC	A eB	AeA	A A
Formaldehyde (50%)	R				R	R	
Formic acid	YARA	Y	A	B B	AR A	RB	C C
Formic acid (90%)				f	e	e	
Freon (TF)	X U		C	A eA	A gA	Cn	A A
Fuel oils	XRU		B	A A	BR B	RB	A A
Fumaric acid	YX U	Y	A	A A	B B	A	A A
Furan	X C		A	U X	U X	A	– X
Furfural	YBLB	Y	A	U nX	BLgC	NXn	U X
Furfuryl alcohol	Y	Y					
Gallic acid	YB*B		A	B* B	B B	RB	A A
Gas blend stocks: alkalate	N						
Gas oil: cracked							
Gasoline, automotive	NXNU–		B	A A	BL B	RC	A A
Gasoline, aviation	N						
Gasoline, casinghead	N						
Gasoline: straight run	N	Y					
Gasoline, white				e	n	p	
Glutaraldehyde, solution							
Glycerine	YARA	Y	A	A eA	AReA	*Ae	A A

CHEMICAL NAME	BUTYL	PC	CPE	NBR	NEO	PVC	VIT
Glycerol	R				R	*	
Glycidyl ethers	A		A			I	
Glycidyl methacralate	Y	Y					
Glycols	ARA		A	A A	AR A	*A	A A
Glyoxal (40% solution)							
Grease (petroleum base)	X		B			B	
Halogens	L				L	*	
Heptachlor	Y	Y					
Heptane	N	Y					
Heptanol	Y	Y					
Heptene (1)	N						
Hexaldehyde (n)	YB		C	X	A	C	I
Hexamethylenediamine	Y	N					
Hexamethylenetetramine	Y L	Y			N	*	
Hexane	NXNU		A	A eA	BNeB	NBn	A A
Hexanol	Y	Y					
Hexene (1)	NX U		I	B B	B B	I	A A
Hexylene glycol	YA	Y	A	B	B	B	A
Hydraulic oil (petroleum)	X U		C	A A	B B	C	A A
Hydrazine	YALA	Y	X	B C	BN C	*I	– I
Hydrazine (65%) [C]				e	e	e	
Hydrobromic acid	A A		A	U X	A B	A	A A
Hydrochloric acid	YAL E		A	E	R	*A	E
Hydrochloric acid (10%)				e	e	e	
Hydrochloric acid (37%) cold	A A		A	B B	B B	A	A A
Hydrochloric acid (38%)				e	e	e	
Hydrocyanic acid	A A		A	B B	B B	B	A A
Hydrofluoric acid (cold)	NBL E		A	E X	R B	*A	EA
Hydrofluoric acid (30%)	*				R	R	
Hydrofluoric acid (48%)				e	e	g	
Hydrofluoric acid (50%)	*				R	L	
Hydrofluoric acid, conc.	B			U	B		A
Hydrogen bromide	Y						
Hydrogen chloride	Y						
Hydrogen cyanide	Y						
Hydrogen fluoride	NC		A			A	
Hydrogen peroxide	YC	Y	A			A	
Hydrogen peroxide (30%)	L			e	Le	* e	
Hydrogen peroxide (90%)	C C		A	U X	– X	A	B B
Hydrogen sulfide (cold)	YA A	Y	A	U X	A A	A	U X
Hydrogen, liquefied	Y	Y					
Hydroquinone	NXL–		A	C C	–L X	RA	U X
Hydroquinone saturated				e	e	e	
Inorganic salts	R				R	*	
Iodine	BL		A	B	L X	NA	A
Isoamyl alcohol	Y	Y					
Isobutane	Y	Y					
Isobutyl acetate	Y	N					

CHEMICAL NAME	BUTYL	PC	CPE	NBR	NEO	PVC	VIT
Isobutyl alcohol [C]	YA A	Y	A	B eB	A eA	Bf	A A
Isobutylamine	N N	N		N			P
Isobutylene	Y	Y					
Isobutyric acid	Y	Y					
Isobutyronitrile							
Isoctaldehyde	Y	N					
Isodecaldehyde	Y E	N		E			E
Isodecyl alcohol	Y	Y					
Isohexane	N						
Isooctane	XNU		A	A eA	BReB	*Bp	A A
Isooctyl alcohol	Y	Y					
Isopentane	N						
Isophorone	YA A	Y	C	U X	– X	X	U X
Isophthalic acid	Y	Y					
Isoprene	N						
Isopropanol	R				R	*	
Isopropyl acetate	YB A	N	C	U X	U X	X	U X
Isopropyl alcohol	YARA	Y	A	B eB	AReB	RAg	A A
Isopropylamine	Y	N					
Isopropyl ether	YX U	Y	A	B B	– C	A	U X
Isopropyl mercaptan	Y	Y					
Isopropyl percarbonate	Y	Y					
Isovaleraldehyde	Y	N					
Jet fuel, JP-1 (kerosene)	N	Y					
Jet fuel, JP-3	NX	Y	A	A	X	C	A
Jet fuel, JP-4	NX	Y	A	A	X	C	A
Jet fuel, JP-5	N	Y					
Kepone	NA	Y	A			A	
Kerosene	NXNU	Y	A	A eA	CReC	RBf	A A
Ketones	CR		C		L	NX	
Lacquer solvents	XNU		X	U X	UL X	NX	U X
Lactic acid	YARA	Y	A	A A	AR A	LA	A A
Lactic acid (25%)	R				R	R	
Lactic acid (85%)				e	e	e	
Latex, liquid synthetic	Y	Y					
Lauric acid	R				R	L	
Lauric acid (36%)				e	e	f	
Lauroyl peroxide	Y	Y					
Lauryl mercaptan	Y	Y					
Lead acetate	YA A	Y	A	B B	B B	A	– A
Lead arsenate	Y	Y					
Lead chloride							
Lead fluoborate							
Lead fluoride	Y	Y					
Lead, inorganic	A		A			A	
Lead iodide	Y	Y					
Lead nitrate	YA A	Y	A	A A	A A	A	– A
Lead salts	A		A			A	
Lead stearate							

CHEMICAL NAME	BUTYL	PC	CPE	NBR	NEO	PVC	VIT
Lead sulfate	Y						
Lead tetraacetate	Y	Y					
Lead thiocyanate	Y	Y					
Lead thiosulfate							
Lead tungstate							
Lime	A		A			A	
Lime sulfur	A A		A	U X	A A	A	A A
Linear alcohols	Y	Y					
Linoleic acid	XNU		A	B eB	UReX	*Ag	B B
Linseed oil	CNB		A	A eA	BReB	RBe	A A
Liquefied natural gas	Y	Y					
Liquefied petroleum gas	YX U	Y	A	A A	B B	A	A A
Litharge							
Lithium aluminum hydride	Y	Y					
Lithium bichromate							
Lithium chromate							
Lithium hydride							
Lithium metal	Y	Y					
Lube oils, petroleum	X*U		A	A A	BR B	RB	A A
Lye	A A		A	B B	B B	A	B B
Magnesium	Y	Y					
Magnesium hydroxide	A*A		A	B B	AR A	RA	A A
Magnesium nitrate	*				R	R	
Magnesium perchlorate	Y	Y					
Magnesium sulfate	A*A		A	A A	AR A	RA	A A
Magnesium salts	A		A			A	
Malathion	YX	Y	C	B	B	B	I
Maleic acid	YCRC	Y	A	– X	–R X	LA	A A
Maleic acid saturated				e	e	g	
Maleic anhydride	YC C	Y	A	– X	– X	A	A A
Maleic hydrazine							
Malic acid	X U		A	A A	B B	A	A A
Mercuric acetate	Y	Y					
Mercuric ammoniumchloride	Y	Y					
Mercuric chloride	YARA	Y	A	A A	AL A	RX	A A
Mercuric cyanide	Y	Y					
Mercuric iodide	Y	Y					
Mercuric nitrate	Y	Y					
Mercuric oxide	Y	Y					
Mercuric sulfate		Y					
Mercuric sulfide	Y	Y					
Mercuric thiocyanate							
Mercurous chloride	Y	Y					
Mercurous nitrate	Y	Y					
Mercury	YARA	Y	A	A A	AL A	RA	A A
Mercury, inorganic	A		A			A	
Mesityl oxide	YB B	Y	C	U X	U X	X	U X
Methallyl chloride	N						

CHEMICAL NAME	BUTYL	PC	CPE	NBR	NEO	PVC	VIT
Methane	YXRU	Y	A	A A	BR B	*A	A A
Methanol	AR		A	A	R A	LA	C
Methoxychlor	Y	Y					
Methyl acetate	NBLB	N	A	U X	BL B	*X	U X
Methyl acrylate	YB B	Y	C	U X	B B	C	U X
Methylacrylic acid	B			–	B		B
Methyl alcohol	YARAE	Y	A	AEeA	AReA	*Ag	CPC
Methylamine	Y *	N		e	Ng	N e	
Methyl amyl acetate	Y	N					
Methyl amyl alcohol	Y	Y					
Methylaniline (n)	Y	Y					
Methyl bromide	NXN–		I	B B	UL X	NI	A A
Methyl butyl ketone	YA A	N	I	U X	U X	I	U X
Methyl cellosolve	BLB		A	– fB	BReB	RA–	U X
Methyl chloride	YCNC	N	X	U X	UN X	NC	A A
Methyl chloroformate	N						
Methylcyclopentane	N U	Y		–	C		A
Methyldichlorosilane	Y	N					
Methylene (bis, 4,4–)	I		C			I	
Methylene bromide	L			n	Nn	* n	
Methylene chloride [C]	XLU		C	U nX	UNnX	NXn	B B
Methyl ethyl ketone {MEK}	YARAF	N	C	UNnX	ULpX	NXn	UNX
Methylethylpyradine	Y	N					
Methyl formal							
Methyl formate	YBLB	Y	B	U X	BL B	*B	– I
Methyl hydrazine	N						
Methyl isobutyl carbinol	Y	Y					
Methyl isobutyl ketone	CRC		C	U pX	ULnX	NXn	U X
Methyl mercaptan	Y	Y					
Methyl methacrylate [C]	YXRU	Y	C	U pX	ULnX	NXn	U X
Methyl parathion	YA	Y	C			C	
Methylstyrene, alpha	Y	Y					
Methyltrichlorosilane	Y	N					
Methyl vinyl ketone	N						
Mineral oil	XNU		A	A A	BR B	RA	A A
Mineral spirits	N						
Mineral spirits, Rule 66				e	g	f	
Mirex							
Molybdic trioxide	Y	Y					
Monochloroacetic acid	Y	Y					
Monochlorobenzene	XNU		C	U X	UN X	*C	A A
Monochlorodifluoromethane	Y	Y					
Monoethanolamine	YBRB	N	I	U eX	UReX	LIe	U X
Monoisopropanolamine	Y	N					
Morpholine	Y R	N		n	Rp	* n	
Motor fuel antiknock cmpd	Y	N					
Muriatic acid	R				R	*	
Mustard gas	A A		A	– A	A A	A	– I
Nabam	Y	Y					
Naled							

CHEMICAL NAME	BUTYL	PC	CPE	NBR	NEO	PVC	VIT
Naphtha	NX*U	Y	A	C C	CR C	NC	A A
Naphtha (VM&P)				e	g	f	
Naphthalene	NXNU		A	U X	UN X	NX	A A
Naphthas, aliphatic	N				R	*	
Naphthas, aromatic	N				L	*	
Naphthenic acid	N U			B	–		A
Naphthylamine	Y	Y					
Neohexane							
Nickel acetate	YA*A	Y	A	B B	BR B	RA	U X
Nickel ammonium sulfate	Y	Y					
Nickel bromide	Y	Y					
Nickel carbonyl	YA	Y	A			A	
Nickel chloride	YA*A	Y	A	A A	AR A	RA	A A
Nickel cyanide	Y	Y					
Nickel fluoborate							
Nickel formate	Y	Y					
Nickel hydroxide	Y	Y					
Nickel, inorganic & cmpds	A		A			A	
Nickel nitrate	Y *	Y			R	R	
Nickel salts	A		A			A	
Nickel sulfate	YA*A	Y	A	A A	AR A	RA	A A
Nicotine	Y	Y					
Nicotine sulfate	Y	Y					
Nitralin							
Nitric acid	NIN E		X	N	L	NI	E
Nitric acid (10%)				e	e	g	
Nitric acid (70%)				n	g	f	
Nitric acid, conc.	C C		X	U X	C X	X	A A
Nitric acid, dilute	B B		A	U X	A A	B	A A
Nitric acid, red fuming	XNU		X	U nX	UNnX	*Xp	C C
Nitric acid, white fuming	N			n	Nn	* P	
Nitric oxide	Y	Y					
Nitriles	C		C			C	
Nitroaniline (2)	Y	Y					
Nitroaniline (4)							
Nitrobenzene	NCNU	N	C	U nX	UNnX	NXn	B B
Nitroethane	YBNB	Y	A	U X	CN C	*X	U X
Nitrogen	YARA	Y	A	A A	AR A	*A	A A
Nitrogen, oxides	A		A			A	
Nitrogen tetroxide	C C		C	U X	U X	C	U X
Nitroglycerin	A		A			I	
Nitromethane	Y NB	Y		U	CN	N	U
Nitromethane (95.5%)				f	e	p	
Nitrophenol	Y	Y					
Nitrophenol (2)	Y	Y					
Nitrophenol (4)	Y	Y					
Nitropropane (2)	Y N	Y			N	N	
Nitropropane (95.5%)				n	g	n	
Nitrosyl chloride	Y	Y					
Nitrotoluene (o)							

CHEMICAL NAME	BUTYL	PC	CPE	NBR	NEO	PVC	VIT
Nitrotoluene (p)							
Nitrous oxide	Y L	Y			L	R	
Nonane	Y	Y					
Nonanol	Y	Y					
Nonene	Y	Y					
Nonene (1)	Y	Y					
Nonyl phenol	Y	N					
Octachlorotoluene	U			U	U		A
Octane	NX U	Y	A	– A	– B	C	A A
Octanol	Y	Y					
Octene (1)	Y	Y					
Octyl alcohol	BRA		A	B eB	AReA	NAf	A A
Octyl epoxy tallate							
Oil: crude	NX		A			B	
Oil: diesel	N						
Oil, edible: castor	Y	Y					
Oil, edible: coconut	Y	Y					
Oil, edible: cottonseed	Y	Y					
Oil, edible: fish	Y	Y					
Oil, edible: lard	Y	Y					
Oil, edible: olive	Y	Y					
Oil, edible: palm	Y	Y					
Oil, edible: peanut	Y	Y					
Oil, edible: safflower	Y	Y					
Oil, edible: soya bean	Y	Y					
Oil, edible: tucum	Y	Y					
Oil, edible: vegetable	YX	Y	A			B	
Oil, fuel: #1 (kerosene)	N	Y					
Oil, fuel: #1D	N	Y					
Oil, fuel: #2	N	Y					
Oil, fuel: #2D	N	Y					
Oil, fuel: #4	N	Y					
Oil, fuel: #5	N	Y					
Oil, fuel: #6	N	Y					
Oil, misc: absorption	N						
Oil, misc: clarified	Y	Y					
Oil, misc: coal tar	N	Y					
Oil, misc: croton	N						
Oil, misc: linseed	Y	Y					
Oil, misc: lubricating	NX	Y	A			B	
Oil, misc: mineral	N	Y					
Oil, misc: mineral seal	N						
Oil, misc: motor							
Oil, misc: neatsfoot	Y	Y					
Oil, misc: penetrating	N						
Oil, misc: range	N						
Oil, misc: resin	Y	Y					
Oil, misc: road	N						
Oil, misc: rosin	Y	Y					
Oil, misc: sperm	Y	Y					

CHEMICAL NAME	BUTYL	PC	CPE	NBR	NEO	PVC	VIT
Oil, misc: spindle	N						
Oil, misc: spray	N						
Oil, misc: transformer	N						
Oil, misc: turbine	N						
Oleic acid	YBLB	Y	A	C eC	CReC	NAf	B B
Oleic acid, potassium salt	Y	Y					
Oleic acid, sodium salt	Y	Y					
Oleum	N						
Oleum (30% Free SO$_3$)	G			G			E
Organotin compounds	C		C			C	
Oxalic acid	YARA	Y	A	B B	BR B	NA	A A
Oxalic acid, saturated				e	e	e	
Oxygen, liquid	Y N	Y			N	*	
Ozone	BLB		A	U X	BL C	RB	A A
Paint thinners	N				L	N	
Paint & varnish removers	N				L	N	
Palmitic acid	BRB		A	A A	BR B	LA	A A
Palmitic acid, saturated				g	e	g	
Paraffin wax	B		A			A	
Paraformaldehyde	Y	Y					
Paraldehyde							
Parathion	YAN	Y	C		N	*C	
Pentaborane	N N				N	*	
Pentachlorophenol [C]	Y L	Y		e	Ne	*f	
Pentadecanol	Y	Y					
Pentaerythritol	Y	Y					
Pentane	N L			e	Re	R n	
Pentene (1)	N						
Peracetic acid	Y	Y					
Perchloric acid	YBLB	Y	I	– X	AR B	NI	A A
Perchloric acid (60%)				e	e	e	
Perchloroethylene [C]	XNU		C	C gC	UNnX	NXn	A A
Perchloromethyl mercaptan	Y	N					
Pesticide manufacturing	C		C			C	
Petrolatum	Y	Y					
Petroleum naphtha	N	Y					
Petroleum spirits	N				R	N	
Phenol [C]	YBRBE	Y	A	–EnX	CReC	LCg	AEA
Phenyldichloroarsine, liq	Y	N					
Phenylhydrazine	X C		I	U X	C X	I	A I
Phosdrin							
Phosgene	YA	Y	A			A	
Phosphoric acid	Y R	Y			L	L	
Phosphoric acid, conc.	B		A	e	e	g	
Phosphoric acid, dil.	A		A			A	
Phosphoric acid (20%)	B A		A	B B	B B	A	A A
Phosphoric acid (45%)	B B		A	U X	B B	A	A A
Phosphorus, black							
Phosphorus oxychloride	Y	Y					
Phosphorus pentasulfide	Y	Y					

CHEMICAL NAME	BUTYL	PC	CPE	NBR	NEO	PVC	VIT
Phosphorus, red	Y	Y					
Phosphorus tribromide	N						
Phosphorus trichloride	YA A	Y	A	U X	U X	A	A A
Phosphorus, white	Y	Y					
Phthalic anhydride	Y	Y					
Pickling solution	CLC		A	– X	–R X	NA	B B
Picric acid	BLB		A	B B	AR A	NA	A A
Picric acid saturated				e	e	e	
Pine oil	XNU		B	B B	UR X	LB	A A
Pinene	X U		B	B B	B C	B	A A
Piperazine	Y	Y					
Plating solutions	RA			A	–R	R	A
Polybutene	Y	Y					
Polychlorinated biphenyls	YA	Y	C			C	
Polyphosphoric acid	Y	Y					
Polypropylene	Y	Y					
Polypropylene glycol	Y	Y					
Potassium	N						
Potassium alum							
Potassium arsenate	Y	Y					
Potassium arsenite							
Potassium binoxalate	Y	Y					
Potassium bisulfate	*				R	R	
Potassium bromide	R				L	R	
Potassium carbonate	*				R	R	
Potassium chlorate	Y	Y					
Potassium chloride	A*A		A	A A	AR A	RA	A A
Potassium chromate	Y	Y					
Potassium chrome alum	R				L	*	
Potassium cyanide	YA*A	Y	A	A A	AR A	RA	A A
Potassium dichromate	YANA	Y	A	A A	AL A	RA	A A
Potassium ferrocyanide	R				L	R	
Potassium hydroxide	YARA	N	A	B B	AR B	RA	B B
Potassium hydroxide (50%)	B		A	e	e	e	
Potassium iodide	Y –	Y			R	*	
Potassiusm nitrate	A*A		A	A A	AR A	RA	A A
Potassium oxalate	Y	Y					
Potassium permanganate	Y	Y					
Potassium peroxide	Y	Y					
Potassium salts, general	A		A			A	
Potassium sulfate	A*A		A	A A	AR A	RA	A A
Printing inks	L				R	*	
Propane	YXRU	Y	C	A A	AR B	*C	A A
Propiolacetone, beta	Y	Y					
Propionaldehyde	Y	N					
Propionic acid	Y	Y					
Propionic anhydride	Y	Y					
Propyl acetate (n)	YBLA	N	C	U fX	–LpX	NCn	U X
Propyl alcohol (n) [C]	YARA	Y	A	A eA	AReA	LAf	A A
Propylamine (n)							

CHEMICAL NAME	BUTYL	PC	CPE	NBR	NEO	PVC	VIT
Propylene	YXRU	Y	A	U X	UR X	*A	A A
Propylene glycol	Y	Y					
Propylene oxide [C]	YB B	Y	I	– nX	U nX	In	– I
Propylene tetramer	Y	Y					
Propyleneimine (inhibited)	Y	N					
Propyl mercaptan	Y	Y					
Pyrethrins	Y R				R	*	
Pyridine	YB B	N	C	U X	U X	C	U X
Pyrogallic acid	Y	Y					
Quinoline							
Red oil	X U		A	A A	B B	B	A A
Refined petroleum solvent	X		C			C	
Resorcinol	Y	Y					
Rubber solvent				e	g	n	
Salicylic acid	YA	Y	A	B	A	A	A
Selenium dioxide							
Selenium trioxide	Y	Y					
Silica, crystalline	A		A			A	
Silicon etch				n	g	f	
Silicone tetrachloride	Y	Y					
Silver acetate	Y	Y					
Silver alkyl sulfates							
Silver carbonate	Y	Y					
Silver fluoride	Y	Y					
Silver iodate	Y	Y					
Silver nitrate	YARA	Y	A	B B	AR A	RA	A A
Silver oxide	Y	Y					
Silver sulfate	Y	Y					
Soda ash	A A		A	A A	A A	A	A A
Sodium acetate	A A		A	B B	B B	A	U I
Sodium amide	Y	Y					
Sodium arsenate	Y	Y					
Sodium arsenite	Y	Y					
Sodium azide	Y	Y					
Sodium bicarbonate	A A		A	A A	A A	A	A A
Sodium bifluoride	Y	Y					
Sodium bisulfite	YA A		A	A A	A A	A	A A
Sodium borate	YA A	Y	A	A A	A A	A	A A
Sodium borohydride	Y	Y					
Sodium cacodylate	Y	Y					
Sodium chlorate	Y	Y					
Sodium chromate	Y	Y					
Sodium cyanide	YA A	Y	A	A A	A A	A	A A
Sodium dechromate	Y	Y					
Sodium ferrocyanide	Y	Y					
Sodium fluoride	Y	Y					
Sodium hydride	Y	Y					
Sodium hydrosulfide sol.	Y	Y					
Sodium hydroxide	YARA	N	A	B B	AR A	LB	B B
Sodium hydroxide (50%)	B E		A	Ee	e	g	E

CHEMICAL NAME	BUTYL	PC	CPE	NBR	NEO	PVC	VIT
Sodium hypochlorite	YB*B	Y	A	B B	BN B	RB	A A
Sodium, metallic	N						
Sodium methylate	Y	Y					
Sodium nitrate	YA A	Y	A	B B	A B	A	– A
Sodium nitrite	Y	Y					
Sodium oxalate	Y	Y					
Sodium phosphate	YA A	Y	A	A A	A A	A	A A
Sodium salts, general	A		A			A	
Sodium senenite							
Sodium silicate	YA A	Y	A	A A	A A	A	A A
Sodium silicofluoride	Y	Y					
Sodium sulfide	Y	Y					
Sodium sulfite	YAR	Y	A	A	L A	RA	A
Sodium thiocyanate	Y	Y					
Sodium thiosulfide	R				L	*	
Sorbitol	Y	Y					
Soybean oil	C C		A	A A	B B	A	A A
Stannous fluoride							
Stearic acid	YBRB	Y	A	B B	BL B	RA	– A
Stoddard solvent	XLU		A	A eA	CLeC	NCf	A A
Strontium chromate							
Strychnine							
Styrene [C]	NXNU		C	U nX	UNnX	NXn	B B
Sucrose	YA A	Y	A	A A	A B	A	– A
Sulfolane	Y	Y					
Sulfur (liquid)	Y	Y					
Sulfur dioxide	YB B	Y	A	U X	C C	A	A A
Sulfuric acid	NI E		X	E		I	E
Sulfuric acid (80%)	R				L	L	
Sulfuric acid (94%)	R				N	N	
Sulfuric acid (95%)				n	f	g	
Sulfuric acid, fuming	X		X			X	
Sulfuric acid, spent	Y	Y					
Sulfuric acid, to 50%	B		A				
Sulfur monochloride	N	N					
Sulfurous acid, conc.	B		A	B	B	B	A
Sulfurous acid, dil.	B		A			B	
Sulfur trioxide	*				N	R	
Sulfuryl chloride	Y	N					
Tallow	Y	Y					
Tallow fatty alcohol	Y	Y					
Tannic acid	YARA	Y	A	A A	AR A	RA	A A
Tannic acid (65%) [C]				e	e	e	
Tar, bituminous	X U		A	B B	C C	A	A A
Tetrabutyl titanate	YB B		A	B B	A B	A	A A
Tetrachloroethane	YC	N	X			I	
Tetrachloroethylene	YX U	N	C	U C	– X	C	A A
Tetradecanol	Y	Y					
Tetradecene (1)	Y	Y					
Tetradecylbenzene							

CHEMICAL NAME	BUTYL	PC	CPE	NBR	NEO	PVC	VIT
Tetraethyl lead	N L	N			R	R	
Tetraethyl pyrophosphate	Y	Y					
Tetraethylene glycol	Y	Y					
Tetraethylenepentamine							
Tetrafluoroethylene	Y	Y					
Tetrahydrofuran {THF}	YBNBN	N	C	−NnX	−LnX	NXn	UNX
Tetrahydronaphthalene	Y	Y					
Tetramethyl lead {TML}	N						
Thiols: N-alkane Mono,	C		C			C	
Thiols: cyclohexane, benzen	C		C			C	
Thiophosgene	Y	Y					
Thiram	Y	Y					
Thorium nitrate							
Titanium tetrachloride	NX U		A	C C	U X	A	A A
Toluene	NXNU	N	C	fC	UNnX	NXn	A A
Toluene diisocyanate {TDI}	BlA		C	−nX	UNnX	NCp	− B
Toluene 2, 4-diisocyanide	Y	Y					
Toluene (p) sulfonic acid							
Toluidine (o)	N						
Toxaphene	Y	Y					
Transformer oil	U			A	B		A
Transmission fluid type A	U			A	B		A
Triacetin	A A		A	B B	B B	A	U X
Tributyl phosphate	A A		C	U X	U X	X	U X
Trichlorfon							
Trichloroacetic acid	BLB		A	B B	BL X	*A	C C
Trichloroethane	YXNU	N	C	U X	UN X	*C	A A
Trichloroethane (1,1,1 −)	C		C			C	
Trichloroethylene {TCE}	NXNU	N	C	C n	ULn	NXn	A
Trichlorofluoromethane	N						
Trichlorophenol	Y	Y					
Trichlorosilane	Y	N					
Trichloro-s-triazinetrion	Y	Y					
Tricresyl phosphate {TCP}	YANA	N	A	U eX	CLfC	NCf	B B
Tridecanol	Y	Y					
Tridecene (1)	N						
Tridecyl alcohol	N				L	*	
Triethanolamine	YBL	N	A	C	R A	LB	X
Triethanolamine (85%) {TEA}				e	e	e	
Triethylaluminum	NC.−−		A	− X	− X	A	B B
Triethylamine	Y	N					
Triethylbenzene	Y	Y					
Triethylene glycol	Y	Y					
Triethylenetetramine	Y	N					
Trifluorochloroethylene							
Trifluralin							
Triisobutyaluminum	Y	Y					
Trimethylamine	Y	Y					

CHEMICAL NAME	BUTYL	PC	CPE	NBR	NEO	PVC	VIT
Trimethylchlorosilane							
Trinitrotoluene {TNT}	XNU		A	U X	BR B	*C	B B
Tripropylene glycol							
Tung oil	CNC		A	A eA	BReB	NCf	A A
Tungsten	A		A			A	
Tungsten carbide, cemented	A		A			A	
Turbine oil	X U		A	B B	B X	C	A A
Turpentine	YXNU	Y	B	A eA	UNnX	LCp	A A
Ultraviolet radiation	I		B			I	
Undecanol (1–)	Y	Y					
Undecene (1)	Y	Y					
Undecylbenzene (n)							
Unsymmetdimethylhydrazine	A			B	B		U
Uranium peroxide							
Uranyl acetate	Y	Y					
Uranyl nitrate	Y	Y					
Uranyl sulfate	Y	Y					
Urea	YA	Y	A	A	A	A	A
Urea peroxide	Y	Y					
Valeraldehyde	Y	N					
Vanadium	A		A			A	
Vanadium oxitrichloride	Y	N					
Vanadium pentoxide	Y	Y					
Vanadyl sulfate	Y	Y					
Varnish	X U		A	B B	C X	C	A A
Vegetable oils	CLA		A	A A	BR C	*C	A A
Vinegar	A A		A	B B	A B	A	A A
Vinyl acetate	YC	N	A	A	A	X	A
Vinyl chloride	YX –	Y	B	– X	U X	X	A A
Vinyl cyanide	I		A			I	
Vinyl ethyl ether							
Vinyl fluoride (inhibited)							
Vinyl halides	C		C	C	C	X	C
Vinylidenechloride	Y	N					
Vinyl styrene	I		X			I	
Vinyl toluene	NI		C			I	
Vinyl trichloride	I		C			I	
Vinyl trichlorosilane	Y	N					
Waste anesthetic gas/vapor	A		A			A	
Wax: carnuba	Y	Y					
Wax: paraffin	Y	Y					
Wood alcohol	R				R	L	
Xylene	NXNU	N	X	U gX	UNnX	NXn	A A
Xylenol	Y	Y					
Xylidenes	X U		C	C C	U X	X	U X
Xylidine	X		C			X	
Zectran							

CHEMICAL NAME	BUTYL	PC	CPE	NBR	NEO	PVC	VIT
Zinc acetate	YA A	Y	A	B B	B B	A	U X
Zinc ammonium chloride							
Zinc arsenate	Y	Y					
Zinc bichromate							
Zinc borate	Y	Y					
Zinc bromide	Y	Y					
Zinc carbonate							
Zinc chloride	YARA	Y	A	A A	AR A	RA	A A
Zinc chromate	Y	Y					
Zinc cyanide							
Zinc fluoborate							
Zinc fluoride							
Zinc formate							
Zinc hydrosulfite							
Zinc nitrate	Y	Y					
Zinc oxide	A		A	A	A	A	A
Zinc phenolsulfonate	Y	Y					
Zinc phosphide	Y	Y					
Zinc potassium chromate							
Zinc salts, general	A		A			A	
Zinc silicofluoride	Y	Y					
Zinc sulfate	YA A	Y	A	A A	A A	A	A A
Zirconium nitrate	Y	Y					
Zirconium oxychloride	Y	Y					
Zirconium sulfate	Y	Y					
Zirconium tetrachloride							

Guidelines for
Decontamination of Fire
Fighters and Their
Equipment Following
Hazardous Materials
Incidents

Canadian Association of Fire Chiefs, Inc., Dangerous Goods Sub-Committee, 1590-7 Liverpool Court, Ottawa, Ontario, Canada K1B 4LZ.

The following supplement is a reprint, with minor editing, of a very useful publication from the Canadian Association of Fire Chiefs. This document includes guidelines on decontamination of Fire Fighters and their equipment after exposure to a variety of levels of hazardous materials, as well as pre-incident planning for decontamination procedures.

We are deeply grateful for the opportunity to include this detailed, practical guide as a supplement to the *Hazardous Materials Response Handbook*.

Guidelines for
DECONTAMINATION
of Fire Fighters and their Equipment
following Hazardous Materials Incidents

Prepared by the Dangerous Goods Sub-Committee
of the CANADIAN ASSOCIATION OF FIRE CHIEFS

Note to Readers

The contents of this booklet are based on information and advice believed to be accurate and reliable. The Canadian Association of Fire Chiefs Inc., its officers and members jointly and severally, make no guarantee and assume no liability in connection with this booklet. Moreover, it should not be assumed that every acceptable procedure is included or that special circumstances may not warrant modified or additional procedures.

The user should be aware that changing technology or regulations may require a change in the recommended procedures contained herein. Appropriate steps should be taken that the information is current when used.

The suggested procedures should not be confused with any federal, provincial, state, municipal, or insurance requirements, or with national safety codes.

Introduction

The number of hazardous materials incidents to which the fire service is called increases year by year. At each of these incidents, there is a good chance that the responding fire fighters may become contaminated with the hazardous material. Frequently, however, the matter of decontamination is never thought of, or is only performed cursorily.

The Dangerous Goods Sub-Committee of the Canadian Association of Fire Chiefs has therefore undertaken to prepare a series of guidelines for decontamination, to be adopted by any fire department that wishes to do so, either as printed here or with local variations due to their own circumstances. Note that these are indeed *guidelines*, *not standards*.

The procedures listed are designed so they can be carried out by any department, rural or urban, volunteer or full-time, with a minimum of investment in special equipment.

The I.A.F.C. Hazardous Materials Committee (among many others) has reviewed this booklet and endorses the philosophies contained in it. However, they share the disclaimer caution on page 523.

Background of the Study and Rationale for Its Conclusions

In 1986 the Dangerous Goods Sub-Committee carried out an extensive study of fire services across the world to investigate the different approaches to decontamination of fire fighters following hazardous materials incidents.

Procedures were reviewed in detail from North America (Phoenix, San Francisco, Colorado, Metropolitan Toronto area) and England (Hampshire, Cambridgeshire, Greater Manchester, London, and the Home Office Guidelines). In addition, information was requested from Hong Kong, New Zealand, Australia, the People's Republic of China, France, Germany, Switzerland, Italy, Sweden, and the Netherlands. From these latter countries no formal replies were received, however, and indications are that decontamination procedures are either limited or absent. Japanese fire officials wrote back to indicate that they were studying various options but had not yet finalized any procedures.

Many magazine articles from the various periodicals published for the European and North American fire service market were reviewed. Furthermore, members consulted with chemical manufacturers, nuclear medicine physicists, hazardous waste disposal companies, industrial hygienists, toxicologists, and various jurisdictional agencies such as the Atomic Energy Control Board and provincial and federal Ministries of Health, Labour, and Environment.

All this research led to the conclusion that there were three basic philosophies in existence:

1. Wet and Dry Procedures;

2. Dispose or Retain Run-off;

3. Severity and Type of the Material.

The idea of a dry procedure made sense because of easier containment and no reaction with water. A single wet procedure, however, was not deemed to be sufficiently comprehensive; on the other hand some wet procedures called for making up solutions of a variety of chemicals, and these were deemed to be too complicated for use by every fire department.

The dispose or retain philosophy is usually considered a wet procedure; however, to base one's procedures solely on the concern about run-off appeared to be overly simplistic. Note, however, that concerns about run-off are addressed later in this document.

The third philosophy was examined in more detail. The methods of determining the severity and type of material were defined by various fire departments along the following lines:

1. By UN class

2. By effects on the environment

3. By chemical characteristics

4. By physiological effects on people

5. By broad groups.

The first alternative was deemed unsuitable because there are too many variables within a class (e.g., some flammable liquids are highly toxic, others are not). The second alternative was deemed to be of secondary importance to fire fighter safety (see the section on Environmental Considerations later in this document). It was found that alternatives three and four could in fact be used to arrive at alternative five, and this was the route taken to define the procedures in this document.

The procedures that were developed are therefore broken down as follows:

—Three general procedures for light, medium, and extreme hazards,

—Two specific procedures for substances that do not fit into the three general groups above (although they share many common factors), and

—One initial routine performed in some cases prior to the start of one of the other procedures.

The Sub-Committee realizes that if the procedures listed here are to be completed thoroughly, a number of decontamination operatives and a Decontamination Officer to oversee them are needed. Typically, this will require the services of at least one fire fighting company. Attention to detail and careful execution of all steps should lead to successful and safe completion of fire fighter decontamination.

Decontamination — General Observations

Six levels of decontamination are outlined. The incident commander will determine which level is applicable for the substance involved, using any reference sources that may state the applicable level. In the absence of such sources, advice should be sought from experts such as toxicologists, chemical company representatives, CANUTEC, CHEMTREC, etc.

The levels are:

A — for light hazards

B – for medium hazards

C – for extreme hazards

D – dry contamination for water-reactive and certain dry substances

E – for etiologic agents and certain dry pesticides and poisons

R – for radioactive materials.

Note that A-level decontamination, the most common, need only be done at the station. However, other levels need to be started at the incident scene as well as being continued on return to the station.

C-level decontamination, the most stringent level for the most toxic substances, may involve the destruction of all clothes worn.

In a few cases, scrubbing of clothes must be done while wearing SCBA as vapours released during cleaning may be harmful.

D-level decontamination is almost always followed by one of the other levels of decontamination, which will be dependent on the substance involved.

The procedures should be initiated if personnel are known or suspected to have been directly exposed to the chemical or its vapors, products of combustion, etc.

Officers should be aware of any cuts, wounds, lesions, or abrasions that their crews may have. If the apparatus is sent to an incident involving hazardous materials, such personnel should wherever possible exercise special care to avoid the chance of contamination through such wounds. Chemicals absorbed through the skin will be absorbed much faster if the skin is cut or abraded, thereby presenting a serious health hazard.

Adequate awareness is necessary to realize when decontamination will be required, so that early action can be taken to bring to the scene the equipment and manpower resources needed to set up and staff the decontamination area.

DECONTAMINATION PROCEDURE

LEVEL "A"

for light hazards

— — — — — — — — — — — —

On Return to Station

1. Wash down all fire fighting clothes with a mild (1 to 2%) trisodium phosphate solution. Rinse with water.

2. Wash down SCBA cylinders and harnesses with a mild trisodium phosphate solution. Take care to wipe, not scrub, around regulator assembly. Rinse with clean water. If damage is suspected to any part of the unit, ensure it is sent for service.

3. Scrub hands and face with soap and water.

NOTE: Where the scrubbing of the fire clothes may release harmful vapors caught in the fibers, it may be necessary to wear breathing apparatus while washing down fire clothes. In these cases, monitor the atmosphere around the washing area. Release of vapors may indicate commercial cleaning is required.

DECONTAMINATION PROCEDURE

LEVEL "B"

for medium hazards

— — — — — — — — — — — —

At the Scene

1. Do not remove SCBA facepiece. Place helmet on back of neck.

2. Assistant to flush fire fighter downwards from head to toe with copious amounts of low pressure water from open end of firehose. Include inside and outside of helmet, mask, harness, boots down from the top, and inside of coat-wrists to the cuff.

3. Do not smoke, eat, drink, or touch face.

On Return to Station

4. Place apparatus temporarily out of service.

5. Remove all fire fighting clothes (coat, belt, boots, helmet, etc.). If possible, remove liner from helmet. Scrub all items, including the helmet liner, inside and out with a mild (1 to 2%) trisodium phosphate solution. Then flush copiously with water.

NOTE: Where the scrubbing of the fire clothes may release harmful vapors caught in the fibers, it may be necessary to wear breathing apparatus while washing down fire clothes. In these cases, monitor the atmosphere around the washing area. Release of vapors may indicate commercial cleaning is required.

6. Scrub all other protective gear such as gloves and breathing apparatus items likewise. Be sure to flush out gloves with water. If SCBA is stored in its case while returning from incident, scrub the case also.

7. Remove all clothing worn at the scene, including underwear, and place in garbage bag for laundering and/or dry cleaning (preferably the latter). Take all garbage bags with contaminated clothing to a place where they can be cleaned separately from other garments.

8. Shower, scrubbing all of the body with soap and water, with particular emphasis on areas around the mouth and nostrils and under fingernails. Shampoo hair and thoroughly clean mustache if you have one.

9. Do not smoke, drink, eat, touch face, or void until step ♡8 completed.

10. Put on clean clothes.

11. Do not put apparatus back in service until clean-up completed.

To Change SCBA Cylinders at the Scene

Flush empty cylinder and surrounding area of fire fighter's back with copious amounts of low pressure water from open end of firehose. Also flush facepiece and breathing tube to prevent inhalation of harmful materials when regulator is disconnected.

Wear gauntlet-type rubber gloves, such as those used by linemen, when changing cylinders. Flush gloves after use before removing them.

DECONTAMINATION PROCEDURE

LEVEL "C"

for extreme hazards

— — — — — — — — — — — —

At the Scene

1. Do not remove SCBA facepiece. Place helmet on back of neck.

2. Assistant, wearing turnout gear and SCBA (plus disposable chemical suit wherever possible), to flush fire fighter downwards from head to toe with copious amounts of low pressure water from open end of firehose. Include inside and outside of helmet, mask, harness, boots down from the top, and inside of coat-wrists to the cuff.

3. Do not smoke, eat, drink, or touch face.

4. Put SCBA, used cylinders, and any equipment (including hoses and tarps) suspected or known to be contaminated in garbage bags. Seal bags and return them to the station. Where circumstances permit, remove and bag firegear also.

On Return to Station

5. Put bags returned from incident scene in exterior cordoned-off area away from public access. Place apparatus out of service.

6. Strip completely. Place all clothing (firegear and personal clothing) in plastic garbage bags. Place portable radios in a separate bag. Seal bags, place in exterior cordoned-off area.

7. Arrange for the supply of a number of steel drums. Upon their arrival, seal garbage bags with contaminated items into drums. Mark drums and place in exterior cordoned-off area, minimum 5-meter radius.

8. Arrange for the drums to be picked up and the contents analyzed. Some or all items may be destroyed, some may be able to be decontaminated and returned.

9. Shower, scrubbing all of the body with soap and water, with particular emphasis on areas around the mouth and nostrils and under fingernails. Shampoo hair. Thoroughly clean mustache if you have one.

Special Attention for Radioactive Incidents:

After showering, scan entire body with a radiation contamination monitor, paying special attention to hair, hands, and fingernails. Hold monitor approximately 3 cm from body. If any reading beyond normal background level is detected, the fire fighter should shower again, scrubbing with more soap than before.

10. Do not smoke, drink, eat, touch face, or void until step #9 is completed.

11. Put on clean clothes.

12. Report to hospital for medical examination. Inform physician which hazardous material was involved.

To Change SCBA Cylinders at the Scene

Flush empty cylinder and surrounding area of fire fighter's back with copious amounts of low pressure water from open end of firehose. Also flush facepiece and breathing tube to prevent inhalation of harmful material when regulator is disconnected.

Wear gauntlet-type rubber gloves, such as those used by linemen, when changing cylinders. Flush gloves after use before removing them.

Place empty cylinder in black plastic garbage bag and seal for subsequent decontamination.

The person doing the flushing and cylinder-changing must wear turnout gear and SCBA, plus a disposable chemical suit if available.

Special Note

Where circumstances, local climate, and available resources permit, the performance of *all* steps at the scene (instead of performing steps 5-11 at the station) is preferable. The procedure is outlined as shown, however, in recognition of the fact that for many departments this will usually be impossible to achieve.

DECONTAMINATION PROCEDURE

LEVEL "D"

for water-reactive hazards

— — — — — — — — — — — —

At the Scene

1. Set up a suitable vacuum cleaner with power supply. Provide a dry brush and a containment capture method for materials falling off the contaminated personnel. Assistants to don full turnout gear and SCBA, plus disposable chemical suits if available and appropriate.

2. If this is a radiation incident:

The fire fighters suspected of being contaminated will be scanned carefully with a radiation monitor suitable for detecting surface contamination. All parts of their clothing and personal equipment will be scanned, including the soles of the boots. If no readings are found, the personnel that have been checked can leave the decontamination area.

3. If not a radiation incident, or if the fire fighter was found to be radioactively contaminated:

Stand fire fighter in center of containment area, clean helmet and place on back of neck, then clean inside of helmet.

4. Commence cleaning from head downwards. Include all external areas. Slacken SCBA harness to allow cleaning behind straps and backplate. Likewise, loosen the hose-key belt and clean behind it.

5. When the fire fighter has been fully vacuumed or brushed off, he will step out of the containment area. As he does so, his boots, including the soles, must be cleaned off so any contaminant will remain within the containment area.

6. Procedures will then continue as follows:

—Radioactive incident — go to Level "R" routine

—Etiological or dry pesticide incident — go to Level "E" routine

—Other incidents — go to Level "B" routine (unless advice is received that Level "C" is more appropriate).

7. All used filters and collected waste are to be placed in a garbage bag, sealed and tagged, and disposed of in a manner acceptable to the agency having jurisdiction.

DECONTAMINATION PROCEDURE

LEVEL "E"

for etiologic hazards
— — — — — — — — — — — —

Special Equipment Required

A presentation spray can (such as used for pesticide spraying), bleach concentrate, orange garbage bags, black garbage bags, sterilization bags as used by hospital laundries, and a box of surgical masks.

At the Scene

1. Make up a 5% to 6% bleach solution in the spray can. Take note of the bleach concentrate percentage when calculating the make-up of the solution. Many brands as purchased in the store are already 6%.

2. Flush the fire fighter downwards from head to toe with low pressure *water* from a firehose. SCBA facepiece can now be removed. Place helmets in black plastic garbage bag(s) and seal. Place surgical mask on fire fighter.

3. Spray the fire fighter's boots (but not their bunker gear) and any tools, hoses, and other equipment used (except for portable radios) with the *bleach* solution in the spray can. Leave for 10 minutes, then flush with water.

4. Remove SCBA Place in black plastic garbage bag and seal. Remove fire fighter's firecoat and gloves. Place in orange plastic garbage bag and seal. Remove any portable radio worn. Place in black plastic garbage bag and seal. Discard surgical masks.

5. Do not smoke, eat, drink, or touch face.

6. Before leaving the scene, a fire fighter wearing SCBA should attempt to spray as much of the ground exposed to the material and the wash-down water as possible with the bleach solution. Then flush the outside of the spray can with clean water.

7. Before leaving the scene, seal the orange garbage bags into the sterilization bags.

On Return to Station

8. Place apparatus temporarily out of service.

9. One fire fighter should dress in firegear and SCBA, and in an outside area perform the following tasks:

–Open the black plastic garbage bags, wipe all helmets, portable radios, SCBA sets, and used cylinders with a rag lightly dampened with a 6% bleach solution. After 10 minutes, wipe these items again with a rag dampened with clean water.

–Seal all used black garbage bags and rags into another bag and put out for normal garbage pick-up. Empty the spray can and flush out to remove bleach residue.

10. Remove all clothing worn at the scene, including underwear, and place in garbage bag for laundering and/or dry cleaning (preferably the latter). Take all garbage bags with contaminated clothing to a place where they can be cleaned separately from other garments.

11. All personnel should shower, scrubbing all of the body with soap and water, with particular emphasis on areas around the mouth and nostrils and under fingernails. Shampoo hair and thoroughly clean moustache if you have one.

12. Do not smoke, eat, drink, touch face, or void until step #11 is completed.

13. Put on clean clothes. Place apparatus back in service when decontamination is completed.

14. Have cleaned firehose and SCBA checked by competent personnel before placing it back in service.

15. Arrange for the sterilization bags to be taken to a hospital laundry facility for cleaning and sterilization of the firecoats, gloves, and any other garments sent in.

Reminder

Black garbage bags are to be used for items retained at the station. Orange bags are for items sent away for sterilization.

To Change SCBA Cylinders at the Scene

Flush empty cylinder and surrounding area of fire fighter's back with copious amounts of low pressure water from open end of firehose. Also flush facepiece and breathing tube to prevent inhalation of harmful material when regulator is disconnected.

Wear gauntlet-type rubber gloves, such as those used by linemen, when changing cylinders. Flush gloves after use before removing them.

Place empty cylinder in black plastic garbage bag and seal for subsequent decontamination.

The person doing the flushing and cylinder-changing must wear turnout gear and SCBA

DECONTAMINATION PROCEDURE

LEVEL "R"

for radioactive hazards

– – – – – – – – – – – –

At the Scene

1. *Preparation*

   A) Mark off a decontamination area with two parts.

   B) Make up a solution of detergent and water. Obtain scrub brushes.

   C) Set out a reserve air supply, preferably with a workline unit or otherwise with a spare SCBA.

   D) In the first part of the decontamination area, set up a runoff capturing method, either with wading pools or through the use of tarpaulins.

E) If appropriate, a "walkway" of polyethylene sheeting (weighted down if necessary) can be placed from the exit from the incident scene to the decontamination area, to prevent possible contamination of the ground.

2. The decontamination crew will don SCBA and, where available, disposable chemical suits.

3. The fire fighters suspected of being contaminated will be scanned carefully with a radiation monitor suitable for detecting surface contamination. All parts of their clothing and personal equipment will be scanned, including the soles of the boots. If no readings are found, the personnel that have been checked can leave the decontamination area.

4. Personnel found to be contaminated will be scrubbed down thoroughly with the detergent solution by the decontamination crew. This is followed by a flushing off using low pressure water. Efforts should be made to capture the runoff.

5. The fire fighters will then move to the second part of the decontamination area, where they will be scanned again with the radiation monitor. If any readings are found, they will return to the first part of the decontamination area and step 4 will be repeated.

6. When all personnel have been cleaned of contamination, the decontamination crew themselves will be hosed down. The matter of the captured runoff water will be discussed with environmental authorities and disposal arranged in a manner acceptable to them.

7. In the event fire fighters being decontaminated run out of breathing air, the reserve supply set out in step 1 will be passed to them. They should hold their breath while changing facepieces.

8. In the event that, despite repeated scrubbing, any firefighters cannot be decontaminated, they will remove as much of their clothing as possible in the second part of the decontamination area, and don clean or spare clothing. The clothing that has been taken off will be sealed into garbage bags and returned to the station. This evolution must be executed in such a manner as not to contaminate the clean clothing.

9. Any equipment suspected or known to be contaminated will be sealed into garbage bags and returned to the station.

On Return to Station

Follow the Level "C" procedure steps 5 to 12 for those fire fighters who were found to be contaminated in step 3 above, and for any contaminated equipment.

To Change SCBA Cylinders at the Scene

Personnel emerging from the incident to have their breathing apparatus cylinder changed will be scanned with a radiation contamination monitor in a manner identical to step 3 above.

If no readings are found, the fire fighter can proceed to the SCBA cylinder change area and may then return to the incident with a fresh cylinder.

Personnel found to be contaminated may not return to the incident. They will be put through the full Level "R" decontamination procedure, and other fire fighters will be sent in to the incident to replace the fire fighters withdrawn.

Before the replacement fire fighters go in, they should attempt to obtain information as to where the other personnel might have received their contamination, in order to allow them to take the necessary caution when approaching that area.

NOTE: Steps 1 and 2 of the Level "R" procedure must be in place by the time the first fire fighter emerges from the incident. If circumstances permit, these preparations should be made before personnel even enter the incident area for the first time.

Decontamination — Specific Observations

Pre-incident Planning

— Review the procedures and, if they are suitable for your location, assemble the equipment necessary into an easily transported container. Some departments, for instance, have all the special items needed for etiologic decontamination carried in a "Level E Decontamination Kit."

— Many departments will have infrequent need to use these procedures. To prevent skill decay, and to prevent certain critical steps in the procedures being accidentally left out, it is suggested that a copy of the procedures be available at the scene and that regular training in the procedures take place. Executing these procedures accurately is not as easy as it would seem.

— The time when you have twenty garbage bags with contaminated clothing sitting on your apparatus floor is not the time to start looking for a laundry that will clean them. Most commercial cleaning companies will not be interested in handling contaminated clothing.

— Furthermore, it should be recognized that at some incidents the nature or extent of the contamination may be such that full decontamination is beyond the resources of the fire department (especially with Levels C, E, and R) and will require specialist treatment. With these three levels, consideration should be given to the destruction of all permeable items in case of serious exposure.

—You should therefore make prior arrangements for the following:

• Obtaining steel drums at any time of the day or night. The drums must be clean and must have a removable lid—not just a bung and vent-hole.

• Analysis and expert decontamination of equipment and clothing contaminated by severely hazardous substances. This is needed for Level C and Level R, although different companies are likely to be needed for the two levels.

• Acceptable methods of disposal for items that cannot be cleaned, or that would be uneconomic to attempt to clean, for Level C, E, and R contaminants.

• The use of a hospital laundry service to perform Level E decontamination on fire gear. This laundry should be approached for the loan of a number of sterilization bags, which are typically used in the hospitals to put dirty laundry in for shipment to the laundry service. Check that the hospital laundry service can take firecoats—in some cases the buckles may bash the inside of their machines too much.

• Check the availability of replacement fire gear and equipment that can be used while the original items are out being decontaminated under Levels C, E, or R.

—You may want to establish a policy regarding personal items such as rings, wallets, watches, etc. Many of these, especially leather items, cannot be decontaminated and may have to be destroyed. Fire fighters should be aware of their department's policy with regard to recompense or replacement.

—One further item of preplanning will always stand you in good stead; note the names and contact numbers of any local experts who could assist and advise you during the incident and its subsequent decontamination.

Decontamination Area Layout

—When choosing the location of the decontamination area, consider the following:

• Prevailing weather conditions (temperature, precipitation, etc.),

• Wind direction,

• Slope of the ground,

• Surface material and porosity (grass, gravel, asphalt, etc.),

• Availability of water,

• Availability of power and lighting,

• Proximity of the incident,

- Location of drains, sewers, and watercourses.

   —When setting up the area, provide the following features:

- Containment of wash-down water if that is necessary,

- Spare supply of breathing air (extra SCBA, extra cylinders, or workline units),

- A supply of industrial-strength garbage bags, double- or triple-bagged if necessary,

- Clearly marked boundaries, not just a rope lying on the ground,

- Clearly marked entry and exit points with the exit upwind, away from the incident and its contaminated area,

- A waiting location at the entry point where contaminated personnel can await their turn without spreading contamination further,

- Access to triage and other medical aid upon exit if necessary,

- Protection of personnel from adverse weather conditions,

- Security and control from the setting up of the area to final clean-up of the site.

Environmental Considerations

   —One fundamental concept forms the basis for these decontamination procedures:

   "The human being comes before the environment."

   —Notwithstanding the above, where containment of run-off is called for, genuine attempts must be made if only to avoid possible legal consequences. Examples of containment basins are:

- Children's wading pools,

- Portable tanks (as used in rural fire fighting),

- Tarps laid over a square formed by hard suction hose or small ground ladders,

- Diking with earth, sandbags, etc. covered with tarps.

   Fire fighters stepping out of a containment basin should lift one foot, have it rinsed off so the water falls inside the basin, step out with that foot, and repeat for the other foot.

When the containment basin is full, it should be able to be siphoned or pumped off into drums or into a vacuum truck for controlled disposal in a manner acceptable to the authority having jurisdiction.

— Any run-off that is not contained will eventually enter sewers and water-courses, or if it sinks into the ground will ultimately reach the water-table. The Department of Mechanical and Fluid Engineering at Leeds (U.K.) University has determined that provided a chemical is diluted with water at the rate of approximately 2000:1, pollution of water-courses will be significantly reduced.

— There is also a change in attitude coming with the environmental authorities, whereby they recognize that the small amount of chemical likely to be washed off contaminated fire fighters with adequate dilution will result in minimum damage to the environment, especially when compared to the results of the spill that generally led to the personnel contamination in the first place.

— Any substances that enter sewers and water-courses should be reported to environmental authorities and to the sewage treatment plant likely to receive it. If necessary, advise water authorities downstream from the decontamination area of actual or potential pollution.

— The most appropriate decontamination for materials that have a severe effect on the environment will usually be to use minimal amounts of water, with run-off containment. Other substances should be deluged off personnel with the 2000:1 factor as a minimum guideline.

Weather

— If decontamination is done indoors because of bad weather, ensure that the drains go into a holding tank and not directly into the sewers.

— If the hazardous material involved requires Level D decontamination, and it is raining or snowing, protect fire fighters from the precipitation until they have been processed.

— Take care when using instruments in wet weather. Extreme cold may affect the operating effectiveness of instruments, especially delicate ones originally designed for use in laboratory environments.

— Under extreme weather conditions (heat or cold), decontamination personnel must be rotated more frequently.

— These decontamination procedures should be reviewed in light of your local climate and adapted if necessary where your area's weather conditions dictate.

Fluid Replacement

—At hazardous materials incidents, especially when chemical suits are worn, serious dehydration can occur in fire fighters. Replacement of fluids should only be permitted if at least gross decontamination is performed — a washdown especially around the head and upper body.

—The preferable method of consuming liquids is by means of drinking boxes with straws (the straw inserted by someone with uncontaminated hands), or by means of a squeeze bottle with an attached drinking tube as used by athletes.

Chemical Suit Decontamination

—When a chemical suit is taken off its wearer, a suitably protected assistant should roll it in on itself in order to keep the outside of the suit from coming into contact with the wearer.

—Because of the inherent smoothness and impermeability of chemical suits, it is usually only required that the on-scene washdown part of fire fighter decontamination is performed. Upon return to the station, instead of doing the steps listed in the appropriate procedure, fire fighters should wash and rinse the chemical suits and examine them carefully for damage caused at the incident. Zippers should be lubricated with their special lubricant.

—Follow-up communication with the suit manufacturer as to the exposure, as well as follow-up from the exposing chemical's manufacturer, is useful in determining long-term effect of exposure to chemical-protective ensembles. Any questionable or unusual findings anywhere in the decontamination or testing process should be immediately referred to the manufacturers; the clothing should be placed out of service until it can be repaired or reevaluated.

Vacuum Cleaners for Level D

When selecting a vacuum cleaner, the following points should be taken into consideration:

—Can it operate off a generator, or is it unforgiving so far as voltage fluctuations are concerned?

—Will it operate safely in an area where it might get wet?

—How effective are the filters?

—Can the unit be safely cleaned out itself?

—Are replacement hoses easy to come by?

Although you won't want to operate your vacuum cleaner under water, it might accidentally get splashed so some basic water protection will be of benefit.

The degree of filtering achieved is important. Most wet/dry industrial type vacuums will achieve a reasonable effectiveness. Some specialized cleaners, equipped with HEPA (High Efficiency Particulate Air) filters will go down to 0.3 microns, but they are expensive. The small, cigarette-lighter plug powered car interior cleaners are not suitable as they filter very little, instead blowing most particulate they pick up back out through their exhaust ports.

Easy removal of contaminated filters will help, as will good access to the machine's insides for its own decontamination. You will usually find that it is impossible to guarantee the effectiveness of cleaning of the accordion-style hose, and you should consider replacing these if they become contaminated.

Remember not to operate a vacuum cleaner in a flammable or explosive atmosphere, unless yours is intrinsically safe.

Bleach

— Bleach, as shown on the bottle's label, is corrosive. Do not spray bleach on fire fighters' skin — it hurts! It is also reactive — do not let it come in contact with fuels and solvents, as a heat-generating reaction (and possible fire) will result.

— Do not spray bleach on bunker gear as it deteriorates and discolours the garment. It also impairs its fire retardancy.

— Regular bleach containers are often made of a hard plastic and are liable to crack or leak around the cap when carried on a vehicle. Consider transferring the bleach to a container such as a new, unused plastic gasoline can, which is far sturdier, but then be sure to label the gas can as containing bleach. Never put bleach in a metal container as it will react.

— At room temperature, bleach shielded from sunlight will degrade about 1% per year, i.e., from 6% to 5% (faster in warmer temperatures, slower at colder temperatures). You should therefore replace the bleach at the appropriate time with a fresh supply, as its strength will have decreased with time.

Record Keeping

A member of the crew responsible for performing the decontamination should maintain written records of the following:

- Fire fighter's name, material involved, length of exposure,

- Level of decontamination performed,

- Any ill effects observed,

- Where fire fighter went, i.e.:

  — returned to station

  — sent to rest area

  — removed to hospital

  — reassigned to other duties at the scene

  — etc.

— At the station, entries should be made on the fire fighters' medical records of the incident date, material involved, and decontamination performed, where exposure is known or suspected. This will assist both in tracing future sickness through synergistic effects of chemicals in the body and with support of any later injury or sickness claims.

— If appropriate, records should also be kept of the length of time each chemical suit was exposed, and what substance it was exposed to. This will permit the tracking of cumulative degradation of the suit material due to exposure to a variety of chemicals or due to repeated exposure to one particular substance.

Contamination of Vehicles

Any vehicle driven through a contaminated area must be washed down, including the undercarriage, chassis, and cab. Air filters on vehicle (and, where appropriate, generator) engines must be replaced. Porous items such as wooden hose beds, wooden equipment handles, seats, and cotton jacketed hose may be difficult to clean completely and may have to be discarded.

It is therefore better to take the "uphill and upwind" approach and keep vehicles at a suitable distance from incidents.

One Final Observation

The entire foregoing contents of this document have probably made you realize by now that it is much more desirable to handle hazardous materials incidents with chemical suits than with regular fire fighting turnouts. The cost of disposable suits is relatively cheap; even for a small department, throwing away a few hundred dollars worth of disposable suits after one single use will be cheaper than replacing turnouts or paying for commercial cleaning. In many jurisdictions, the fire service is permitted to recover the cost of destroyed equipment from the party responsible for the incident's occurrence, and the cost of disposable chemical suits can thus be recovered.

Always remember: if the emergency response crew is not equipped with gear suitable for entry into a hazardous or toxic atmosphere, then the option of "no go" should be considered the most appropriate tactic.

Speed — a Case for Exception?

Decontamination should emphasize thoroughness, not speed. Under non-critical conditions certain commonsense actions should be taken, such as decontaminating the fire fighter with the lowest air reserve first.

Speed is only important where a victim is involved and even then decontamination should be as thorough as is practicable.

Circumstances may dictate that emergency decontamination becomes necessary, examples of such situations being where a protective suit has become split or damaged, or when a fire fighter is injured. Emergency decontamination may also be applicable when contaminated civilians or other emergency workers (police, ambulance, etc.) are involved.

Emergency Decontamination Procedure

Paragraphs 1 to 6 below, although arranged in a basic chronological order, do not necessarily have to be undertaken in the exact sequence outlined. The officer-in-charge should act in the most expedient manner appropriate without worsening the situation.

The procedure outlined should be carried out as quickly as possible.

To protect the ambulance crew and hospital staff as well as the victim, every attempt must be made to perform at least this emergency procedure prior to transporting the victim to the hospital.

1. Remove the victim from the contaminated area into the decontamination zone and ensure he is supplied with uncontaminated air or oxygen.

2. Remove fire helmet if worn and immediately wash with flooding quantities of water any exposed parts of the body that may have been contaminated.

3. If the victim is wearing SCBA, release the harness and remove the set leaving the face mask in position.

4. Remove contaminated fire gear or clothing (if necessary by cutting it off the victim) ensuring where practicable that the victim does not come into further contact with any contaminant. Maintain the washing of the victim while the clothing removal is taking place.

5. Remove the victim to a clean area. Render first aid as required, but do not apply mouth-to-mouth resuscitation. Send victim for medical treatment as soon as this emergency decontamination procedure has been completed.

6. Ensure hospital/ambulance personnel are informed of the contaminant involved.

Index

Additional haz mats resources help you be prepared!

FIRE PROTECTION HANDBOOK

Only one comprehensive sourcebook answers the full range of today's fire protection and prevention questions — the *Fire Protection Handbook*. With new fire technologies, new materials, and constantly changing fire hazards, the very nature of fire protection grows more complicated every day. The *Fire Protection Handbook* covers:

- hazardous wastes and materials
- venting
- emergency response
- fire hazards of robotics
- aerosol charging
- mining safety and emergency intervention.

It's all here, and more, in one comprehensive volume.

Order your copy today!

Fire Protection Guide on Hazardous Materials

Get all the haz-mats codes you need — at a savings! The *Fire Protection Guide on Hazardous Materials* includes the complete texts of four NFPA codes which classify most common hazardous materials:

- NFPA 325M: Fire Hazard Properties of Flammable Liquids, Gases, and Volatile Solids (1984)
- NFPA 49: Hazardous Chemicals Data (1975)
- NFPA 491M: Manual of Hazardous Chemical Reactions (1985)

- NFPA 704: Identification of the Fire Hazards of Materials (1985)

This essential guide is critical in dealing effectively with a haz-mat fire, spill, or accident. Also covers storage and handling procedures (488 pp., 1986).

Be prepared . . . be sure you have a copy of the *Fire Protection Guide on Hazardous Materials!*

Order your copy today!

NO POSTAGE
NECESSARY
IF MAILED
IN THE
UNITED STATES

BUSINESS REPLY MAIL
FIRST CLASS   PERMIT NO. 3376   QUINCY, MA

POSTAGE WILL BE PAID BY ADDRESSEE

NATIONAL FIRE PROTECTION ASSOCIATION
1 BATTERYMARCH PARK
PO BOX 9101
QUINCY MA 02269-9904

NO POSTAGE
NECESSARY
IF MAILED
IN THE
UNITED STATES

BUSINESS REPLY MAIL
FIRST CLASS   PERMIT NO. 3376   QUINCY, MA

POSTAGE WILL BE PAID BY ADDRESSEE

NATIONAL FIRE PROTECTION ASSOCIATION
1 BATTERYMARCH PARK
PO BOX 9101
QUINCY MA 02269-9904